HAZARDOUS WASTE INCINERATION

Evaluating the Human Health
and Environmental Risks

HAZARDOUS WASTE INCINERATION

Evaluating the Human Health
and Environmental Risks

Edited by

Stephen M. Roberts, Ph.D.
University of Florida
Gainesville, FL

Christopher M. Teaf, Ph.D.
Florida State University
Tallahassee, FL

Judy A. Bean, Ph.D.
Children's Hospital Medical Center
Cincinnati, OH

LEWIS PUBLISHERS
Boca Raton London New York Washington, D.C.

Library of Congress Cataloging-in-Publication Data

Roberts, Stephen M., 1950–
    Hazardous waste incineration: evaluating the human health and
environmental risks / Stephen M. Roberts, Christopher M. Teaf, Judy
A. Bean.
       p.   cm.
    Includes bibliographical references and index.
    ISBN 1-56670-250-X (alk. paper)
    1. Hazardous wastes—Incineration—Health aspects—United States.
2. Hazardous wastes—Incineration—Environmental aspects—United
States.   3. Incineration—Waste disposal—Environmental aspects—
United States.   4. Incineration—Waste disposal—Health aspects—
United States.   5. Health risk assessment—United States.
    I. Teaf, Christopher M.   II. Bean, Judy A.   III. Title.
    RA578.H38R64 1998
    363.72'8—dc21

RA
578
.H38
H38
1999

98-28649
CIP

This book contains information obtained from authentic and highly regarded sources. Reprinted material is quoted with permission, and sources are indicated. A wide variety of references are listed. Reasonable efforts have been made to publish reliable data and information, but the author and the publisher cannot assume responsibility for the validity of all materials or for the consequences of their use.

© 1999 by CRC Press LLC.
Lewis Publishers is an imprint of CRC Press LLC

No claim to original U.S. Government works
International Standard Book Number 1-56670-250-X
Library of Congress Card Number 98-28649
Printed in the United States of America 1 2 3 4 5 6 7 8 9 0
Printed on acid-free paper

About the Editors

Stephen M. Roberts is Director of the Center for Environmental and Human Toxicology at the University of Florida, Gainesville, and is an associate professor with joint appointments in the Department of Physiological Sciences in the College of Veterinary Medicine and the Department of Pharmacology and Therapeutics in the College of Medicine.

He received his Ph.D. from the University of Utah College of Medicine in 1977, and subsequently completed a National Institutes of Health (NIH) individual postdoctoral fellowship at the State University of New York at Buffalo. He has served on the faculties of the College of Pharmacy at the University of Cincinnati and the College of Medicine at the University of Arkansas for Medical Sciences.

Dr. Roberts has an active research program funded by the NIH to examine mechanisms of toxicity and has published more than 60 articles on toxicology in humans and animals. His teaching responsibilities at the University of Florida include graduate courses in both general and advanced toxicology, risk assessment, and issues in the responsible conduct of research.

Dr. Roberts served as chairman of the Florida Risk-Based Priority Council and currently advises the Florida Department of Environmental Protection on issues pertaining to toxicology and risk assessment.

Christopher M. Teaf is Associate Director of the Center for Biomedical and Toxicological Research at Florida State University, Tallahassee, and President and Director of Toxicology for Hazardous Substance and Waste Management Research, Inc.

He received his B.S. in biology from Pennsylvania State University in 1975, his M.S. in biological science from Florida State University in 1980, and his Ph.D. in toxicology from the University of Arkansas for Medical Sciences in 1985.

For more than 15 years, Dr. Teaf has conducted or reviewed risk-based evaluations under federal or state occupational and environmental requirements. His principal areas of interest include performance and evaluation of risk assessments under the requirements of CERCLA, SARA, RCRA,

TSCA, and related legislation; potential human health impacts of exposure to environmental contaminants; and development of risk-based targets to guide remedial decisions.

He has served as toxicologist to the state Financial and Technical Advisory Committee and is chairman of the Department of Labor and Employment Security's Toxic Substances Advisory Council. He served on the steering committees for the 1992, 1994, and 1996 International Symposia on Contamination in Central and Eastern Europe and is currently risk assessment advisor with the Institute for Environmental Studies of Industrial Areas in Poland.

Judy A. Bean is Director of the Biostatistics Program, Children's Hospital Medical Center, Cincinnati, Ohio, and has served as Director of Biostatistics at the University of Miami in Florida.

She received her B.A. in mathematics and biology from Murray State College in 1963, an M.P.H. in biostatistics from the University of Michigan in 1965, and a Ph.D. in biostatistics from the University of Texas at Houston in 1973.

Dr. Bean has been involved in statistical consulting and development of statistical methods for more than 20 years. Her research interests include survey methodology, environmental health epidemiology, cancer epidemiology and statistical methods in epidemiology. She is a member of the Scientific Advisory Board for the U.S. EPA Drinking Water Committee and has served on several committees for EPA, FDA, National Institutes of Health, National Institute for Environmental Health and Sciences, and National Cancer Institute. She is a reviewer for numerous epidemiology, toxicology, occupational health, and statistics journals.

Contributors

Judy A. Bean, Ph.D.
Children's Hospital Medical Center
Cincinnati, Ohio

Stuart M. Brooks, M.D.
Department of Environmental and
  Occupational Health
College of Public Health
University of South Florida
Tampa

C. David Cooper, Ph.D., P.E.
College of Engineering
Department of Civil and
  Environmental Engineering
University of Central Florida
Orlando

Robert P. DeMott, Ph.D.
ATRA Occupational and
  Environmental Services, Inc.
Tallahassee, Florida, and
University of South Florida
Tampa

Frank J. Dombrowski
ChemRisk Division
McLaren/Hart
Alameda, California

**Lora E. Fleming, M.D., Ph.D.,
  M.P.H., M.Sc.**
Associate Professor
Department of Epidemiology and
  Public Health
University of Miami
Miami, Florida

Julia Gill, M.P.H., Ph.D.
Department of Environmental and
  Occupational Health
College of Public Health
University of South Forida
Tampa

Charles R. Harman
ChemRisk Division
McLaren/Hart
Alameda, California

Richard C. Hertzberg
U.S. Environmental Protection
  Agency
National Center for Environmental
  Assessment
Office of Research and
  Development
Atlanta, Georgia

J. Michael Kuperberg
Center for Biomedical and
  Toxicological Research and Waste
  Management
Florida State University
Tallahassee

Christopher E. Mackay, Ph.D.
Exponent
Bellevue, Washington

William Thomas Marsh
Environmental Assessment and
  Toxicology
QST Environmental, Inc.
Gainesville, Florida

Thomas Mason, Ph.D.
Department of Epidemiology and
  Biostatistics
College of Public Health
University of South Florida
Tampa

Philip A. Moffat
Center for Biomedical and
  Toxicological Research and Waste
  Management
Florida State University
Tallahassee

Glenn Rice
U.S. Environmental Protection
  Agency
National Center for Environmental
  Assessment
Office of Research and
  Development
Cincinnati, Ohio

Stephen M. Roberts, Ph.D.
Center for Environmental and
  Human Toxicology
University of Florida
Gainesville

John Schert, Ph.D.
Florida Center for Solid and
  Hazardous Waste Management
University of Florida
Gainesville

Patrick J. Sheehan, Ph.D.
ChemRisk Division
McLaren/Hart
Alameda, California

Ann Bergquist Shortelle
Environmental Assessment and
  Toxicology
QST Environmental, Inc.
Gainesville, Florida

Isabel K. Stabile, M.D., Ph.D.
Center for Prevention and Early
  Intervention Policy
Florida State University
Tallahassee

Christopher M. Teaf, Ph.D.
Center for Biomedical and
  Toxicological Research and Waste
  Management
Florida State University
Tallahassee

Linda K. Teuschler
U.S. Environmental Protection
  Agency
National Center for Environmental
  Assessment
Office of Research and
  Development
Cincinnati, Ohio

Michael J. Ungs
ChemRisk Division
McLaren/Hart
Alameda, California

Contents

Preface

Despite considerable technical and regulatory efforts to minimize the production of hazardous wastes, hundreds of millions of tons of these materials continue to be produced annually in the United States. Some of this waste is an unavoidable consequence of the creation of products that modern society values, while other waste is generated from the ongoing cleanup of contaminated sites. While the potential sources of hazardous waste are many, the established options for disposal are few. One historically common approach is to place the hazardous waste in a landfill, essentially storing it until such time as a more permanent solution can be found. Another common method is incineration. Hazardous waste incineration offers the advantage of waste volume reduction and the permanent destruction of organic wastes. With these advantages come some limitations as well.

Probably the largest single obstacle to the use of incineration as part of an overall hazardous waste management strategy is concern about potential environmental impacts, both in terms of human health and effects on ecosystems. This is an important issue not only for industry and regulatory agencies but one in which the public has taken considerable interest. Few things are as guaranteed to generate a heated and impassioned public debate as a proposal to site a hazardous waste incinerator in or near a community. Reasoned decisions regarding the acceptability of an existing or proposed incinerator facility require a careful evaluation of the potential health risks it may pose, and there may be several facets to this evaluation. One aspect is the collection of knowledge and experience regarding other hazardous waste incinerators. What is known about the emissions from existing hazardous waste incinerators and their possible health effects? What human health impacts and ecosystem effects have been plausibly attributed to hazardous waste incinerators in the past? Are these facilities prone to accidental releases that may have health consequences? For a proposed hazardous waste incinerator — one that exists only on paper — a second aspect is the ability to effectively predict the health impacts of a completed and operational facility. This requires a way to predict the nature and rate of emissions as well as their dispersion in the environment. There must also be a framework for identifying exposures of potentially affected species and estimating the health risks associated with these. A third aspect pertains to existing facilities:

how to identify and measure health impacts. What studies could be performed to determine whether a hazardous waste incinerator is having an adverse effect on human or ecosystem health? Conceivably, this aspect could be relevant to proposed facilities as well, if approval was contingent upon development and implementation of an ongoing health monitoring program.

Faced with these issues, the Florida Department of Environmental Protection tasked a consortium of individuals from public and private universities to prepare a report concerning potential health impacts of hazardous waste incineration. The analysis in the report focused on four key questions:

- What is known about existing hazardous waste incinerators and their impacts on human health?
- Can the impacts of a proposed facility be evaluated before it is built, and, if so, how?
- What is the regulatory compliance record of existing commercial hazardous waste incinerators?
- What methods can be used to monitor a facility's impacts after it is built?

This volume is based on that report, with updated and expanded information in key areas.

Chapters 1 to 3 provide a background on hazardous waste incineration and the emissions that may be associated with it. A discussion of methodology for modeling atmospheric dispersion of hazardous waste incinerator emissions is also included in Chapter 4. Chapter 5 deals with the difficult issue of human health risk assessment for a complex combustion mixture, and Chapters 6 through 9 summarize existing knowledge of ecological impacts of hazardous waste incineration as well as provide guidance and examples of ecological risk assessment and environmental monitoring procedures. Chapters 10 through 12 discuss the state of knowledge regarding human health impacts of incineration and present strategies for health monitoring studies. Chapter 13 covers the issue of uncertainty in risk assessments, and Chapter 14 presents the results of a study of regulatory compliance within the hazardous waste industry.

This book represents a comprehensive review of information pertinent to the evaluation of human health and ecological risks of hazardous waste incineration. It is our sincere hope that this book will be useful to regulatory agencies, industry, environmental groups, and citizen groups who must grapple with deciding whether incineration is the right choice for dealing with hazardous waste under their particular circumstances. We are grateful to each of the authors for contributing their time and expertise to this volume and to the Florida Department of Environmental Protection for initiating this project and for their support and encouragement.

Stephen M. Roberts, Ph.D.
Christopher M. Teaf, Ph.D.
Judy A. Bean, Ph.D.

chapter one

Fundamentals of hazardous waste incineration

C. David Cooper

Contents

Introduction

Hazardous wastes are some of the inevitable unwanted by-products of our industrialized society. They take on many forms and are generated by many different industries and commercial activities. Hazardous wastes are even generated in small quantities by individuals in their own households. Several years ago, the U.S. Environmental Protection Agency (EPA) estimated the total Resource Conservation and Recovery Act (RCRA) hazardous waste generation rate to be 180 million tons per year. At that same time, another estimate by the EPA's Office of Technology put the amount at about 280 million tons of hazardous wastes per year, but this included waste oils and other materials not considered RCRA hazardous wastes.

1-56670-250-X/99/$0.00+$.50
© 1999 by CRC Press LLC

1

Many physical, chemical, and biological processes can be used to treat hazardous wastes. The treatments can destroy, convert, or detoxify hazardous wastes or can simply concentrate and isolate them for disposal. Recently much emphasis has been placed on *prevention* and *waste minimization* (making changes in the industrial process itself to produce fewer or smaller amounts of hazardous wastes). Pollution prevention is always an important part of the overall treatment hierarchy for any waste.

Physical processes for hazardous waste treatment include separation and concentration to reduce hazardous waste volume and make it easier or more economical to treat. Filtration, centrifugation, and flotation are all physical separation processes that operate on heterogeneous mixtures. Ion exchange, reverse osmosis, ultrafiltration, carbon adsorption, liquid absorption, and air stripping are examples of physical processes that operate on an ionic or molecular scale.

Chemical treatments include thermal processes (incineration, calcination, volatilization, catalytic oxidation, etc.), wet processes (chlorinolysis, hydrolysis, neutralization, electrolysis, chemical oxidation/reduction), and precipitation processes [adding one or more chemicals to the waste mixture to remove the hazardous component (meanwhile forming a potentially hazardous sludge)]. Biological processes include activated sludge, soil vapor aeration and extraction, in situ bioremediation, aerated lagoons, anaerobic digestion, composting, and others. These are usually best applied to very dilute water streams or to contaminated underground sites that cannot be easily excavated.

After one or more of these treatment processes, there is always the need to dispose of the residual material that remains. The residuals may be solid, liquid, or gaseous. They may still be somewhat hazardous or may be innocuous. The final disposal option may be to store the residuals in an isolated area (hazardous waste landfill, deep well injection) or to disperse the residuals into the environment (exhaust gases from an incinerator). All of the above-mentioned processes for dealing with hazardous wastes have advantages and disadvantages, but incineration is one of the more widely used processes in the United States and in many other countries. Also, for certain types of wastes, incineration may be the only viable option. Therefore, it behooves us to understand this process better.

What is hazardous waste incineration?

Incineration can be viewed as the flame-initiated, high temperature, air oxidation of organic matter. Incineration is currently practiced to some extent on municipal solid waste (MSW), medical waste (MW), and hazardous waste (HW). HW incineration is an important part of the nation's overall capabilities for treating certain classes of HW. Incineration can only destroy the organic compounds in HW; it cannot destroy mineral compounds (which end up as residual ash). Emissions from hazardous waste incinerators

(HWIs) that are of concern to the general public include unburned organic wastes, products of incomplete combustion or by-products of combustion, heavy metals, acid gases, ash, and others. It is noted that MSW and MW incinerators emit these same pollutants. Emissions of these pollutants typically can be controlled to very low rates by modern air pollution control equipment.

Incineration has several disadvantages as well as advantages when compared with other methods of treatment, so it is not always the preferred choice. However, the key advantage of incineration is that it destroys organic wastes permanently. To be acceptable, however, HWIs must meet strict emission limits not only for various hazardous pollutants (including organics not combusted completely and toxic metals not capable of destruction) but also for other pollutants, such as hydrogen chloride (HCl), carbon monoxide (CO), and nitrogen dioxide (usually combined with nitric oxide and referred to as NO_x).

Combustion chemistry

Since incineration is high-temperature oxidation, the first step toward understanding the whole process is to review the basic chemistry involved. Very simply, the oxygen in the air combines with the elements in the waste compounds to create oxidized products. Problems arise from incomplete combustion, from combustion products that are themselves pollutants, from side reactions that form pollutants, or from emissions of noncombustible contaminants. Because the waste must be oxidized nearly completely [99.99% destruction and removal efficiency (DRE) is required under HWI regulations], a large excess of air is used to ensure sufficient oxygen to do the job.

An idealized oxidation (combustion) reaction is shown below using an assumed formula for the organic waste material (with no ash) and using stoichiometry that assumes complete combustion in 50% excess air. Note that the reaction shown in Equation 1 does not occur as written and is used only to depict the overall stoichiometry of the idealized reaction.

$$C_{1914}H_{2963}O_{518}N_{36}Cl_{51}S_{17} + 3600\ O_2 + 13543\ N_2 \rightarrow$$

$$1914\ CO_2 + 51\ HCl + 1456\ H_2O + 1200\ O_2 + 13561\ N_2 + 17\ SO_2 \quad (1)$$

The *desired* end products of incineration are shown in Equation 1. It is hoped that all of the carbon in the waste will be converted to carbon dioxide, all of the hydrogen will go to water, all of the nitrogen will come out as N_2, and all of the chlorine will appear as HCl [which can then be scrubbed out easily with appropriate air pollution control (APC) equipment].

For this assumed waste (which is relatively low in chlorine content), and this assumed combustion reaction, for every 5,000 g of waste burned, 138 g of HCl is created. This idealized reaction also assumes that none of the nitrogen in the waste is oxidized to NO_x, which is not accurate, as discussed below.

Also, note the large amount of air (oxygen and nitrogen) that must be supplied to the incinerator. For this assumed waste and this reaction, for every 5,000 lb/hour of waste burned, and assuming that the gases exit the incinerator at 1900°F, the gaseous flow rate leaving the incinerator is about 85,000 actual cubic feet per minute (acfm).

In the real case, it is understood that not all the waste is destroyed (at 99.99% DRE, 1 g of waste is not burned for every 9,999 g that are burned). Thus, a small fraction of the original waste is emitted from the furnace, which may either be collected by APC equipment or emitted out the stack. In addition, most wastes contain some noncombustibles (ash) that are not represented in Equation 1. These materials (sand particles, minerals, metal compounds) may be present in large quantities in a mixed waste stream and must be collected by APC equipment and disposed of properly.

In addition to the above considerations, some of the organic waste that is changed from its original molecular form may not be *completely* oxidized, as shown in Equation 1. Relatively stable, smaller organic compounds are always formed as part of the process by which the parent compound is burned. These products of incomplete combustion (PICs) can sometimes resist further oxidation; many PICs are hazardous. Also, PICs may participate in recombination reactions to form dioxins and/or furans — extremely toxic organic by-products.

Furthermore, at the very high temperatures characteristic of flames, oxygen and nitrogen molecules present in air dissociate into oxygen and nitrogen atoms then recombine to form NO (the main emitted component of NO_x), as shown by the two main reactions of the well-known Zeldovich mechanism:

$$N_2 + O \rightarrow NO + N \qquad (2)$$

$$N + O_2 \rightarrow NO + O \qquad (3)$$

NO_x emissions may be an important issue in some locations.

Finally, the actual burning process is made up of hundreds of elementary reactions, which occur both sequentially and simultaneously in an extremely complex system of chain-branching chemical reactions that proceed very rapidly. The detailed mechanisms that have been proposed to explain the combustion of even simple pure compounds (such as heptane or chlorobenzene) involve over 200 chemical species and more than 100 reactions.

Types and prevalence of hazardous waste incinerators

Of the total amount of HW generated each year, about 20% contain organics and can be incinerated, but the EPA estimated that only about 10 to 15% of these were incinerated in 1987 (ASME, 1988). In 1987, Oppelt reported that at that time there were over 200 HWIs in use, including 42 rotary kilns, 95 liquid injection units, 25 fume incinerators, 32 hearth-type units, and 14 fluidized bed incinerators. Those figures included all incinerators, both commercial and captive. As of mid-1994, there were 19 commercial HWIs operating in the United States.

A large furnace or combustor cannot exist and operate by itself; a schematic block diagram showing the major subsystems of an HWI facility is presented in Figure 1.1. The HWI facility is in essence a complex industrial processing plant that includes waste receiving, storage, handling and preparation areas, one or more combustors, auxiliary equipment (fans, pumps, valves, instruments, etc.), ash disposal facilities, APC facilities, residue treatment areas, an exhaust stack, operations control buildings, a laboratory, and maintenance buildings. Its efficient operation requires well-trained employees, adequate utilities (water, gas, electricity), good communications, and a reliable and safe transportation system.

The main types of HWIs are liquid injection chambers, rotary kilns, fixed hearth furnaces, and fluidized bed incinerators. The first three of these types are shown in Figures 1.2 through 1.4. Each type of incinerator is designed to handle specific waste types. A liquid waste (e.g., a mixture of hydrocarbons or chlorinated solvents) can best be burned in a liquid injection-type incinerator (Figure 1.2). Some liquid wastes with good heating values can be burned in a main burner and can sustain a stable flame. Aqueous wastes that cannot sustain a flame must be injected into the flame or even downstream of the flame zone.

A rotary kiln with an afterburner (Figure 1.3) is very versatile and can handle bulky solid waste mixtures (such as contaminated soils, sludges, and even whole drums of waste). These kilns can be over 100 feet long and tilted slightly from one end to the other. The whole kiln slowly rotates to permit the solid wastes to tumble slowly down toward the ash disposal end. Liquid wastes can be injected in the afterburner, which always follows a rotary kiln to ensure that all the off-gases from the kiln are treated (subjected to high temperatures for the required residence time).

A fixed-hearth incinerator (Figure 1.4) handles solid wastes that are fairly dry. This is the type most commonly found incinerating MSW and MW. Typically, it is much smaller and cheaper than a rotary kiln. The wastes are fed into the incinerator one batch at a time with a ram feeder. In the main chamber, the solid wastes burn and gasify (produce a variety of organic gases that are not fully combusted). These gases flow into an afterburner where more fuel is burned and more air is added to create a highly oxidizing environment. The remaining organic compounds in the gas stream are oxidized nearly to completion.

All HWIs are designed to meet a DRE requirement of 99.99% for principal organic hazardous compounds (POHCs). At 99.99% DRE, for every 10,000 lb of waste that enters the incinerator, 9,999 lb of waste are burned,

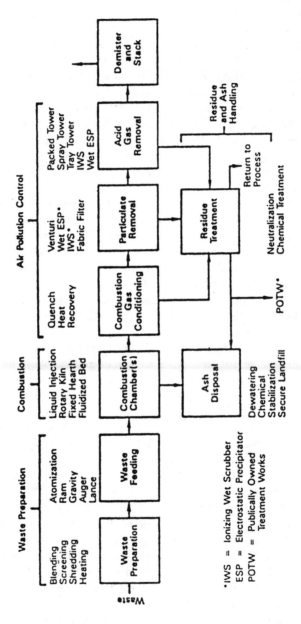

Figure 1.1 General orientation of incineration subsystems and typical process component options. (From Oppelt, E. T. *J. Air Pollut. Control Assoc.*, 37(5), May 1987. With permission.)

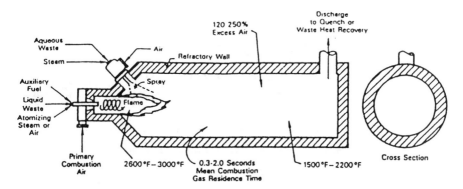

Figure 1.2 Typical liquid injection combustion chamber. (From Oppelt, E. T. *J. Air Pollut. Control Assoc.*, 37(5), May 1987. With permission.)

while 1 lb is not. Thus, a small fraction of the original waste is emitted as a pollutant. In addition, the noncombustible ash (sand particles, minerals, and metal compounds), which may be present in large quantities in a mixed waste stream, and the noncombusted waste residues must be collected in APC equipment and disposed of properly.

In addition to the above considerations, some of that 99.99% of the organic waste that is destroyed may not be oxidized *completely* to CO_2 and H_2O, as was shown in Reaction 1. Relatively stable, smaller organic compounds are

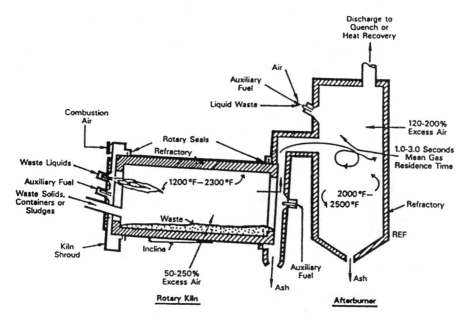

Figure 1.3 Typical rotary kiln/afterburner combustion chamber. (From Oppelt, E. T. *J. Air Pollut. Control Assoc.*, 37(5), May 1987. With permission.)

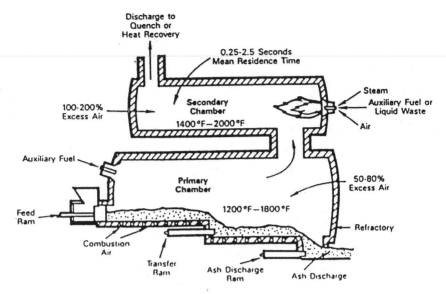

Figure 1.4 Typical fixed-hearth combustion chamber. (From Oppelt, E. T. *J. Air Pollut. Control Assoc.*, 37(5), May 1987. With permission.)

always formed as part of the process by which the parent compound is burned. These PICs can sometimes resist further oxidation and be emitted. Many PICs are hazardous themselves. Also, PICs may participate in other reactions to form dioxins and/or furans — toxic organic by-products that are often associated with incineration. While not ignoring the contribution of dioxins and furans from HWIs, it is noted that ordinary MSW and MW incinerators may emit more of these compounds than HWIs (U.S. EPA, 1994).

Simplified description of a hazardous waste incinerator

A schematic flow diagram of an HWI is presented in Figure 1.5, reference to which (as well as to Figure 1.1) may help the reader during the following description of operations. Wastes are brought to an HWI by truck or rail; then they are checked, sampled, and unloaded into the proper storage areas. Wastes are blended as appropriate for feeding into the incinerator. It is *very* desirable that the burning process be as smooth and continuous as possible and that upsets and sudden changes be avoided. The facility must have a good system of instruments, alarms, and controls to ensure optimum operations.

The wastes are burned using auxiliary fuel to achieve the desired high temperatures. The proper design of the combustor is critical to achieve the required good mixing of air and waste and the desired length of residence time of the gases. In addition, the systems must be operated and maintained by well-trained personnel. Typically, the equipment and personnel at an HWI are more sophisticated than at, say, an MSW incinerator.

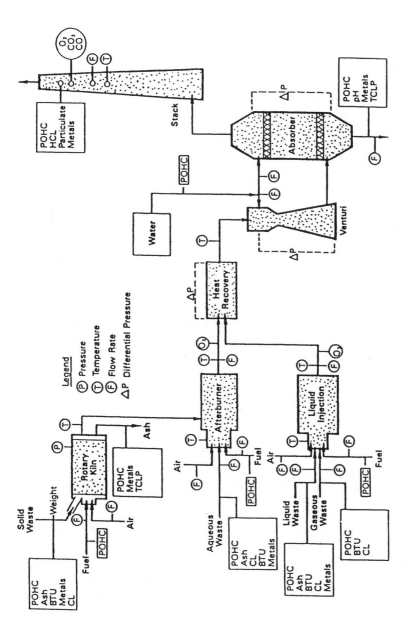

Figure 1.5   Potential sampling points for assessing incinerator performance. (From Oppelt, E. T. *J. Air Pollut. Control Assoc.*, 37(5), May 1987. With permission.)

The end products of the fuel and waste combustion are ash and gases. Some ash is removed from the combustor and disposed of properly, while the gases and the remaining ash (in the form of very small particles) flow to the APC equipment. The APC train is designed to remove almost all of the particles and noxious gases that exit from the incinerator. During this process, solid and/or liquid residues may be created that must be tested and then discharged appropriately. The cleaned exhaust gases flow out a tall stack into the atmosphere.

Throughout the process operation, a number of key variables must be monitored continuously (such as certain temperatures, flow rates, and concentrations) to ensure good efficient operations. Other variables are sampled on a regular basis to monitor performance. As with any industrial facility, the routine inspection and maintenance of the equipment, the training of operators, and the regular calibration and tuning of instruments are all very important to the continuing proper operation of an HWI. In addition, frequent inspections by government regulators and routine reporting by management of the HWI to government agencies appear to enhance the performance of these units.

Air pollution and pollution control equipment

The rate of air pollutant emissions from an HWI is dependent upon many factors. Among the most important are waste feed rate, waste composition, furnace design, combustor operating conditions, instrumentation and controls, maintenance practices, and APC equipment (type, size, design, operation, and maintenance). Many of these factors (and thus emissions rates) can vary significantly over the life of the facility. The literature on emissions from an HWI is predominantly focused on stack emissions from the incinerator. The exhaust stack is downstream of all the APC equipment that may be installed to capture and remove the pollutants that exit from the incinerator itself.

Because the pollutants are emitted continuously from the stack, and then are transported and dispersed in the atmosphere to locations that may be far away from the site, more attention (properly) is paid to them. However, fugitive emissions (those pollutants that escape through leaks from valves, pumps, storage tanks, unloading stations, or leaking trucks, for example) may also be important and are less well documented. A third category of emissions — accidental releases due to an unplanned emergency situation at the HWI facility — should really be considered as a very low-probability, high-consequences event. What is the risk of a large emissions event that might happen sometime during the life of the facility? The quantitative assessment of this situation is very difficult and essentially has been ignored in most regulatory reviews.

The emissions from an HWI can be classified broadly by type as (1) toxic metals, (2) hazardous organic compounds, and (3) other pollutants. The

Table 1.1   U.S. EPA-Proposed Hazardous
Waste Incineration Standards

Destruction and removal	99.9999% most dioxin-listed wastes
efficiency (DRE)	99.99% all other wastes
Particulate matter	0.08 gr/dscf @ 7% O_2
Carbon monoxide (tier I)	100 ppmv (d) @ 7% O_2
Hydrocarbons (tier II)	20 ppmv (d) @ 7% O_2
Continuous emissions monitoring	CO, O_2, HC

**Tier III reference air concentrations
(annual limits, μg/m³**

Hydrogen chloride	0.7	Free chlorine	0.4
Carcinogenic metals		**Noncarcinogenic metals**	
Arsenic	2.3×10^{-3}	Antimony	0.3
Beryllium	4.1×10^{-3}	Barium	50
Cadmium	5.5×10^{-3}	Lead	0.09
Chromium	8.3×10^{-4}	Mercury	0.3
		Silver	3
		Thallium	0.3

Source: From Donnelly, J. R. and Brown, W. A. *Hazardous Waste Incineration Air Toxics Emissions and Controls,* presented at the 86th Annual Meeting and Exhibition of the Air and Waste Management Association, Denver, 1993.

metals may be emitted as fine particulate matter or in some cases as vapors and may be in various forms or chemical composition (elements, oxides, chlorides, or mixtures as part of the ash). The organics may be traces of unburned waste (called POHCs), may be partially oxidized fragments from the original waste (called PICs), or may in fact have been formed de novo in or downstream of the incinerator (such as dioxins and furans). The category of "other" includes the criteria pollutants (CO, NO_x, SO_x, PM-10) and any other compounds (such as HCl, the emission of which, because of the prevalence of chlorine in hazardous wastes, is regulated) that do not fall into either of the first two categories.

Stack emissions standards continue to tighten as we learn more about the types and amounts of emissions, about how the waste characteristics and operating factors influence them, and as we develop better technology to prevent or control them. The recently proposed U.S. standards for HWIs are presented in Table 1.1. Table 1.2 presents some data from various HWIs, their emissions, their DREs, and the collection efficiencies of specific types of APC for various pollutants (Christiansen and Brown, 1992; Donnelly and Brown, 1993; Rigo, 1994; Walker and Cooper, 1992).

Air pollution control equipment

APC devices for HWIs fall into two broad categories: those that remove particles and those that remove gases. Particulate matter control devices

Table 1.2 Incinerator Performance and Stack Data (Data Reported as Averages for each Facility)

Facility type	O_2 (%)	CO (ppm)	TUHC[a] (ppm)	DRE (%)	Particulate[b] (mg/m³)	HCl control (%)
Commercial rotary kiln liquid incinerator	10.5	6.2	1.0	99.999	152	99.4
Commercial fixed-hearth, two-stage incinerator	11.4	6.9	1.0	99.994	400	98.3
On-site two-stage liquid incinerator	8.1	9.4	6.0	99.994	143	99.7
Commercial fixed-hearth, two-stage incinerator	11.0	327.7	18.7	99.997	60	—[c]
On-site liquid injection incinerator	13.2	11.9	1.0	99.999	186	—[c]
Commercial two-stage incinerator	10.2	1.1	1.3	99.998	902	—[c]
On-site rotary kiln incinerator	9.7	554.0	61.7	99.999	23	99.9
Commercial two-stage fixed-hearth incinerator	13.4	26.8	1.8	99.996	168	98.3
On-site rotary kiln	—[d]	794.5	NA[e]	99.998	184	99.7
On-site liquid injection incinerator	9.7	66.3	7.8	99.994	95	—[c]
On-site rotary kiln incinerator	10.7	5.8	NA	99.996	404	99.9
On-site rotary kiln incinerator	14.1	323.0	NA	99.996	NA	99.8
On-site liquid injection incinerator	12.4	31.9	1.9	99.999	163	98.6

On-site liquid injection incinerator	9.3	1.0	NA	99.996	40	—[c]
On-site fluidized bed incinerator	3.6	67.4	NA	99.996	259	—[c]
On-site fixed-hearth incinerator	12.9	ND[f]	NA	99.999	93	—[c]
On-site liquid injection incinerator	4.5	358.0	NA	99.995	99	—[c]
On-site liquid injection incinerator	3.6	28.4	NA	99.998	12	99.3
Commercial rotary kiln incinerator	9.4	8.0	0.5	99.999	172	99.9
On-site liquid injection incinerator	3.1	779.3	NA	99.999	88	99.6
On-site liquid furnace incinerator	6.4	56.3	NA	99.999	4	99.9
On-site fixed-hearth incinerator	13.5	5.0	NA	99.999	150	98.4

[a] Total unburned hydrocarbons.
[b] RCRA requires 180 mg/m^3.
[c] HCl emissions <4 lb/h.
[d] Reported only as a range (3.1–16.7%).
[e] NA — not available.
[f] Not detected.

Source: ASME. *Hazardous Waste Incineration,* American Society of Mechanical Engineers, New York. 1988.

include electrostatic precipitators, baghouses, and scrubbers. In the following few paragraphs we give a brief description of each device and then compare their efficiencies for removing particulate matter.

An electrostatic precipitator removes particulate matter from a gas stream by creating a high-voltage drop between electrodes. As the gas with particles passes between the electrodes, gas molecules are ionized, the resulting ions stick to the particles, and the particles acquire a charge. The charged particles are attracted to and collected on the oppositely charged plates while the cleaned gas flows through the device. During the operation of the device, the plates are rapped periodically to shake off the layer of dust that builds up. The dust is collected and disposed of properly depending on its hazardous characteristics.

A baghouse can be thought of as a giant multiple-bag vacuum cleaner. Gas containing the particulates is made to flow through cloth filter bags. The dust is filtered from the gas stream, while the gas passes through the cloth and is exhausted to the atmosphere. The bags are periodically cleaned (two methods are by shaking the bags or by blowing clean air backward through them) to knock the dust down to the bottom hoppers from where it can be removed to be either recycled or disposed.

Scrubbers operate on the principle of collision between particles and water droplets, collecting the particles in the larger, heavier water drops. The water falls through the upward-flowing gases, colliding with and removing particles, and accumulates in the bottom of the scrubber. Later, the "dirty" water can be treated to remove the solids.

Another major use of scrubbers besides removing particles is to absorb a pollutant gas from a mixture of gases. The rate and extent of absorption is commonly assisted by chemical reaction in the absorbing medium. A widespread example is the scrubbing of HCl from HWI combustion gases by an alkaline solution, which not only absorbs the HCl gas but also neutralizes the acid.

Dry scrubbing is a recent innovation in which powdered lime (CaO) is injected into the stream of exhaust gases and then is captured by a baghouse, forming a layer of lime dust on the bags. As the gases flow through the baghouse and the layer of lime dust, the acid gases (mainly HCl) are neutralized by the lime, forming water (which evaporates) and calcium chloride. Dry scrubbing can be very effective and is used in a number of MW incinerators and some HWIs.

Gases such as HCl can be scrubbed out of the exhaust gases with alkaline aqueous solutions or with alkaline powders, and this is currently being done at a number of HWIs throughout the world. However, a gas such as NO_2 or NO, which is relatively insoluble, is very difficult to remove once it has been formed. Therefore, the principal control mechanism for NO_x in the United States has been to minimize its formation by the proper design and operation of burners and furnaces. However, catalytic reduction is also used and is popular in Japan.

Finally, it must be mentioned that dioxin/furan formation has been shown to be closely related to the temperature–time profiles experienced by the gases in the APC equipment. A rapid quench to below about 300°F is a good method for minimizing formation of these hazardous compounds.

Emissions of metals

Keep in mind that an incinerator cannot destroy metals, only change their physical or chemical form. Much of the metal content of HW is from non-hazardous metals such as iron or aluminum, but a substantial fraction is often one or more of the following: lead, mercury, cadmium, chromium, arsenic, beryllium, nickel, vanadium, zinc, and others. Most metals accumulate in the bottom ash from an incinerator, but substantial fractions are found in the collected fly ash, and much smaller fractions escape as vapors or fumes.

The emissions of metals into the *atmosphere* are almost entirely via the route of stack emissions. The rate of emissions of each metal from the stack is highly dependent upon the rate of waste feed, the individual content of each metal in the waste, the partitioning of the metal among bottom ash, fly ash, and vapors, and the capture efficiencies (for each metal) of all APC equipment in the pollution control train (Abbruzzese, 1992; Chandler et al., 1994; Guest, 1992; Morency et al., 1994; Rigo, 1994; Roth, 1992; Srinivasachar et al., 1992).

With the notable exception of mercury, most of the emissions of metals from the stack are in the form of small particulate matter (PM). Hence better control of PM means better control of metals emission rates. Therefore, any discussion of emissions of metals must include a discussion of the collection efficiencies of different APC equipment, the two main types being dry scrubbers/baghouses or wet scrubbers. Of course, once the particulate matter has been collected, it must be disposed of in such a manner as to prevent the release of toxic metals into the environment. Typically, the ash is stabilized and placed into a secure landfill, so the risk of re-release is minimal. It is noted that incineration of mercury-containing wastes is now prohibited by law in Florida.

Recent literature suggests that better metals control is achieved with a dry scrubber/baghouse system, which includes lime injection and carbon or coke injection, than with wet scrubber systems (Blumbach and Nethe, 1992; Brinckman; 1992; Carroll et al., 1994; Christiansen and Brown, 1992; Feldt, 1991; Hartenstein, 1992; Kreindl and Brinckman, 1991; Livengood et al., 1994; Murowchick and Rice, 1992). The type of fabric used is very important because it must be able to capture the extremely fine PM on which much of the mass of metals is adsorbed. Lime injection helps neutralize acid gases and provides better PM removal by building up a denser filtering medium supported on the fabric. Mercury can exist as a vapor as low as 80°F. Addition of carbon or coke injection helps adsorb mercury very effectively as demonstrated in Finland, Denmark, Sweden, and Germany (Christiansen and Brown, 1992). As discussed in the sections on organics control, the mixture of lime and carbon or coke also helps control chlorinated organics, especially dioxins and furans (as long as the baghouse is not operated too hot). In a hot baghouse, dioxins and furans that have been collected on the filter cake may revolatilize and be emitted to the air (Williamson, 1994).

Hazardous organic compound emissions

There are hundreds of organic chemicals that have been designated as hazardous. Chlorinated organic compounds are found frequently in hazardous wastes and many are extremely difficult to destroy in an incinerator. Perhaps the most infamous are those known as dioxins and furans. Dioxins belong to the family of compounds known chemically as polychlorinated dibenzo-*p*-dioxins and exist as a series of 75 related congeners. Furans are in a family of 135 compounds known as polychlorinated dibenzofurans. The general chemical structures of dioxins and furans are shown in Figure 1.6. Dioxins and furans are found in small quantities in the exhaust gases of almost all incinerators regardless of waste type. They are extremely difficult to detect and quantify, and the tests are very costly.

It is now widely accepted that the main source of dioxins and furans in incinerator exhaust gases is the de novo formation from simple precursors such as chlorine and unrelated trace amounts of organics that escape destruction in the combustor (Williamson, 1994). Copper compounds, iron compounds, and even fly ash have been shown to be effective catalysts for the formation reactions at 500 to 650°F. "Oxygen and HCl concentrations are

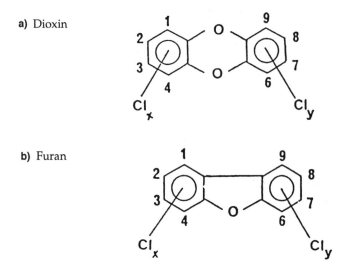

a) Dioxin

b) Furan

Figure 1.6 Schematic structural diagrams of dioxins and furans.

important, but, under normal hazardous waste incinerator operating conditions, sufficient amounts will be present to ensure that their concentration is not limiting and the temperature becomes the critical factor" (Williamson, 1994). Temperatures of 1,800°F or more in the combustor along with residence times of more than 1.5 seconds are well above the critical level for dioxin and furan destruction, but it is important to destroy as much of the precursors as possible.

Formation of these compounds can be minimized with good combustion of precursors in the incinerator, but proper operating conditions of the downstream equipment is critical to prevent formation of the tiny amounts that would cause problems in meeting the proposed standards. Thus, cooling of the gases in the downstream equipment is extremely important. A rapid (1 to 2 seconds) adiabatic quench is much preferred over waste heat recovery in a boiler that lets the gases linger at formation temperatures for 5 to 10 seconds. If a waste heat boiler is employed, it should be cleaned (by soot-blowing) frequently. Once formed in the downstream equipment, dioxins/furans can be captured by either wet scrubbing or spray drying/fabric filtration. Current thinking is that, when supplemented with 50 to 200% excess lime injection and operated below about 275°F with an air-to-cloth ratio of 5:1 or less, a fabric filtration system is superior (Williamson, 1994).

Evidence in the literature suggests that dioxins and furans are more of a problem in MSW and MW incinerators than in hazardous waste incinerators (Oppelt, 1987; Williamson, 1994). By averaging data reported by Walker and Cooper (1992), a gross estimate of the dioxin concentrations in stack gases from existing MW incinerators can be made. That estimate is about 550 ng/m^3. In more modern MSW and MW incineration facilities, it has been shown that lime injection (supplemented with carbon injection) in concert with a fabric filtration system resulted in very low emissions of dioxins and furans (Christiansen and Brown, 1992; Licata et al., 1994).

Although the main focus of the draft CETRED (U.S. EPA, 1994) document was on PM and dioxin/furan emissions from HWIs, the CETRED presented some limited data on MSW incinerators that showed dioxins from modern MSW incinerators equipped with ESPs as their main APC device emitted at concentrations about 400 ng/dry standard cubic meter (dscm), whereas, if they were equipped with a lime injection spray dryer with fabric filtration, the concentrations were only about 10 ng/dscm. In that document, dioxins from cement kilns and HWIs equipped with ESPs or fabric filtration systems averaged about 400 to 500 ng/dscm, but the levels from HWIs equipped with rapid quench and wet scrubbers averaged about 4.5 ng/dscm, illustrating the large difference that can occur with different thermal treatment of the postflame combustion gases. It is noted that many MSW and MW incinerators are equipped with waste heat boilers (which have been implicated as prime spots for the formation of dioxins). The CETRED also suggests that an appropriate dioxin level for the best controlled HWIs should be about 5 to 10 ng/dscm based on observed performance of the best several facilities. EPA emission guidelines for dioxins/furans from large and very large existing MSW incinerators are 60 to 125 ng/dscm (Fricilli, 1991).

Much data have been published on emissions of organics from incinerators (for example, see ASME, 1988; Christiansen and Brown, 1992; Donnelly and Brown, 1993; Feldt, 1991; Licata et al., 1994). In several of these studies, it was reported that, when properly designed and operated APC equipment was taken into account, the control efficiencies for POHCs and other organics were 99.99%, while the control efficiencies for metals were above 99%. In a 1988 review of existing plant test data (ASME, 1988), a correlation was found between emissions of organics and their waste feed concentrations but further showed that most of the emissions fell within a two-orders-of-magnitude range regardless of feed concentration (see Figure 1.7).

Other pollutants

There are several other pollutants that are emitted and frequently regulated for HWIs. Four that will be mentioned are PM, HCl, CO, and NO_x. PM is a

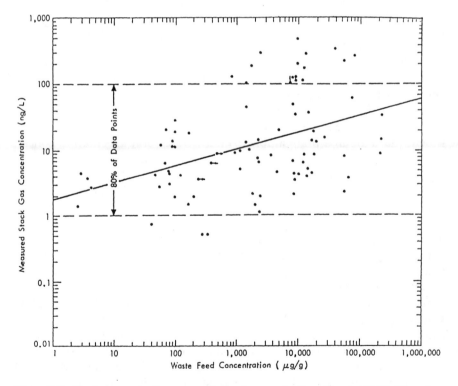

Figure 1.7 Stack concentration vs. waste feed concentration. (From ASME. *Hazardous Waste Incineration*, American Society of Mechanical Engineers, New York, 1988. With permission.)

class of pollution rather than a particular substance. It represents uncollected fine particles of ash that usually contains mostly silicon and aluminum oxides (sand) and other common metal oxides. However, it may also contain some toxic metals, either embedded or adsorbed.

HCl is a thermodynamically stable end product of combustion of any wastes that contain chlorine. It is easily scrubbed with a wet scrubber and can also be removed with great effectiveness with a dry scrubber/baghouse. It must be removed because, if emitted in large quantities, it could cause severe local corrosion and respiratory problems.

CO is a criteria pollutant and a ubiquitous end product of not-quite-complete combustion. It is a stable intermediate compound in the chain of reactions that we know as incineration, which requires high temperatures and long residence times to completely destroy. In a well-operated combustion system, the exit concentrations of CO will be low. It is more important as an indicator of poor combustion and is often monitored continuously to help incinerator operators spot deviations from good combustion.

NO_x is a criteria pollutant and is an unavoidable end product of any combustion operation. It is formed by the high-temperature reactions between the nitrogen and the oxygen in the air used for combustion. It can be minimized with modern low-NO_x burners but once formed is difficult to remove without great expense. It is more usually thought of as a problem of fossil fuel power plants and very large industrial furnaces, but in certain regions a large HWI might add enough NO_x to the regional emissions that it could be of concern to air quality managers.

Summary

In summary, incineration is the high-temperature oxidation of wastes in facilities specifically designed to handle them. There are a number of different types of HWIs located in various states in the U.S., each of which is equipped with different types of sophisticated APC equipment. An HWI typically is a highly regulated, complex facility with a number of interacting subprocesses. The types of pollutants emitted from HWIs include toxic metals, hazardous organic compounds, acid gases, and others. Emission rates depend on the types and amounts of wastes being burned, the style, operation and maintenance of the incinerator, and the design and operation of the APC equipment. Providing a detailed description of the process and reliable estimates of the emission rates is the first step in the regulatory analysis of a proposed facility.

References

M. J. Abbruzzese. The Fate of Arsenic in Cement Kilns, presented at the 85th Annual Meeting and Exhibition of the Air and Waste Management Association, Kansas City, 1992.

ASME. *Hazardous Waste Incineration*, sponsored by the ASME Research Committee on Industrial and Municipal Wastes, cosponsored by the Air Pollution Control Association, the American Institute of Chemical Engineers, and the U.S. Environmental Protection Agency, The American Society of Mechanical Engineers, New York, 1988.

J. Blumbach and L. P. Nethe. SORBALIT — A New Economic Approach Reducing Mercury and Dioxin Emissions, presented at the 85th Annual Meeting and Exhibition of the Air and Waste Management Association, Kansas City, 1992.

G. A. Brinckman. Microporous Gore-Tex Membrane Filter Media Can Help Control Submicron Particulate Matter, Heavy Metal, and Dioxin/Furan Emissions, presented at the 85th Annual Meeting and Exhibition of the Air and Waste Management Association, Kansas City, 1992.

C. R. Brunner. *Incineration Systems — Selection and Design*, Incinerator Consultants Incorporated, Reston, VA, 1984.

A. Carpi, D. W. Ditz, L. H. Weinstein, J. M. Waldman, and A. Greenberg. Biological Monitoring Around a Municipal Solid Waste Incinerator, presented at the 85th Annual Meeting and Exhibition of the Air and Waste Management Association, Kansas City, 1992.

G. J. Carroll, R. C. Thurnau, and D. J. Fournier, Jr. Control of Mercury Emissions from Hazardous Waste Incineration, presented at the 87th Annual Meeting and Exhibition of the Air and Waste Management Association, Cincinnati, 1994.

A. J. Chandler, H. G. Rigo, and S. E. Sawell. Effect of Lead Acid Battery and Cadmium Spiking on Incinerator Emissions, presented at the 87th Annual Meeting and Exhibition of the Air and Waste Management Association, Cincinnati, 1994.

O. B. Christiansen and B. Brown. Control of Heavy Metals and Dioxins from Hazardous Waste Incinerators by Spray Dryer Absorption Systems and Activated Carbon Injection, presented at the 85th Annual Meeting and Exhibition of the Air and Waste Management Association, Kansas City, 1992.

C. R. Dempsey. A Comparison: Organic Emissions from Hazardous Waste Incinerators Versus the 1990 Toxics Release Inventory Air Releases, presented at the 85th Annual Meeting and Exhibition of the Air and Waste Management Association, Kansas City, 1992.

J. R. Donnelly and W. A. Brown. Hazardous Waste Incineration Air Toxics Emissions and Controls, presented at the 86th Annual Meeting and Exhibition of the Air and Waste Management Association, Denver, 1993.

K. G. Feldt. Electrostatic Precipitator vs. Fabric Filter for Dioxin/Furan Emission Control on Hazardous Waste Incinerator at Sakab, Sweden, presented at the 84th Annual Meeting and Exhibition of the Air and Waste Management Association, Vancouver, BC, 1991.

P. W. Fricilli. Impact of EPA's Air Pollution Emission Standards and Guidelines on Municipal Waste Combustion Units, presented at the 84th Annual Meeting and Exhibition of the Air and Waste Management Association, Vancouver, BC, 1991.

T. L. Guest. Mercury Retention in Fly Ash Using Activated Carbon Absorption, presented at the 85th Annual Meeting and Exhibition of the Air and Waste Management Association, Kansas City, 1992.

Guidance on Setting Permit Conditions and Reporting Trial Burn Results, a handbook published by the Office of Solid Waste and Emergency Response, U.S. Environmental Protection Agency, Washington, DC, 1989.

H. U. Hartenstein. A Fixed Bed Activated Coke/Carbon Filter as a Final Gas Cleaning Stage Retrofitted for a Hazardous Waste Incineration Plant — The First 6 Months of Operating Experience, presented at the 85th Annual Meeting and Exhibition of the Air and Waste Management Association, Kansas City, 1992.

F. Hasselriis. Optimization of Combustion Conditions to Minimize Dioxin Emissions, presented at the ISWA-WHO-DAKOFA specialized seminar, Copenhagen, 1987.

L. C. Holcomb and J. F. Pedelty. Exposure to Emissions from Cement Kilns Burning Waste Derived Fuels Compared to Health Based Standards, presented at the 85th Annual Meeting and Exhibition of the Air and Waste Management Association, Kansas City, 1992.

T. Kreindl and G. A. Brinckman. Case Studies of Particulate Matter, Heavy Metal, and Total Equivalent Dioxin/Furan Collection Using PTFE Membrane Filter Media, presented at the 84th Annual Meeting and Exhibition of the Air and Waste Management Association, Vancouver, BC, 1991.

P. J. Kroll and R. C. Chang. Unsteady-State Model of Incinerator ESV Emissions, presented at the 85th Annual Meeting and Exhibition of the Air and Waste Management Association, Kansas City, 1992.

K. Lee. Research Areas for Improved Incineration System Performance, presented at the Engineering Foundation Conference on Hazardous Waste Management Technologies, South Charleston, WV, 1988.

Y. J. Lee. *Flue Gas Cleanup with Hydroxyl Radical Reactions*, Pittsburgh Energy Technology Center, Pittsburgh, 1990.

A. Licata, M. Babu, and L. P. Nethe. An Economic Alternative to Controlling Acid Gases, Mercury and Dioxin from MWCs, presented at the 87th Annual Meeting and Exhibition of the Air and Waste Management Association, Cincinnati, 1994.

C. D. Livengood, H. S. Huang, and J. M. Wu. Experimental Evaluation of Sorbents for the Capture of Mercury in Flue Gases, presented at the 87th Annual Meeting and Exhibition of the Air and Waste Management Association, Cincinnati, 1994.

C. Lodi and L. E. Clark. MSW Incinerator Air Pollution Control Device Retrofit Experience at the Pittsfield Resource Recovery Facility, presented at the 86th Annual Meeting and Exhibition of the Air and Waste Management Association, Denver, 1993.

Medical Waste Incineration and Pollution Prevention, Green, A. E. S., ed., Van Nostrand Reinhold, New York, 1992.

J. R. Morency, S. Srinivasachar, P. M. Lemieux, J. V. Ryan, and M. C. Meckes. Control of Trace Metal Species in Combustion Systems, presented at the 87th Annual Meeting and Exhibition of the Air and Waste Management Association, Cincinnati, 1994.

P. S. Murowchick and B. L. Rice. Metals Control Efficiency Test at a Dry Scrubber and Fabric Filter Equipped Hazardous Waste Incinerator, presented at the 85th Annual Meeting and Exhibition of the Air and Waste Management Association, Kansas City, 1992.

E. T. Oppelt. Incineration of hazardous waste — a critical review, *J. Air Pollut. Control Assoc.*, 37(No. 5), May 1987.

PEI Associates, Inc. and JACA Corporation. *Permit Writer's Guide To Test Burn Data — Hazardous Waste Incineration*, Office of Research and Development, U.S. Environmental Protection Agency, Cincinnati, 1986.

A. A. Pope and C. R. Blackley. EPA's FIRE (Factor Information Retrieval) — The System and its Contents, presented at the 87th Annual Meeting and Exhibition of the Air and Waste Management Association, Cincinnati, 1994.

H. G. Rigo. Effect of Metals in Waste Components on Incinerator Emissions, presented at the 87th Annual Meeting and Exhibition of the Air and Waste Management Association, Cincinnati, 1994.

H. G. Rigo, F. A. Ferraro, and M. Wilson. Effect of Nitrogen in Waste Components on Incinerator Emissions, presented at the 87th Annual Meeting and Exhibition of the Air and Waste Management Association, Cincinnati, 1994.

R. G. Rizeq, W. Clark, and W. R. Seeker. Engineering Analysis of Metals Emissions from Hazardous Waste Incinerators, presented at the 85th Annual Meeting and Exhibition of the Air and Waste Management Association, Kansas City, 1992.

A. J. Roth. BIF Rule Metals Stack Emissions, Removal Efficiencies, and Fate in a Conditioned Preheater-type Cement Kiln, presented at the 85th Annual Meeting and Exhibition of the Air and Waste Management Association, Kansas City, 1992.

R. A. Rothstein, P. Billig, and G. W. Siple. Implications of EPA's Draft Combustion Strategy on Permitting and Retrofit of Hazardous Waste Incinerators and BIFs, presented at the 87th Annual Meeting and Exhibition of the Air and Waste Management Association, Cincinnati, 1994.

S. Srinivasachar, J. R. Morency, and B. E. Wyslouzil. Heavy Metal Transformations and Capture During Incineration, presented at the 85th Meeting and Exhibition of the Air and Waste Management Association, Kansas City, 1992.

L. Theodore and J. Reynolds. *Introduction to Hazardous Waste Incineration*, John Wiley & Sons, New York, 1987.

A. R. Trenholm, D. W. Kapella, and G. D. Hinshaw. Organic Products of Incomplete Combustion from Hazardous Waste Combustion, presented at the 85th Annual Meeting and Exhibition of the Air and Waste Management Association, Kansas City, 1992.

W. L. Troxler, G. D. Smith, C. B. Henke, J. D. Lauber, J. J. Czapla, and J. J. Segada. Ionizing Wet Scrubber Pilot Test Results for Controlling Emissions from a Liquid Hazardous Waste Incinerator, presented at the 85th Annual Meeting and Exhibition of the Air and Waste Management Association, Kansas City, 1992.

T. Tsui and D. Fickling. Performance Testing at Hennepin County Resource Recovery Facility, presented at the 85th Annual Meeting and Exhibition of the Air and Waste Management Association, Kansas City, 1992.

K. Tsuji, I. Shiraishi, and D. G. Olson. Simultaneous Reduction of Sox, Nox, Mercury and VOCs from Combustion and Incineration Sources Using the GE-Mitsui-BF Activated Coke Process, presented at the 87th Annual Meeting and Exhibition of the Air and Waste Management Association, Cincinnati, 1994.

U.S. Environmental Protection Agency, Office of Solid Waste and Emergency Response. Combustion Emissions Technical Resource Document (CETRED) — DRAFT, EPA530-R-94-014, May 1994.

B. L. Walker and C. D. Cooper. Air pollution emission factors for medical waste incinerators, *J. Air Waste Manage. Assoc.*, 42(6), June 1992.

P. Williamson. Production and Control of Polychlorinated Dibenzo-p-Dioxins and Dibenzofurans in Incineration Systems: A Review, presented at the 87th Annual Meeting and Exhibition of the Air and Waste Management Association, Cincinnati, 1994.

W. H. Ziegler and G. Miller. Physical and Chemical Processes Relevant to the Formation of Chlorinated Dioxins in Municipal Solid Waste Incinerators, presented at the 85th Annual Meeting and Exhibition of the Air and Waste Management Association, Kansas City, 1992.

chapter two

Characterization of potential emissions from hazardous waste incinerators and related facilities

Christopher M. Teaf, Isabel K. Stabile, and Philip A. Moffat

Contents

1-56670-250-X/99/$0.00+$.50
© 1999 by CRC Press LLC

Introduction and overview

Factors that influence the composition of incinerator emissions

Under ideal conditions, the complete combustion of organic materials, including wastes, produces primarily carbon dioxide, water, and inert ash. In practice, however, variable waste compositions, equipment performance, and fluctuating combustion parameters may lead to the formation of a multitude of ancillary products. These products, in addition to any uncombusted residual components that were present in the initial waste stream, represent the aggregate emissions from the incinerator.

The main emission pathways from an incineration system include air emissions, wastewater discharges from scrubber systems, and ash residues. The composition of these emissions is heavily influenced by the composition of the waste stream as well as combustion performance parameters including the magnitude and consistency of the temperature in the combustion chamber(s), the residence time (or "dwell time") of the waste in the combustion chamber(s), and the degree of mixing of the waste that occurs during the combustion process. Therefore, available data concerning the characterization of hazardous waste incinerator emissions are highly site and time specific. However, a number of consistent features apply to incineration facilities. This section describes and summarizes relevant information concerning incinerator emissions, focusing principally on airborne emissions, since these typically attract greatest interest in terms of potential human or environmental exposures in the vicinity of hazardous waste incinerators and related facilities.

The final two tables in this chapter (Table 2.8 and Table 2.9) summarize over 200 organic and inorganic analytes that have been identified in studies of hazardous waste incinerator emissions or that have been projected for such emission sources on the basis of structural similarity to other analytes that were detected (e.g., isomers of the chlorinated dioxins and furans). Based on the available information, these analytes provide reasonable representations for the evaluation of potential risks that may be associated with such facilities.

Background and historical data sources

An accurate assessment of hazardous waste incineration emissions requires knowledge of the feed materials, engineering knowledge of the facility, careful sampling, and extensive analysis. Prior to the 1980s, limited data were available concerning pollutant emissions from what were at that time termed "hazardous waste thermal destruction devices" (i.e., incinerators). During and subsequent to the early 1980s, however, several groups, including the U.S. Environmental Protection Agency (U.S. EPA), conducted a number of facility-specific site evaluations (termed "test burns") that were designed to estimate the magnitude and characteristics of potential facility releases and, thus, to estimate the potential environmental impacts of these incinerator operations. These test burns were designed to provide information on the ability of such facilities to destroy organic substances of a variety of types and the ability to control the level of facility-specific emissions. The following reports contain detailed information concerning these test facilities, the test procedures employed, and the results of the sampling and analyses for these facilities.

- *Performance Evaluation of Full-Scale Hazardous Waste Incinerators*[1] presents the results from a series of extensive tests at eight incinerators, prepared in response to the U.S. EPA's need to conduct a regulatory impact analysis (RIA). Different organic waste streams (non-PCB containing) were combusted in each of the eight trial burns. Products of incomplete combustion (PICs) were defined for purposes of this report as any RCRA Appendix VIII analyte emitted in the stack gas but that was not present in the initial waste stream in concentrations greater than 100 mg/kg.
- *Total Mass Emissions from a Hazardous Waste Incinerator*[2] presents results from emissions testing at the Dow Chemical plant at Plaquemine, Louisiana. Measurements were made for both RCRA Appendix VIII and non-Appendix VIII analytes in all effluent media (i.e., stack gas, scrubber water, and ash) using a wide array of sampling and analytical techniques.
- *Characterization of Hazardous Waste Incineration Residuals*[3] provides data on solid and liquid emissions from ten incinerator facilities. All inlet and outlet liquid and solid waste streams were sampled and analyzed for organic and inorganic pollutants.
- *Incineration of Hazardous Waste: A Critical Review Update*[4] provides a detailed review of the state of the science regarding hazardous waste incinerator facilities.
- *Risk Assessment for the Waste Technologies Industries (WTI) Hazardous Waste Incineration Facility (East Liverpool, Ohio) volume III: External Review Draft*[5] provides a detailed characterization of the nature and magnitude of all relevant emissions from this contemporary facility.

Many additional relevant studies were used to augment and to expand the list of analytes that are presented in Table 2.8 and Table 2.9. These studies and analytes are discussed in subsequent parts of this chapter.

Organic analytes from waste feed detected in stack emissions and residuals

Introduction

Organic compounds that are emitted from hazardous waste incinerators and related facilities generally are classified in one of two categories:

- Principal organic hazardous constituents (POHCs), representing components of the input waste stream
- Products of incomplete combustion (PICs), representing compounds that are formed during the combustion process and that may or may not also be present in the original waste stream

POHCs are defined in the Resource Conservation and Recovery Act (RCRA) as those Appendix VIII constituents that are detected in emissions *and* that are identified in the incoming waste stream at concentrations greater than 100 mg/kg (100 ppm).

A consensus regulatory definition of PICs has not yet been achieved due to the broad range of chemicals included and the number of factors which affect their generation. In the U.S. EPA's test program reported by Travis and Cook,[6] however, compounds were considered to be PICs if they were regulated organic compounds (i.e., listed in Appendix VIII of CFR 40 Part 261 under RCRA) *and* were detected in stack emissions *but* were not detected at a concentration greater than 100 mg/kg in the incoming waste stream. By that convention then, an analyte that is not included on the RCRA Appendix VIII list cannot be classified as a PIC or as a POHC. This implies that an unspecified number of stack emission analytes (non-POHCs and non-PICs) may remain unclassified from a regulatory perspective.[7] Nevertheless, in the selection of analytes for inclusion in Table 2.8 and Table 2.9, all reported detections were considered, regardless of regulatory status. Such an approach has become more common than a constrained regulatory definition.

In terms of helping to define emissions, the "incinerability," or destruction characteristics, of several organic analytes was investigated and ranked by Mournighan et al.[8] with the objective of determining potentially appropriate surrogates for incinerator performance evaluations. For that purpose, toluene was judged to be more stable than tetrachloroethene, chlorobenzene, or pentachlorobenzene. The authors concluded that sulfur hexafluoride, which previously had been proposed for use as an indicator compound (or "surrogate analyte") during emissions evaluations,[9] was not well suited to this purpose.

Organic constituents in stack gas emissions

One primary measure of incinerator performance is the characteristic known as the "destruction and removal efficiency" (DRE) for POHCs. The explicit RCRA performance standard requires the achievement of a 99.99% DRE ("four nines") for non-PCB and nondioxin wastes. PCB-containing material and dioxin-containing wastes require incineration facilities that are capable of achieving 99.9999% DRE ("six nines"). Data from U.S. EPA studies demonstrated that many of the incinerators that were tested by the agency can achieve the 99.99% level of performance for analyte POHCs that are present in concentrations greater than 1,200 ppm in the waste feed. However, organic analytes that comprise smaller proportions (e.g., between 200 and 1,200 ppm in the waste feed) frequently may not be destroyed to a 99.99% DRE, and compounds that are present at less than 200 ppm in the waste feed typically do not achieve the RCRA DRE limit.[4] Table 2.7 presents organic emission results from this study. Analysis of pooled data by Oppelt[10] resulted in the conclusion that no single or absolute combination of combustion temperature, residence time, or carbon monoxide emission concentration was correlated consistently with achievement of 99.99% DRE for incineration facilities.

The observation has been made that facilities that do not achieve a 99.99% DRE rate principally exhibit two conditions. The first condition is low concentration of the POHC in the waste feed (i.e., <100 to 200 ppm). The second condition involves compounds that commonly are identified as PICs (especially chloroform, methylene chloride, benzene, and naphthalene), but that *also* may be present in the incoming waste stream. The formation and persistence of these particular PIC compounds during the incineration of chlorinated organics increase their concentration in the stack gas, thereby functionally resulting in a lower calculated DRE for the incinerator.[1] It generally is not possible to distinguish between an analyte that was present in the incoming waste stream vs. that same analyte that may be formed as a PIC.

Halogenated analytes

Halogenated organics in the incoming waste stream correlate closely with the quantity and concentration of halogenated analytes in stack gas emissions. Halogenated hydrocarbons, including simple alkyl and aromatic (aryl) halides, alcohols, and ethers (Figure 2.1) represent the major categories of compounds associated with stack emissions when halogenated materials are incinerated. Many potential sources may account for the presence of halogenated material in the incinerator waste feed. Several of these sources include PCB-contaminated media, refuse from wood preservative processing, pesticides, herbicides, and residuals from commercial solvent production and usage.[6,11-13]

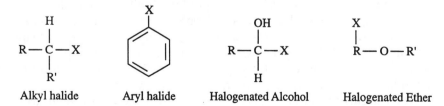

Alkyl halide   Aryl halide   Halogenated Alcohol   Halogenated Ether

Figure 2.1 Example structures of halogenated hydrocarbons associated with emissions from hazardous waste incinerators and related facilities. (X = halide group such as chlorine, fluorine, and bromine; R = other substituent groups.)

Halogenated compounds containing chlorine (as opposed to bromine, fluorine, or iodine) are frequently the focus of concern when evaluating the human health risks associated with incinerator emissions. In the Dempsey and Oppelt study,[4] some of the principal chlorinated analytes that were identified as potential fugitive emissions included chloroform, methylene chloride, trichloroethylene, and chlorobenzene. Each of these is volatile and is classified as a priority pollutant by the EPA. In fact, of the 37 reasonable worst-case emissions that were specified in that 1993 study, 25 relate to chlorinated analytes. In light of these and related findings as well as the associated putative health risks, many environmental groups are calling for a reduction in the amount of chlorinated material incinerated at any one time.

Halogenated emissions, at sufficient exposure levels, may be associated with a wide range of human health effects, including liver dysfunction, immune system alterations, cancer, neurotoxicological changes, and developmental defects.[4,14-16] A number of these analytes are environmentally persistent and subsequently may be detected in air, soils, water, or sediments after release. Therefore, knowledge of the waste feed composition, coupled with a thorough analysis and identification of the emitted halogenated analytes, is necessary for a proper human health risk characterization. Hence, every risk characterization is inherently site-specific. A list of many halogenated organic analytes reported for hazardous waste incinerator emissions is presented in Table 2.8.

Nonhalogenated analytes

A variety of nonhalogenated analyte types require consideration when characterizing the human health risks that may be associated with emissions from hazardous waste incinerators and related facilities. Some of these analyte classes include alcohols (e.g., phenol), ketones, nitrogen and sulfur heterocyclic amines, and polycyclic aromatic hydrocarbons (PAHs). When listing potentially significant analyte emissions from hazardous waste incinerators, Dempsey and Oppelt[4] projected the presence of several volatile and semivolatile compounds. These included benzene, toluene, naphthalene, methyl ethyl ketone, and several phthalate compounds. These compounds find their way into hazardous waste incinerator feeds as residuals from numerous industrial processes, including but not limited to manufacture of the following commercial products: solvents, plastics, synthetic rubbers, fuels, paints, dyes, adhesives, resins, cleaners, hydraulic fuels, liquid soaps, and varnishes.[17] Of the nonhalogenated analytes, the PAHs typically are among the groups that attract the greatest interest when considering the human health risks associated with incinerator stack emissions.

At sufficient concentrations, PAHs may be associated with a wide variety of potential human health effects, including myelotoxic effects (destruction of bone marrow), cancer, CNS depression, liver dysfunction, reproductive behavioral anomalies, and even death. Benzene, toluene, and the xylenes (all volatile and structurally among the simplest of the aromatic hydrocarbons) have some of the most acutely toxic effects. Inhalation is the primary means of exposure to these analytes. Generally, the larger PAHs are less volatile and acutely toxic. The larger PAH compounds (4 to 5 rings), for example benzo-a-pyrene and dibenzo(a)anthracene, demonstrate greater environmental persistence, thus having the potential for chronic exposures and associated toxicological effects. Some of the PAH compounds in this size class are considered potential human carcinogens. Since these are nonvolatile and insoluble in water, the primary routes of exposure include not only particulate inhalation but also dermal absorption and ingestion.[17]

Table 2.8 presents many of the reported nonhalogenated organic analytes from hazardous waste incinerator emissions.

Organic constituents in incinerator ash residuals

In the study of Trenholm et al.,[1] incinerator ash from several hazardous waste incinerators was analyzed for organic constituents. Only two facilities generated ash that exhibited concentrations of organic compounds greater than 35 μg/g (equivalent to 35 mg/kg). When organic compounds were detected, they frequently included toluene, phenol, or naphthalene at concentrations less than 10 mg/kg. The results of a separate 10-incinerator test program[3] generally were similar to those of Trenholm et al.[1] In the latter study, more organic compounds were detected across all facilities (19 volatile and 24 semivolatile). The concentrations in ash typically were equal to or less than 30 mg/kg. Table 2.1 lists the volatile and semivolatile organic compounds that were detected in ash samples from the 10 incinerator sites. The U.S. EPA risk assessment[5] for the WTI incineration facility in Ohio found no detectable concentrations of organics in the facility's ash residuals. However, only fly ash residuals were analyzed.

Organic constituents in wastewater discharges

Other residual materials, such as ash and process water from incinerator facilities, have not received the same level of scrutiny as have air emissions. Most of the aqueous emissions from hazardous waste incinerator facilities are represented by air pollution control device (APCD) effluents such as scrubber water. As with ash residues discussed previously, the components of interest are organic compounds and metals. The results of the study by Trenholm et al.[1] indicated that similar organic analytes were detected as in the case of ash. When detected, these organic analytes typically were represented by the more water-soluble analytes, such as toluene, phenol, or naphthalene, at concentrations less than 20 μg/L. Van Buren et al.[3] reported a greater number of compounds in scrubber waters across the 10 facilities in their study, in comparison to the study by Trenholm et al.,[1] and reported generally higher concentrations. Semivolatiles ranged from below detection limits to 100 μg/L, while volatile compounds ranged from 0 to 32 mg/L).

Table 2.1   Volatile and Semivolatile Organics Detected
in Incinerator Ash from 10 Facilities

Volatile organics	Semivolatile organics
Acetone	Acenaphthene
Benzene	Aniline
Bromomethane	Anthracene
2-Butanone	Benzo(b)fluoranthene
Carbon disulfide	Benzoic acid
Chlorobenzene	Benzyl butyl phthalate
Chloromethane	Bis(2-ethylhexyl)phthalate
Chloroform	Chrysene
1,2-Dichloroethane	Diethyl phthalate
trans-1,2-Dichloroethene	Dimethyl phthalate
Ethylbenzene	Di-n-butyl phthalate
Methylene chloride	Fluoranthene
4-Methyl-2-pentanone	Isopherone
Styrene	2-Methylnaphthalene
Tetrachloroethene	2-Methylphenol
Toluene	Naphthalene
1,1,1-Trichloroethane	2-Nitrophenol
Trichloroethene	4-Nitrophenol
Xylenes	N-Nitrosodiphenylamine
	Phenanthrene
	Phenol
	Pyrene
	1,2,4-Trichlorobenzene

Source: Van Buren, D. et al., *Characterization of Hazardous Waste Incinerator Residuals,* Acurex Corporation, EPA-600.2-87/017, NTIS PB87-168159, 1987.

Organic products of incomplete combustion and other process reactions

Introduction

Analysis of incinerator stack gas samples often gives clear evidence of the emission of organics that originally were not detected in the waste stream. These detected substances may include PICs from the hazardous waste incineration process. This concept has been widely accepted for some time.[4,6,18-21]

Strictly speaking, PICs represent any organic analytes that are present in the emissions from the incineration process but that were not present or detectable in the waste fuel or the air which is fed to the incinerator. In general regulatory practice, compounds are considered to be PICs if they represent RCRA Appendix VIII organic compounds that are detected in stack emissions but that are not detected in the incoming waste feed at concentrations that exceed 100 mg/kg. This 100 mg/kg regulatory value is distinct from the empirical 200 mg/kg value published by Dempsey and Oppelt.[4]

Incineration facilities that exhibit a high degree of DRE for POHCs typically also have low PIC emissions.[10]

Three potential explanations[1,6] for the presence of PICs (however they may be defined) in emissions are as follows:

- The compounds were actually present at low concentrations (<100 mg/kg) in the waste feed and subsequently were destroyed by the facility at a relatively low DRE.
- The compounds were introduced to the system from some source other than the waste, such as ambient air or scrubber water.
- The compounds were formed as products of the combustion reactions and represent the recombination of molecular fragments that were formed during the thermal degradation process.

From an exposure point of view and a potential risk perspective, it is not important whether the PICs detected in stack gas emissions are compounds formed during the incineration process or are compounds that originally were present in the waste stream, the scrubber water, or the fuel. Whichever is true, these compounds are available to be released to the atmosphere, or to ash or water, as a result of the incineration process, and their potential human health impact is not otherwise likely to be addressed in considerations directed toward the POHCs. Common examples of PICs are discussed in the following section, including the chlorinated dioxins and furans, which have received a great deal of interest.

Polychlorinated dioxins and furans: recent findings

The term "chlorinated dioxin" refers to one or more members of a class of 75 structurally similar chemical compounds known formally as polychlorinated dibenzo-*p*-dioxins (PCDDs). The term "chlorinated furans" refers to one or more of the 135 related chemical compounds known as polychlorinated dibenzofurans (PCDFs). These compounds are not intentionally produced for any commercial purpose but rather represent chemically predictable by-products that are created during the manufacture of other chemicals such as some chlorinated pesticides/herbicides or as a result of incomplete combustion of mixtures containing certain chlorinated organic compounds.[4,6,21,22]

Initially, the greatest degree of concern was focused on 2,3,7,8-tetrachloridibenzo-*p*-dioxin (2,3,7,8-TCDD). A wide variety of toxicological studies have demonstrated that, at sufficient exposures to typically very low doses, 2,3,7,8-TCDD can produce a variety of toxic effects including cancer, immune dysfunction, and reproductive effects in laboratory animals.[4] More information is needed, however, on the effect of the substance on humans. U.S. EPA recently released[23] a document representing several years of review that reevaluates the toxicity of the dioxins. The agency, working closely with the scientific community, concluded that the primary pathway for human exposure to dioxin and related compounds occurred through the potential ingestion of dioxin-tainted foods. Contamination typically occurs via atmospheric deposition and transferral through the food chain. One school of thought with regard to the dioxins holds that there is an identified series of biological steps common to vertebrates that accounts for many of the biological effects caused by the dioxin compounds. The U.S. EPA report concluded that, in many instances, adequate scientific evidence was available to conclude that, at significant exposures, dioxin and related compounds may have a wide variety of human health effects. These effects are similar to those found in laboratory animals, including certain adverse developmental effects. Several human mortality studies that had been completed since EPA's last review of dioxin's human carcinogenicity in the 1980s suggested a human risk specific-dose estimate (1E-6) of 0.01 pg TEQ/kg body weight/day. Current estimates of human background exposure go as high as 3 to 6 pg TEQ/kg body weight/day for all dioxin-like compounds including dioxin-like PCBs.[23]

These conclusions are considered controversial in the scientific and public health communities. For example, Environ[24] assembled several expert groups to examine the agency's recent findings, arguing that the agency's estimates of dioxin's levels of adverse effect are too high. They concluded that adverse human health effects do not necessarily occur at the currently estimated background burden levels and that there is insufficient scientific evidence to support EPA's controversial conclusions. These opposing positions presently are the subject of intense debate.

Dioxin and furan emissions may have one or all of the following three principal potential sources in the hazardous waste combustion process:[11]

- They may be present in the waste feed and may pass through the combustion process without being completely destroyed.
- They may be formed in the combustion process as PICs.
- They may be formed from elements and compounds that act as precursors in cooler areas downstream of the actual combustion process.

Formation of dioxins, and perhaps PCDFs as well, appears to be influenced by the following factors, as discussed by Travis and Cook,[6] Acharya et al.,[11] Hutzinger and Fiedler,[12] and Rappe et al.[13]

- PCDD in waste feed — The likelihood of finding dioxins in the feed of a combustion process historically reflects the widespread uses of commercial products (e.g., pesticides and wood preservatives) in which PCDDs can be found as trace contaminants.
- Dioxin precursors in waste feed — Several classes of chemical compounds have been isolated and studied as potential precursors of PCDD emissions from the combustion process: chlorinated phenols, chlorinated benzenes, and PCBs. The occurrence, magnitude, and significance of these processes is dependent on many operating parameters of an individual facility.

- Chlorine content of waste feed — Chlorine is a necessary element in the formation of PCDD, and increased chlorine concentrations in the feed can enhance the occurrence of PCDD formation.
- Combustion temperature — Temperatures between 500 and 800°C are observed to promote PCDD formation, while destruction of PCDDs is expected only at temperatures above 800°C. Combustion temperature is a function of fuel Btu content, available oxygen content, and fuel processing operations. Thus, the conditions of municipal waste incinerator combustion (low temperatures, high moisture content, low Btu-value fuel, lack of supplemental fuel, and poor waste mixing during combustion) are conducive to and are considered a significant source of PCDD formation. Hazardous waste incinerators and high efficiency boilers, on the other hand, have combustion design features that are less conducive to poor performance and the formation and emission of dioxins or furans.
- Combustion chamber dwell time — In general, PICs of all kinds are less likely to be formed in incinerators with long dwell times at very high temperatures. The residence times that are necessary for decomposition of PCDDs become shorter as the temperature is increased.
- Oxygen availability — Poor combustion conditions, often leading to the formation of PCDDs, result from low ratios of air-to-fuel and/or poor mixing of air and fuel. However, Huffman and Staley[25] observed increased PIC formation at high excess air values.
- Processing of waste feed — Waste feeds that are processed into small, homogeneous particles enhance the mixing of air and fuel and require shorter residence times to achieve acceptable destruction. Wastes that are composed of larger particles, or poorly mixed batches, may result in poorly mixed oxygen and fuel and require longer residence times. Waste mixes that exhibit high moisture content may decrease combustion efficiency.
- Use of supplemental fuel — Combustion sources that utilize firing of the combustion chamber with supplemental fuel of high Btu content minimize the production of PCDDs. Combustion sources burning low Btu wastes that do not utilize supplemental fuel are high potential sources of PCDDs.
- Availability of catalytic materials (e.g., copper).

Trenholm et al.[1] analyzed samples from six incinerator facilities that were capable of burning non-PCB-containing waste. The authors sought to detect tetrachlorodibenzofurans, pentachlorodibenzofurans, tetrachlorodibenzodioxins, or pentachlorodibenzodioxins. Furans were detected in samples at three of the six facilities, and dioxins were detected in samples from one of the six facilities.

Oppelt[22] reported on dioxin and furan emissions from five hazardous waste incinerators, six industrial boilers, and three calcining kilns that employed hazardous waste (e.g., high Btu solvents) as a fuel source. The author reported that none of the facilities emitted 2,3,7,8-TCDD, although four were emitting other PCDDs and five were emitting PCDFs. By way of comparison, the author also reported that a variety of PCDDs and PCDFs were detected in the emissions from 22 municipal waste incinerators for which complete stack emissions data were available. It is interesting to note that the average emission concentrations of PCDDs and PCDFs (3.3 $\mu g/m^3$ and 2.7 $\mu g/m^3$, respectively) from the municipal waste incinerators were nearly three orders of magnitude (1,000 times) greater than the highest values that were reported for any of the 14 hazardous waste incineration units that were evaluated. This was judged to be due, at least in part, to the fact that municipal waste combustors operate at lower temperatures and lower efficiencies and typically exhibit poorer air-to-fuel mixing characteristics during the combustion process than do well-designed and operated hazardous waste incinerators. The emission concentrations from the Oppelt[22] study are listed in conjunction with the results from the study of Trenholm et al.[1] in Table 2.2.

Table 2.2   Range of Dioxins and Furans Reported
in Stack Samples from 19 Hazardous Waste
Incinerators and Related Facilities

Analyte	Concentration
TCDF	0.0007–0.047
PCDF	0.0003–0.095
TCDD	0–0.021
PCDD	0.00064–0.076

Note: All data expressed in units of ng/L
(equvalent to $\mu g/m^3$).

Source: Trenholm, A. et al., *Performance Evaluation of Full-Scale Hazardous Waste Incinerator*, Vol. I–V, Industrial Environmental Research Laboratory, U.S. Environmental Protection Agency, EPA-600/2-84/181a-e, NTIS PB85-129500, 1984; Oppelt, E. T., Incineration of hazardous waste: a critical review, *J. Air Pollut. Control Assoc.*, 37(5), 558, 1987. With permission.

In the WTI risk assessment, U.S. EPA[5] tested chlorine feed rates ranging from 400 lb/hr to 3,300 lb/hr to determine if the facility's emission concentrations were in compliance under a variety of operating scenarios. EPA concluded that the facility was operating well within the permitted concentration level of 30 ng/dscm (total sum of all the tetra through octa congeners averaged over five test runs per operating scenario) once an enhanced carbon injection system (ECIS) was installed. The WTI risk assessment results for dioxin and furan emissions are not included in Table 2.2 because concentrations were reported as a summation of all the tetra through octa congeners and not the tetra through penta congeners only.

Field studies of incineration facilities that were used in the treatment of dioxin-contaminated soils in the mid-1980s demonstrated the ability to achieve DRE values in excess of 99.9999%[26] for these waste sources.

Hutzinger and Fiedler[12] noted that 2,3,7,8-TCDD was not detected in gaseous emissions following PCB incineration, and the other PCDD and PCDF concentrations were in the low part per trillion range in air.

Other products of incomplete combustion

In addition to the dioxins and furans, a number of other products of incomplete combustion have been identified. These PICs generally can be readily divided into the following three categories:

- Volatiles
- Semivolatiles
- Trihalomethanes (THMs: comprised of the sum of chloroform, bromodichloromethane, dibromochloromethane, and bromoform)

Travis and Cook[6] reported that the PICs that were found in stack effluents at eight U.S. EPA trial burns included representatives from all three of these groups and included benzene, chloroform, bromodichloromethane, dibromochloromethane, bromoform, naphthalene, chlorobenzene, tetrachloroethene, 1,1,1-trichloroethane, hexachlorobenzene, methylene chloride, *o*-nitrophenol, phenol, and toluene. The operating premise is that most PICs are formed as the result of a recombination of molecular fragments from partially degraded compounds, an observation that is supported by the appearance of many of the same organic compounds in almost all trial burn stack gas samples, even considering the highly variable waste feeds.

Table 2.3 presents a list of the RCRA Appendix VIII compounds that were detected in stack emissions from eight incinerators as evaluated by Trenholm et al.[1] The authors note that the detected trihalomethanes may have been contributed by chlorination of the water supplies that are used in the air pollution control systems at some facilities.

Inorganic analytes and related materials in emissions

Metals

Metals are of interest in the context of hazardous waste incineration evaluations, not only because of the possible adverse health effects that may result from exposure to the emissions, but also because they are not destroyed by the combustion process and may accumulate in, for example, the ash residuals. Since metals are not destroyed, one measure of a facility's efficiency in effectively handling and thus preventing the release of metals is its system removal efficiency (SRE), calculated as the ratio of the metal emission rate to metal feed rate. While the incineration process may change the form or the valence state of metals in waste streams, it does not have the capability to destroy the metals. As a result, approximately the same mass of metals that enters the combustion chamber is expected to emerge either in the form of bottom ash or in the form of fly ash that goes up the stack and is typically

Table 2.3 Products of Incomplete Combustion Reported from Stack Sampling at Eight Incinerators

Analyte	Number of facilities	Concentration range ($\mu g/m^3$)
Benzene	6	12–670
Bromochloromethane	1	14
Bromodichloromethane	4	3–32
Bromoform	3	0.2–24
Bromomethane	1	1
Carbon disulfide	1	32
Chlorobenzene	3	1–10
Chloroform	5	1–1,330
Chloromethane	1	3
o-Chlorophenol	1	2–22
Dibromochloromethane	4	1–12
Dichlorobenzene	1	2–4
Diethyl phthalate	1	7
2,4-Dimethyl phenol	1	1–21
Fluoranthene	1	1
Hexachlorobenzene	2	0.5–7
Methyl ethyl ketone	1	3
Methylene bromide	1	18
Methylene chloride	2	2–27
Naphthalene	3	5–100
o-Nitrophenol	2	25–50
Pentachlorophenol	1	6
Phenol	2	4–22
Pyrene	1	1
Tetrachloroethylene	3	0.1–2.5
Toluene	2	2–75
Trichlorobenzene	1	7
1,1,1-Trichloroethane	3	0.1–1.5
2,4,6-Trichlorophenol	1	110

Source: Trenholm, A. et al., *Performance Evaluation of Full-Scale Hazardous Waste Incinerator*, Vol. I–IV, Industrial Environmental Research Laboratory, U.S. Environmental Protection Agency, EPA-600/2-84/181a-3, NTIS PB85-129500, 1984.

trapped by APCDs. Metals generally travel in the form of particles (fly ash) or, to a lesser extent, vapors, in cases such as mercury and selenium.

The U.S. EPA typically identifies the following 12 metals as principal hazardous constituents: antimony, arsenic, barium, beryllium, cadmium, chromium, lead, mercury, nickel, selenium, silver, and thallium. Of this list, only arsenic, cadmium, chromium, and beryllium pose well-defined carcinogenic risks, primarily from the inhalation pathway.[4] While nickel is often classed as an inhalation carcinogen, this is limited to the refinery dust form. Lead also is defined presently by U.S. EPA as a carcinogen, but no data are available relative to its potency in that regard.

Table 2.4 presents estimated feed and emission rates along with a calculation of the system removal efficiency (SRE) for each metal studied at the WTI incinerator facility.[5] For seven metals, i.e., antimony, arsenic, beryllium, cadmium, chromium, lead, and mercury, SREs were calculated using trial burn data. The values for the remaining eight metals listed in Table 2.4 were estimated using surrogates from the trial burn and the appropriate thermodynamic modeling techniques. All calculations of the SREs range from 99.68% to 99.99932%, excluding mercury, which was rarely removed.

Table 2.4 Estimated Average Metal Emission Rates
and SRE for the WTI Hazardous
Waste Incineration Facility

Analyte	SRE % measured (or estimated)	Feed rate (g/sec)	Emission rate (g/sec)
Aluminum	NA (99.99932)[a]	18	2.4 E-04
Antimony	99.986	0.030	4.2 E-06
Arsenic	99.977	0.16	3.7 E-05
Barium	NA (99.977)[b]	0.67	1.5 E-04
Beryllium	99.9907	0.00035	3.33 E-08
Cadmium	99.987	0.12	1.6 E-05
Chromium	99.99932	0.10	7.1 E-07
Copper	NA (99.977)[b]	0.41	9.4 E-05
Lead	99.99	0.44	4.3 E-05
Mercury	0[c]	0.0014	1.4 E-03
Nickel	NA (99.977)[b]	0.022	5.0 E-06
Selenium	99.68	0.15	4.7 E-04
Silver	NA (99.977)[b]	0.065	1.5 E-05
Thallium	NA (99.977)[b]	0.15	3.4 E-05
Zinc	NA (99.977)[b]	0.54	1.2 E-04

Note: System Removal Efficiency (SRE) determined from March 1993 trial burn. NA: measured value not applicable; SRE not determined from trial burn, but estimated.

[a] Estimated based on chromium SRE

[b] Estimated based on arsenic SRE.

[c] Assumed to be zero although very low, non-zero efficiency was measured.

Source: U.S. Environmental Protection Agency, Risk Assessment for the Waste Technologies (WTI) Hazardous Waste Incinerator Facility (East Liverpool, Ohio), Vol. III, EPA/905/D95/0026, 1995a.

Table 2.5 presents the reported range of metal concentrations in waste feeds and the resultant concentration of metal particulate emissions from the five facilities in the study by Trenholm et al.[1] Analysis of the table reveals that the concentrations of antimony (Sb), barium (Ba), lead (Pb), chromium (Cr), cadmium (Cd), and nickel (Ni) in the stack emissions were much greater than their concentrations in the incinerated waste feeds. This phenomenon

Table 2.5 Ranges of Metal Concentrations in Waste
Feeds and Particulate Emissions for Five Hazardous
Waste Incinerators

Concentration analyte	Emission range (mg/kg of particulate)	Waste feed (mg/kg)
Antimony	<1,200–15,200	<0.184–437
Arsenic	ND	<3.5–45.2
Barium	41–3,090	<0.24–1,460
Beryllium	<0.32–6	<0.15–1.8
Cadmium	140–4,300	<0.016–224
Chromium	668–47,500	<0.034–2,170
Lead	3,100–100,000	<0.25–9,830
Mercury	56[a]	<3–50
Nickel	230–49,000	<0.063–6,670
Selenium	<500–61,600	<2–546
Silver	7.6–1,880	<0.03–<2.6
Thallium	—[b]	<4–<23

Note: < Concentration was below the reported detection limit.
ND Not detected in any test.

[a] Detected only one run; others not detected.

[b] Relevant range could not be developed because of elevated
detection limits.

Source: Trenholm, A. et al., *Performance Evaluation of Full-Scale
Hazardous Waste Incinerators*, Vol. I–IV, Industrial Environmental
Research Laboratory, U.S. Environmental Protection Agency,
EPA-600/2-84/181a-e, NTIS PB85-129500, 1984.

is known as "enrichment" and occurs because metals exhibit a tendency to
condense on the surfaces of the finer particles that are difficult to remove
from emission gases by air pollution control equipment. Lead and chromium
were detected in the stack gases at concentrations of 100,000 mg/kg of
particulates and 47,500 mg/kg of particulates, respectively. At these concentrations, lead and chromium comprise 10% and 4.8%, respectively, of the
particulate emissions by mass.

Combustion chamber ash ("bottom ash") also was analyzed for metals
by Trenholm et al.[1] Detected concentrations varied widely and were a function of facility operating conditions, residue processing, and the quantity of
metal in the input waste stream. Metals most frequently detected in the
bottom ash included chromium, zinc, copper, nickel, lead, arsenic, and silver.
The majority of the metals in the toxicity characteristic leaching procedure
(TCLP) extracts of the ash residues were at concentrations below the standards that define a waste as characteristically hazardous. However, disposal
of the ash may be subject to tighter restrictive land disposal restriction (LDR)
standards. Among the metals that most frequently exceeded the LDR limits
were arsenic, nickel, and lead. This suggests that incinerator residues that
are subject to LDR standards may require additional treatment prior to
disposal.

Combustion chamber ash was not analyzed for potential emissions at the WTI East Liverpool facility.[5] However, seven metals were detected in the fly ash samples trapped by the APCD system, and these were used to compile a list of potential fugitive metal emissions. The potential fugitive fly ash emissions included arsenic, barium, cadmium, lead, nickel, selenium, and silver. No concentrations of these metals were reported.

Table 2.6 presents results of the analysis by Van Buren et al.[3] for metals in APCD effluents at eight selected incinerator sites. According to comparisons with TCLP regulatory limits, the effluent for site 1 would be considered hazardous for cadmium (3.5 mg/L), chromium (11 mg/L), and lead (860 mg/L), while the effluent from site 8 would be considered hazardous for cadmium (2.8 mg/L), lead (31 mg/L), and selenium (2.1 mg/L). Sites 4 and 10, which incinerate low metal content wastes, have only 0.4 and 1.4 mg/L, respectively, of priority pollutant metals in their APCD effluents and, thus, would not trigger regulatory action on that basis.

Acid gases

As noted previously, when halogen (e.g., chlorine, bromine, fluorine) compounds are present in waste feed, hydrogen halides may be formed and

Table 2.6 Concentrations of Priority Pollutant Metals in Aqueous Effluent from Hazardous Waste Incinerator Air Pollution Control Devices

Analyte	Site Number							
	1[a]	2[a]	3[a]	4[b]	7[c]	8[a,d]	9[a]	10[a]
Antimony	0.1[e]	<0.01	0.61	<0.01	1.7	4.1	<0.03	0.13
Arsenic	0.2	<0.01	<0.01	<0.01	0.06	0.4	<0.1	<0.1
Beryllium	<0.01	<0.01	<0.01	<0.01	<0.01	0.01	<0.01	<0.01
Cadmium	3.5	<0.01	0.04	<0.01	0.08	2.8	<0.01	<0.01
Chromium	11	<0.05	0.1	0.06	0.28	3.8	0.27	0.28
Copper	550	<0.04	0.26	<0.04	0.64	2.2	0.46	0.05
Lead	860	<0.01	2.6	<0.01	2.6	31	0.38	0.1
Mercury	0.06	0.013	0.013	<0.001	<0.005	<0.005	<0.005	<0.005
Nickel	<0.02	23	0.17	0.05	0.75	1.5	0.07	0.48
Selenium	0.09	<0.01	<0.01	<0.01	0.6	2.1	<0.1	0.2
Silver	<0.01	<0.02	0.04	<0.02	0.05	0.15	0.61	<0.01
Thallium	<0.01	1.3	16	0.02	0.16	1.6	0.31	0.03
Zinc	950	0.02	16	0.27	6.7	1.6	0.16	0.11
Total	2,380	24.3	35.7	0.4	13.6	51.3	2.26	1.38

[a] Rotary kiln with secondary combustor.

[b] Fluidized bed incinerator.

[c] Fixed hearth with secondary combustor.

[d] Highest values reported for aqueous effluent that was recirculated from two cooling ponds.

[e] All data expressed in units of mg/L of effluent liquid.

Source: Adapted from Van Buren, D. et al., *Characterization of Hazardous Waste Incinerator Residuals,* Acurex Corporation, EPA-600-2-87/017, NTIS PB87-168159, 1987.

emitted, including hydrogen chloride (HCl), hydrogen fluoride (HF), and hydrogen iodide (HI). Halide emissions in exhaust gases are a major concern because several of the hydrogen halides are highly corrosive gases. HCl emissions typically are limited by regulatory stricture to the greater of 1.8 kg/hr (approximately 4 lb/hr) of HCl or 1% of the total organic chloride in the waste. Since HCl is not consumed during combustion, other control measures must be employed for incinerators to meet HCl emission standards. Limiting the input waste chlorine and APCDs are two methods commonly used to control HCl emissions. Typically, absorption scrubbing systems using either water or a caustic absorbing medium are used for HCl control.

All facilities in the studies by WTI[5] and Trenholm et al.[1] were in compliance with the HCl control standard of 1.8 kg/hr with the exception of a single run at one facility in the Trenholm study (run 3 at plant B). Two of the three facilities with no HCl control had little chloride in the waste and, thus, were able to comply with the mass emissions standards. At the other facility, which exceeded the limits, almost all of the chlorides in the waste feed were the result of carbon tetrachloride and trichloroethene that had been spiked intentionally into the input waste stream for investigative purposes. Under normal operating conditions the facility would probably have complied with the standard.

Particulate matter

Emitted particulate matter, which may include solids or liquids (e.g., aerosols), comprises a complex category of materials. The size range of interest varies from just over that of large individual molecules (e.g., 0.01 μm) to approximately 500 μm in diameter.[27] The principal sources of particulate emissions from hazardous waste incinerators include the waste feed, auxiliary fuels, and scrubbing liquids used in the air pollution control system. During the combustion process, fly ash from the waste feed or auxiliary fuel can be entrained in the combustion gas exhaust and emitted as particulate matter. Suspended and dissolved solids from water used in the air pollution control system are another potential source of particle emissions. If water droplets are evaporated in a quench chamber or scrubber, solids contained in these droplets can be entrained in the exhaust stream as particles.[1]

A variety of control techniques has been employed to control particulate emissions from hazardous waste incinerators. The most common techniques include Venturi scrubbers and ionizing wet scrubbers. Five of the eight facilities that were presented by Trenholm et al.[1] had some type of particulate control. Two had high-energy wet scrubbers, one had a low-energy wet scrubber (which provides minimal control), and one had an ionizing wet scrubber. The data from Trenholm et al.[1] indicate that four of the eight sites tested failed to meet the particulate emission standard of 180 mg/nm^3, and two of the sites met the standard with less than a 20% margin. One site is equipped with a packed bed scrubber that provides little control for particles less than 5 μm in diameter, and available data suggest that most particles from hazardous waste incinerators are less than 2 μm in diameter.[1] In the WTI risk assessment, particle size was estimated to be under 10 μm with an emission rate averaging 0.07 g/sec. Although nothing was mentioned concerning the type of scrubber system employed, it is assumed one was in operation during the test burns.[5] Thus, scrubber technology has the potential to exert a significant influence on the ability to comply with emission standards.

Other

Other analytes such as carbon monoxide and carbon dioxide are addressed in the literature with respect to hazardous waste incinerators and related facilities.[28] These are not dealt with as specific entities in this chapter, as they are general emissions from combustion facilities, and are not specific to hazardous waste incinerators and related facilities.

Summary

The evaluations that have been conducted for the purpose of characterizing emissions from hazardous waste incinerators and related facilities have demonstrated repeatedly that the waste feed, the operating characteristics, and APCDs are the principal controlling factors in defining emissions. It is these emissions that determine the type and degree of health hazard that may be posed by a specific facility. Some toxicological considerations for selected emissions are presented in the following chapter.

Tables 2.7 through 2.9 and references follow.

Table 2.7 Reasonable Upper Limit Estimates for Emissions of Specific Organics from Incinerators and Related Facilities Burning Hazardous Wastes

Analyte	Classified as carcinogen (Y/N)	Conc. in emission ($\mu g/m^3$)	Percent of total[a]
Acetonitrile	N	0.26	0.00066
Benzene	Y	4,928	12.6
Benzo(a)Anthracene	Y	1.10	0.0028
Bromomethane	N	2.13	0.0054
Butylbenzyl phthalate	N	3	0.0076
C1 Hydrocarbons (unspecified)	N	9,600	24.5
C2 Hydrocarbons (unspecified)	N	17,000	43.3
Carbon tetrachloride	Y	99.5	0.25
Chlorobenzene	N	195	0.50
Chloroform	Y	1,407	3.6
Chloromethane	Y	807	2.1
Dibutyl phthalate	N	3.6	0.0092
o-Dichlorobenzene	N	95	0.24
p-Dichlorobenzene	N	86	0.22
1,1-Dichloroethane	N	3.37	0.0086
1,2-Dichloroethane	Y	714	1.8
1,1-Dichloroethene	Y	31.6	0.081
Dichlorodifluoromethane	N	1.22	0.0031
2,4-Dichlorophenol	N	0.50	0.0013
Diethyl phthalate	N	31	0.079
bis(2-Ethylhexyl) phthalate	Y	77.7	0.20
Formaldehyde	Y	892	2.3
Hexachlorobenzene	Y	8.95	0.023
Methyl ethyl ketone	N	33.2	0.085
Methylene chloride[b]	Y	1,755	4.5
Naphthalene	N	130	0.33
Other Carcinogens	Y	4.6	0.0117
Other Noncarcinogens	N	2.8	0.007
PCDD	Y	0.10246	0.00026
Pentachlorophenol	Y	9.3	0.024
Phenol	N	33.1	0.084
TCDF	Y	0.00141	0.0000036
Tetrachloroethene	Y	297	0.76
Toluene	N	551	1.4
1,2,4-Trichlorobenzene	N	77	0.20
1,1,1-Trichloroethane	N	64	0.16
1,1,2,2-Tetrachloroethane	Y	17	0.043
1,1,2-Trichloroethane	Y	36.7	0.094
Trichloroethene	Y	81.8	0.21
2,4,5-Trichlorophenol	N	144	0.37
Vinyl chloride	Y	14	0.036

[a] Total does not equal 100% due to rounding conventions employed.

[b] Potentially attributable to laboratory contamination.

Source: Adapted from Dempsey, C. R. and Oppelt, E. T., Incineration of hazardous waste: a critical review update, *Air Waste*, 43, 25, 1993.

Table 2.8 Organic Analytes to Be Considered
in Hazardous Waste Incinerator Evaluation

Analyte	CAS no.
Acenaphthene	83-32-9
Acetaldehyde	75-07-0
Acetone	67-64-1
Acetonitrile	75-05-8
Acetophenone	98-86-2
2-Acetylaminofluorene	53-96-3
Acenaphthylene	208-96-8
Acrolein	107-02-8
Acrylonitrile	107-13-1
Aniline	62-53-3
Anthracene	120-12-7
Azobenzene(cis)	103-33-3
Benzaldehyde	100-52-7
Benzo[a]anthracene	56-55-3
Benzo[e]anthracene	NF
Benzene	71-43-2
1,2-Benzenedicarboxylic acid	88-99-3
Benzidine	92-87-5
Benzoic Acid	65-85-0
Benzo[b]fluoranthene	205-99-2
Benzo[j]fluoranthene	205-82-3
Benzo[k]fluoranthene	207-08-9
Benzo[g,h,i]perylene	191-24-2
Benzo[a]pyrene	50-32-8
Benzo[e]pyrene	NF
p-Benzoquinone	192-97-2
Benzotrichloride	98-07-7
Benzo(b)fluoranthene	205-99-2
Benzyl Chloride	100-44-7
1,1-Biphenyl	92-52-4
Bis[2-chloroethoxy]methane	111-91-1
Bis[2-chloroethyl]ether	111-44-4
Bis[2-chloroisopropyl]ether	39638-32-9
Bis[chloromethyl]ether	542-88-1
Bis[2-ethylhexyl]phthalate	117-81-7
Bromochloromethane	74-97-5
Bromodichloromethane	75-27-4
Bromoethane	74-96-4
Bromoform	75-25-2
1,3-Butadiene	106-99-0
Butyl acetate	123-86-4
Butyl benzyl phthalate	85-68-7
Calcium chromate	13765-19-0
Captan	133-06-2
Carbon disulfide	75-15-0

(continues)

Table 2.8   (continued)

Analyte	CAS no.
Carbon tetrachloride	56-23-5
Chlordane	57-74-9
2-Chloroacetophenone	532-27-4
p-Chloroaniline	106-47-8
Chlorobenzene	108-90-7
Chlorobenzilate	510-15-6
Chlorocyclopentadiene	41851-50-7
4-Chlorodiphenylether	7005-72-3
Chloroethane	75-00-3
Chloroform	67-66-3
β-Chloronaphthalene	91-58-7
o-Chlorophenol	95-57-8
2-Chloropropane	75-29-6
Chrysene	218-01-9
Coronene	197-07-1
Creosote	8001-58-9
m-Cresol	108-39-4
o-Cresol	95-48-7
p-Cresol	106-44-5
Crotonaldehyde	4170-30-3
Cumene	98-82-8
Cyclohexane	110-82-7
Cyclohexanone	108-94-1
Dibenzodioxins, chlorinated	NF
Dibenz(a,h)anthracene	53-70-3
Dibenzo[a,e]fluoranthene	5385-75-1
Dibenzo[a,h]fluoranthene	NF
Dibenzofuran	132-64-9
Dibenzofurans, chlorinated	NF
Dibromochloromethane	124-48-1
Dibromomethane	74-95-3
1,2-Dibromo-3-chloropropane	96-12-8
Dibutyl phthalate	84-74-2
1,2-Dichlorobenzene	95-50-1
1,3-Dichlorobenzene	541-73-1
1,4-Dichlorobenzene	106-46-7
3,3′-Dichlorobenzidine	91-94-1
Dichlorobiphenyl	2050-68-2
1,4-Dichloro-2-butene (cis & trans)	264-41-0
Dichlorodifluoromethane	75-71-8
Dichlorofluoromethane	75-43-4
1,1-Dichloroethane	75-34-3
1,2-Dichloroethane	107-06-2
1,1-Dichloroethene	75-35-4
1,2-Dichlorethylene (trans)	156-60-5
2,4-Dichlorophenol	120-83-2

Table 2.8 (continued)

Analyte	CAS no.
1,2-Dichloropropane	78-87-5
1,3-Dichloropropene (cis & trans)	542-75-6
Diethyl phthalate	84-66-2
Dimethyl amine	124-40-3
3,3'-Dimethoxybenzidine	119-90-4
Dimethyl hydrazine	57-14-7
2,4-Dimethylphenol	105-67-9
Dimethyl sulfate	77-78-1
Dimethyl sulfide	75-18-3
Dimethyl phthalate	131-11-3
Di-n-butylphthalate	84-74-2
1,2-Dinitrobenzene	528-29-0
1,3-Dinitrobenzene	99-65-0
1,4-Dinitrobenzene	100-25-4
2,4-Dinitrophenol	51-28-5
2,4-Dinitrotoluene	121-14-2
2,6-Dinitrotoluene	606-20-2
Di-n-octylphthalate	117-84-0
1,4-Dioxane	123-91-1
Epichlorohydrin	106-89-8
2-Ethoxyethanol	110-80-5
Ethylacrylate	140-88-5
Ethylbenzene	100-41-4
Ethyl methacrylate	97-90-5
Ethylene dibromide	106-93-4
Ethylene oxide	75-21-8
Ethylene thiourea	96-45-7
Fluoranthene	106-44-0
Fluorene	86-73-7
Formaldehyde	50-00-0
Formic acid	64-18-6
Furfural	98-01-1
Heptachlor	76-44-8
Heptane	142-82-5
Heptachlorobiphenyl	28655-71-2
Hexachlorobenzene	118-74-1
Hexachlorobiphenyl	26601-64-9
Hexachlorobutadiene	87-68-3
α-Hexachlorocyclohexane	319-84-6
β-Hexachlorocyclohexane	319-85-7
γ-Hexachlorocyclohexane	58-89-9
Hexachlorocyclopentadiene	77-47-4
Hexachloroethane	64-72-1
Hexachlorophene	70-30-4
n-Hexane	110-54-3
2-Hexanone	591-78-6

(continues)

Table 2.8   (continued)

Analyte	CAS no.
3-Hexanone	589-38-8
Hydrazine	302-01-2
Hydrogen sulfide	7783-06-4
Indeno(1,2,3-*cd*)pyrene	193-39-5
Isobutanol	78-83-1
Isophorone	78-59-1
Isopropanol	67-63-0
Isosafrole	120-58-1
Maleic anhydride	108-31-6
Maleic hydrazide	123-33-1
Methanol	67-56-1
Methoxychlor	72-43-5
Methyl bromide (bromomethane)	74-83-9
1-Methyl butadiene	504-60-9
Methyl chloride (chloromethane)	74-87-3
Methylene bromide	74-95-3
Methylene chloride (dichloromethane)	75-09-2
Methyl ethyl ketone	78-93-3
Methyl isobutyl ketone	108-10-1
Methyl mercaptan	74-93-1
Methyl mercury	22967-92-6
Methyl methacrylate	80-62-6
2-Methylnaphthalene	91-57-6
4-Methyl-2-pentanone	108-10-1
Methyl-tert-butyl ether	1634-04-4
Naphthalene	91-20-3
Naphthylamine	25168-10-9
2-Nitroaniline	88-74-4
3-Nitroaniline	99-02-2
4-Nitroaniline	100-01-6
Nitrobenzene	98-95-3
2-Nitrophenol	88-75-5
4-Nitrophenol	100-02-7
o-Nitrophenol	88-75-5
2-Nitropropane	79-46-9
N-nitrosodiethanolamine	1116-54-7
N-nitrosodiethylamine	55-18-5
N-nitroso-di-*n*-propylamine	621-64-7
N-nitrosodiphenylamine	86-30-6
N-nitrosomethylethylamine	10595-95-6
N-nitroso-*n*-dibutylamine	924-16-3
N-nitrosopyrrolidine	930-55-2
Nonachlorobiphenyl	53742-07-7
Octachlorobiphenyl	55722-26-4
Pentachlorobenzene	608-93-5
Pentachlorobiphenyl	25429-29-2

Table 2.8 (continued)

Analyte	CAS no.
Pentachloroethane	76-01-7
Pentachloronitrobenzene	82-68-8
Pentachlorophenol (PCP)	87-86-5
Phenanthrene	85-01-8
Phenol	108-95-2
Phosgene	75-44-5
Phthalic anhydride	85-44-9
2-Picoline	109-06-8
Polychlorinated biphenyls (PCB)	1336-36-3
Propionaldehyde	123-38-6
Pyrene	129-00-0
Pyridine	110-86-1
Quinoline	91-22-5
Quinone	106-51-4
Resorcinol	108-46-3
Safrole	94-59-7
Styrene	100-42-5
1,2,4,5-Tetrachlorobenzene	95-94-3
Tetrachlorobiphenyl	26914-33-0
1,1,1,2-Tetrachloroethane	630-20-6
1,1,2,2-Tetrachloroethane	79-34-5
Tetrachloroethene	127-18-4
2,3,4,6-Tetrachlorophenol	58-90-2
Toluene	108-88-3
Toluenediamine	95-80-7
2,6-Toluenediisocyanate	584-84-9
o-Toluidine	95-53-4
p-Toluidine	106-49-0
1,2,4-Trichlorobenzene	25323-68-6
Trichlorobiphenyl	120-82-1
1,1,1-Trichloroethane	71-55-6
1,1,2-Trichloroethane	79-00-5
Trichloroethene	79-01-6
Trichlorofluoroethane	27154-33-2
Trichlorofluoromethane	75-69-4
2,4,5-Trichlorophenol	95-95-4
2,4,6-Trichlorophenol	88-06-2
Trichloropropane	25735-29-9
1,1,2-Trichloro-1,2,2-Trifluoroethane	76-13-1
Vinyl acetate	108-05-4
Vinyl chloride	75-01-4
m-Xylene	108-38-3
o-Xylene	95-47-6
p-Xylene	106-42-3
Xylenes (total)	1330-20-7

Table 2.9 Inorganic Analytes to Be Considered
in Hazardous Waste Incinerator Evaluation

Analyte	CAS no.
Aluminum	7429-90-5
Antimony	7440-36-0
Arsenic, inorganic	7440-38-2
Barium	7440-39-3
Beryllium	7440-41-7
Cadmium	7440-43-9
Carbon monoxide (Co)	630-08-0
Chlorine	7782-50-5
Chromium (III)	16065-83-1
Chromium (VI)	18540-29-9
Cobalt	7440-48-4
Copper	7440-50-8
Hydrogen chloride (HCl)	7647-01-0
Hydrogen fluoride	7664-39-3
Iron	7439-89-6
Lead	7439-92-1
Magnesium	7439-95-4
Manganese	7439-96-5
Mercury	7439-97-6
Mercuric chloride	7487-94-7
Molybdenum	7439-98-7
Nickel	7440-02-0
Nitrogen dioxide	10102-44-0
Nitrogen monoxide (NO)	10102-43-9
Particulate matter	NA
Scandium	7440-20-2
Selenium (and compounds)	7782-49-2
Silicon	7440-21-3
Silver	7440-22-4
Sulfur dioxide	7446-09-5
Sulfur trioxide	7446-11-9
Tellurium	13494-80-9
Thallium	7440-28-0
Tin	7440-31-5
Titanium	7440-32-6
Vanadium	7440-62-2
Zinc	7440-66-6

References

1. Trenholm, A. et al., *Performance Evaluation of Full-Scale Hazardous Waste Incinerator*, Vol. I–V, Industrial Environmental Research Laboratory, U.S. Environmental Protection Agency, EPA-600/2-84/181a-e, NTIS PB85-129500, 1984.
2. Trenholm, A. et al., *Total Mass Emissions from a Hazardous Waste Incinerator*, Midwest Research Institute, U.S. Environmental Protection Agency, EPA-600/2-87/064, NTIS PB87-228508, 1987.
3. Van Buren, D. et al., *Characterization of Hazardous Waste Incinerator Residuals*, Acurex Corporation, EPA-600.2-87/017, NTIS PB87-168159, 1987.
4. Dempsey, C. R. and Oppelt, E. T. Incineration of hazardous waste: a critical review update, *Air & Waste*, 43, 25, 1993.
5. U.S. Environmental Protection Agency, *Risk Assessment for the Waste Technologies (WTI) Hazardous Waste Incinerator Facility (East Liverpool, Ohio)*, Vol. III, EPA/905/D95/002c, 1995.
6. Travis, C. C. and Cook, S. C., *Hazardous Waste Incineration and Human Health*, CRC Press, Boca Raton, FL, 1989, 1.
7. Trenholm, A. and Lee, C. C., Analysis of PIC and total mass emissions from an incinerator, in *Land Disposal, Remedial Action, Incineration and Treatment of Hazardous Waste*, Proceedings of the Twelfth Annual Research Symposium at Cincinnati, EPA 600/9-86/022, 1986.
8. Mournighan, R. E. et al., Incinerability ranking of hazardous organic compounds, in *Remedial Action, Treatment and Disposal of Hazardous Waste*, Proceedings of the Fifteenth Annual Research Symposium, EPA 600/9-90/006, 1990.
9. Taylor, P. H. and Chadbourne, J. F., Sulfur hexafluoride as a surrogate for monitoring hazardous waste incineration performance, *J. Air Pollut. Control Assoc.*, 37(6), 729, 1987.
10. Oppelt, E. T., Hazardous waste destruction, *Environ. Sci. Technol.*, 20(4), 312, 1986.
11. Acharya, P., DeCicco, S. G., and Novak, R. G., Factors that can influence and control the emissions of dioxins and furans from hazardous waste incinerators, *J. Air Waste Manage. Assoc.*, 41, 1605, 1991.
12. Hutzinger, O. and Fiedler, H., Sources and emissions of PCDD/PCDF, *Chemosphere*, 18(1-6), 23, 1989.
13. Rappe, C., Andersson, R., Bergqvist, P., Brohede, C., Hansson, M., Kjeller, L., Lindstrom, G., Marklund, S., Nygren, M., Swanson, S. E., Tysklind, M., and Wiberg, K., Sources and relative importance of PCDD and PCDF emissions, *Waste Mgmt. Res.*, 5, 225, 1987.
14. Hume, A. S. et al., Toxicity of solvents, in *Basic Environmental Toxicology*, Cockerman, L. G. and Shane, B. S., eds., CRC Press, Boca Raton, FL, 1994, 157.
15. Chambers, J. E., Toxicity of pesticides, in *Basic Environmental Toxicology*, Cockerman, L. G. and Shane, B. S., eds., CRC Press, Boca Raton, FL, 1994, ch.7.
16. Hansen, L. G., Halogenated aromatic compounds, in *Basic Environmental Toxicology*, Cockerman, L. G. and Shane, B. S., eds., CRC Press, Boca Raton, FL, 1994, 199.
17. Overton, E. B. et al., Toxicity of petroleum, in *Basic Environmental Toxicology*, Cockerman, L. G. and Shane, B. S., eds., CRC Press, Boca Raton, FL, 1994, 133.

18. Senser, D. W. et al., PICs — a consequence of stable intermediate formation during hazardous waste incineration (dichloromethane), *Haz. Waste Haz. Materials*, 2(4), 473, 1985.
19. Taylor, P. H. et al., Evaluation of mechanisms of PIC formation in laboratory experiments: implications for PIC formation and control strategies in full-scale incineration systems, in Third International Conference on New Frontiers for Hazardous Waste Management, EPA 600/9-90/072, 1989.
20. Peters, W. A., Solids pyrolysis and volatiles secondary reactions in hazardous waste incineration: implications for toxicants destruction and PIC's generation, *Haz. Waste Haz. Materials*, 7(1), 89, 1990.
21. Tessitore, J. L. et al., *Thermal Destruction of Air Toxics*, Cross/Tessitore & Associates, P.A., Environmental Engineers.
22. Oppelt, E. T., Incineration of hazardous waste: a critical review, *J. Air Pollut. Control Assoc.*, 37(5), 558, 1987.
23. U.S. Environmental Protection Agency, *Estimating Exposure to Dioxin-Like Compounds*. EPA/600/6-88/005C, 1995.
24. Environ, Expert panel discussion summary. *Environ. Sci. Technol.*, 29 (1), 31, 1995.
25. Huffman, G. L. and Staley, L. J., The formation of products of incomplete combustion in research combustors, in *Land Disposal, Remedial Action, Incineration and Treatment of Hazardous Waste*, Proceedings of the Twelfth Annual Research Symposium at Cincinnati, EPA 600/9-86/022, 1986.
26. Freestone, F. et al., Evaluation of on-site incineration for cleanup of dioxin-contaminated materials, in *Land Disposal, Remedial Action, Incineration and Treatment of Hazardous Waste*, Proceedings of the Twelfth Annual Research Symposium at Cincinnati, Ohio. EPA 600/9-86/022, 1986.
27. Brunner, C. R., *Hazardous Waste Incineration*, 2nd ed., McGraw-Hill, New York, 1993.
28. Schreiber, R. J., Jr., Hydrocarbon emissions from cement kilns burning hazardous waste, *Haz. Waste Haz. Materials*, 11(1), 157, 1994.

chapter three

Physical, chemical, and toxicological characteristics of emissions from hazardous waste incinerators and related facilities

Christopher M. Teaf, Isabel K. Stabile, Philip A. Moffat, and J. Michael Kuperberg

Contents

Introduction and historical overview

Any evaluation of the potential for adverse effects on human populations or on the environment as a result of emissions from hazardous waste incinerators and related facilities requires an understanding of the relevant physical properties, chemical characteristics, and toxicological end points. This section presents brief discussions on selected relevant physical/chemical properties and summarizes available values for many of the organic and inorganic analytes that were identified in Chapter 2 as potential monitoring parameters to evaluate the impacts of a hazardous waste incineration facility (see Table 2.8 and Table 2.9). Table 3.1 summarizes these physical/chemical properties for a number of the analytes previously identified. Table 3.2 summarizes the toxicological benchmark values for the same selected analytes that are available from a variety of sources so that these benchmark values may be used in appropriate risk assessment procedures. Table 3.3 summarizes some of the relevant federal water quality criteria that may be applicable in terms of a preliminary evaluation of aquatic ecological impacts from a hazardous waste incinerator or related facility. Many individual states have developed criteria as well. There are not similar values available for soils, and sediment quality criteria are presently undergoing development at the federal level and in many states.

The sources of the summary data from which Tables 3.1 to 3.3 were compiled are identified in the footnotes that accompany the respective table.

Selected physical and chemical properties

Introduction

The purpose of this section is to summarize relevant information concerning the physical and chemical properties of analytes that may be emitted from hazardous waste incinerators. These properties may be used in a number of the calculations that describe the potential risks to humans and to the environment from such emissions.

Table 3.1 Physical and Chemical Data for Selected Analytes Reported for Emissions from Hazardous Waste Incinerators

Chemical Name	CAS Registry No.	Solubility H_2O (mg/L)	Ref.	K_{oc} (L/kg)	Ref.	Molecular Weight (g/mol)	Ref.	Vapor Pressure (mmHg)	Ref.	Henry's Law Constant (atm-m³/mole)	Ref.	Log K_{ow} (dimensionless)	Ref.
Organics													
Acenaphthene	83-32-9	4.24E+00	a	4.90E+03	a	154.21	a	4.47E-03	a	1.55E-04	a	3.92	a
Acenaphthylene	208-96-8	1.60E+01	a	2.50E+03	b	152.00	b	9.10E-04	a	1.10E-04	a	3.60	a
Acetonitrile	75-05-8	1.00E+06	a	1.60E+01	c	41.10	c	9.10E+01	a	3.50E-05	a	-0.34	a
Acrylonitrile	107-13-1	7.90E+04	h	8.50E-01	b	53.06	b	1.00E+02(23°C)	a	8.80E-05	h	0.25	c
Aniline	62-53-3	3.60E+04	c	2.55E+01	c	93.13	c	6.70E-01	c	1.20E-04	c	0.90	c
Anthracene	120-12-7	4.30E-02	a	2.35E+04	a	178.00	a	2.70E-06	a	6.50E-05	a	4.60	a
Azobenzene	103-33-3	6.40E+00	c	2.82E+03	c	182.22	c	3.60E-04	a	1.35E-05	c	3.82	c
Benzaldehyde	100-52-7	3.00E+03	c	9.20E+01	c	106.12	c	1.27E-01	a	2.60E-05	c	1.48	c
Benz[a]anthracene	56-55-3	9.40E-03	a	2.28E+02	a	228.29	a	1.10E-07	a	3.40E-06	a	5.70	a
Benzene	71-43-2	1.80E+03	a	6.20E+01	a	78.10	a	9.50E+01	a	5.60E-03	a	2.10	a
Benzidine	92-87-5	5.20E+02	i	9.12E+02	i,ee	184.24	i,ee	5.00E-04	i	3.00E-07	i	1.34	i
Benz[e]anthracene	NF	NF	—	NF	—	NF	—	NF	—	NF	—	NF	—
Benzo[b]fluoranthene	205-99-2	1.50E-03	a	1.23E+06	a	252.00	a	5.00E-07	a	1.10E-07	a	6.20	a
Benzo[j]fluoranthene	205-82-3	6.76E-03	j	5.95E+04	j	252.31	c,ee	1.50E-08	j	1.00E-06	j	6.12	j
Benzo[k]fluoranthene	207-08-9	8.00E-04	a	1.23E+06	a	252.00	a	2.00E-09	a	8.30E-07	a	6.20	a
Benzo[g,h,i]perylene	191-24-2	2.60E-04	a	1.60E+06	a	276.00	b	1.00E-10	a	1.40E-07	a	6.70	a
Benzo[a]pyrene	50-32-8	1.60E-03	a	9.69E+05	a	252.00	a	5.50E-09	a	1.10E-06	a	6.10	a
Benzo[e]pyrene	192-97-2	6.30E-03	j	7.60E+04	a	252.31	c	5.70E-09	a	3.00E-07	e,c	6.44	a
Benzyl chloride	100-44-7	5.30E+02	a	3.03E+02	c	127.00	c	1.30E+00	a	4.20E-04	a	2.30	c
1,1-Biphenyl	92-52-4	7.48E+00	c	1.45E+03	c	154.21	c	8.93E-03	a	3.00E-04	c	4.01	c
Bis[2-chloro-ethoxy]methane	111-91-1	1.20E+05	a	3.40E+01	a	173.00	c,ee	1.40E-04	a	1.70E-07	a	0.75	a
Bis[2-chloroethyl]ether	111-44-4	1.72E+04	a	7.60E+01	a	143.01	a	1.50E+00(20°C)	a	1.80E-05	a	1.21	a
Bis[chloromethyl]ether	542-88-1	2.20E+04	k	NF	—	114.96	a	3.00E+01(22°C)	a	2.10E-04	k	-0.38	k
Bis[2-ethylhexyl]phthalate	117-81-7	3.40E-01	a	1.11E+05	a	391.00	a	6.40E-06	a	1.00E-07	a	7.30	a
Bromochloromethane	74-97-5	1.67E+04	c	8.00E+01	c	129.38	c	1.55E+02	a	1.50E-03	c	1.14	b
Bromodichloromethane	75-27-4	6.70E+03	a	5.50E+01	a	164.00	a	5.00E+01	a	1.60E-03	a	2.10	a
Bromoform	75-25-2	3.10E+03	a	1.26E+02	a	252.73	a	5.00E+00	a	5.35E-04	a	2.35	a
Bromomethane (methyl bromide)	74-83-9	1.50E+04	a	9.00E+00	a	94.90	a	1.60E-03	a	6.20E-03	a	1.20	a
1,3-Butadiene	106-99-0	7.35E+02	c	2.88E+02	c	54.09	dd	9.10E+02(20°C)	a	6.20E-02	c	1.99	c
Butyl benzyl phthalate	85-68-7	2.69E+00	a	1.37E+04	a	312.36	a	8.60E-06	a	1.26E-06	a	4.84	a
Captan	133-06-2	3.30E+00	c	2.33E+03	c	300.59	c	7.5E-06(20°C)	ff	7.90E-06	c	2.35	c

(continues)

Table 3.1 Physical and Chemical Data for Selected Analytes Reported for Emissions from Hazardous Waste Incinerators (continued)

Chemical Name	CAS Registry No.	Solubility H$_2$O (mg/L)	Ref.	K$_{oc}$ (L/kg)	Ref.	Molecular Weight (g/mol)	Vapor Pressure (mmHg)	Ref.	Henry's Law Constant (atm-m^3/mole)	Ref.	Log K$_{ow}$ (dimensionless)	Ref.
Carbon disulfide	75-15-0	1.19E+03	a	4.57E+01	a	76.13	2.97E+02(20°C)	a	3.03E-02	a	2.00	a
Carbon tetrachloride	56-23-5	7.90E+02	a	1.52E+02	a	154.00	1.20E+02	a	3.00E-02	a	2.70	a
Chlorobenzene	108-90-7	4.70E+02	a	2.24E+02	a	113.00	1.20E-01	a	3.70E-03	a	2.90	a
Chlorocyclopentadiene	41851-50-7	NF	—	5.25E+01	a	100.55	NF	—	NF	—	NF	—
Chloroform	67-66-3	7.90E+03	a	3.88E+02	a	119.00	2.00E+02	a	3.70E-03	a	1.90	a
o-Chlorophenol	95-57-8	2.20E+04	a	3.98E+05	a	128.56	2.20E+00(20°C)	a	3.91E-04	a	2.15	a
Chrysene	218-01-9	1.60E-03	a	3.98E+05	a	228.00	6.20E-09	a	9.50E-05	a	5.70	a
Coronene (hexabenzo-benzene)	191-07-1	1.39E-04	ii	NF	—	300.36	3.75E-13	ii	1.00E-09	ii	NF	—
Cumene	98-82-8	6.10E+01	a	2.82E+03	a	120.00	4.50E+00	a	1.20E+00	a	3.60	a
Dibenzo[a,e]fluoranthene	5385-75-1	NF	—	NF	—	302.37	NF	—	NF	—	NF	—
Dibenzo[a,h]fluoranthene	NF	NF	—	NF	—	NF	NF	—	NF	—	NF	—
Dibenzodioxins, chlorinated												
2,3,7,8-TCDD	1746-01-6	7.90E-06	c	2.45E+07	c	321.97	7.40E-10	c	1.62E-05	c	6.80	c
1,2,3,7,8-PCDD	40321-76-4	1.18E-04	jj	NF	—	356.42	6.60E-10	jj	2.60E-06	jj	9.06	jj,ee
1,2,3,4,6,7,8-HCDD	34465-46-8	4.00E-06	c	2.40E+05	c	390.86	4.40E-11	c	5.70E-06	c	9.795	jj,ee
1,2,3,4,7,8,9-HCDD	37871-00-4	1.90E-03	c	7.80E+5	c	425.28	5.60E-12	jj,ee	2.18E-05	c	10.535	jj,ee
1,2,3,4,6,7,8,9-OCDD	3268-87-9	7.40E-08	c	79432823	c	459.75	8.25E-13	c	6.74E-06	c	11.08	jj,ee
Dibenzofurans, chlorinated												
1,3,7,8-TCDF	57117-35-8	4.20E-04	kk,hh	4.07E+05	kk,hh	305.96	9.21E-07	kk,hh	1.48E-05	kk	5.82	kk,hh
1,2,3,4,8-PCDF	67517-48-0	2.40E-04	kk,hh	NF	—	340.42	2.73E-07	kk,hh	2.63E-05	kk	6.79	kk
1,2,3,4,7,8-HCDF	70648-26-9	1.80E-05	kk	NF	—	374.87	6.07E-08	—	2.78E-05	kk	NF	—
1,2,3,4,6,7,8-HCDF	67562-39-4	1.40E-05	kk	NF	—	409.31	1.68E-08	kk	4.10E-06	kk	7.92	kk
1,2,3,4,6,7,8,9-OCDF	39001-02-0	1.20E-06	kk	3.72E+08	kk	443.76	NF	—	1.70E-06	kk	8.20	kk
Dibenzofuran	132-64-9	1.00E+01	a	9.12E+03	a	168.00	1.80E-04	a	1.30E-05	a	4.20	a
Dibromochloromethane	124-48-1	2.60E+03	a	6.31E+01	a	208.00	4.90E+00	a	7.80E-04	a	2.20	a
1,2-Dibromo-3-choropropane	96-12-8	1.20E+03	b	9.80E+01	b	236.00	5.80E-01	a	1.50E-04	a	2.3	a
Dibutyl phthalate	84-74-2	1.10E+01	a	1.57E+03	a	278.00	7.30E-05	a	9.40E-10	a	4.60	a
1,2-Dichlorobenzene	95-50-1	1.60E+02	a	3.79E+00	a	147.00	1.40E+00	a	1.90E-03	a	3.40	a
1,3-Dichlorobenzene	541-73-1	1.30E+02	e	4.98E+02	a	147.00	2.20E+00	a	3.10E-03	a	3.50	a
1,4-Dichlorobenzene	106-46-7	7.40E+01	a	6.16E+02	a	147.00	1.00E+00	a	2.40E-03	a	3.40	a
3,3'-Dichlorobenzidine	91-94-1	3.11E+00	a	7.24E+02	a	253.13	NF	—	4.00E-09	a	3.51	a

(continues)

Chemical	CAS No.											
3,4-Dichlorobutene	760-23-6	4.20E+02	c	1.60E+02	c	125.00	2.19E+01	a	8.56E-03	c	NF	—
Dichlorodifluoromethane	75-71-8	2.80E+02	a	2.00E+02	c	121.00	4.80E+08	a	3.40E-01	a	2.20	a
1,1-Dichloroethane	75-34-3	5.10E+03	a	5.30E+03	a	99.00	2.30E+02	a	5.60E-03	a	1.80	a
1,2-Dichloroethane	107-06-2	8.50E+03	a	3.80E+01	a	99.00	7.90E+01	a	9.80E-04	a	1.50	a
1,1-Dichloroethene	75-35-4	2.30E+03	a	6.50R+01	a	96.90	6.00E+02	a	2.60E-02	a	2.10	a
2,4-Dichlorophenol	120-83-2	4.50E+03	a	1.59E+02	a	163.00	7.50E-02	a	3.16E-06	a	3.08	a
1,3-Dichloropropene	542-75-6	2.70E+03	a	2.70E+01	a	11.00	3.40E+01	a	1.80E-02	a	2.00	a
1,3-Dichloropropene (cis)	10061-01-5	2.70E+03	a	2.70E+03	c	111.00	3.30E+01	a	NF	—	2.00	a
Diethyl phthalate	84-66-2	1.10E+03	a	8.22E+01	a	222.00	1.70E-03	a	4.50E-07	a	2.50	a
2,4-Dimethyl phenol	105-67-9	7.90E+03	a	2.09E+02	a	122.00	9.80E-02	a	2.00E+00	a	2.40	a
Dimethyl phthalate	131-11-3	4.00E+03	a	4.60E+01	a	194.00	1.70E-03	a	1.10E-07	a	1.60	a
Dimethyl sulfide	75-18-3	2.20E+04	c	4.35E+01	c	62.13	5.02E+02	a	1.69E-03	c	NF	—
1,4-Dioxane	123-91-1	>1.0E+06	c	1.70E+01	c	88.11	3.70E+01	c	4.88E-06	gg	-0.27	c
Epichlorohydrin	106-89-8	NF	—	1.23E+02	e	92.53	1.00E+01(16.6°C)	a	NF	a	0.26	c
Ethylbenzene	100-41-4	1.70E+02	a	2.04E+02	a	106.00	9.60E+00	a	7.90E-03	a	3.10	a
Fluoranthene	206-44-0	2.10E-01	a	4.91E+04	a	202.00	7.80E-06	a	1.60E-05	a	5.10	a
Fluorene	86-73-7	2.00E+00	a	7.71E+03	a	166.00	6.30E-04	a	6.40E-05	1	4.20	a
Formaldehyde	50-00-0	NF	—	2.20E+00	—	30.03	3.88E+03	a	3.27E-07	a	0.35	1
Hexachlorobenzene	118-74-1	6.20E+00	a	8.00E+04	a	284.78	1.09E-05	a	1.32E-03	a	5.89	a
Hexachlorobutadiene	87-68-3	3.20E+00	a	5.37E+04	a	261.00	2.20E-01	a	8.10E-03	a	4.80	a
Hexachlorocyclo-pentadiene	77-47-4	1.80E+00	a	2.00E+05	a	272.77	8.00E-02	a	2.70E-02	a	5.39	a
Hexachloroethane	67-72-1	5.00E+01	a	1.78E+03	a	236.74	2.10E-01(20°C)	a	3.89E-03	a	4.00	a
Hydrogen sulfide	6/4/83	4.10E+03	a	NF	—	34.10	1.60E+04	a	NF	—	NF	—
Methyl bromide (bromomethane)	74-83-9	1.52E+04	a	9.00E+00	a	94.94	1.62E+03	a	6.24E-03	a	1.19	a
Methyl chloride (chloromethane)	74-87-3	5.32E+03	c	5.25E+01	a	50.49	4.30E+03	c	8.82E-02	c	0.91	c
Methyl mercaptan	74-93-1	1.54E+04	c	1.70E+01	c	48.10	1.52E+03	c	1.23E-01	c	NF	—
Methyl mercury	22967-92-6	NF	—	NF	—	215.62	NF	—	NF	—	NF	—
Methylene bromide (dibromomethane)	74-95-3	1.17E+01	c	2.50E+01	c	173.83	4.00E+01(23.3°C)	a	8.88E-04	a	NF	—
Methylene chloride (dichloromethane)	75-09-2	1.30E+04	a	1.00E+01	a	84.90	4.30E+02	a	2.20E-03	a	1.30	a
Methyl ethyl ketone	78-93-3	2.20E+05	a	3.40E+01	c	72.10	9.50E+01	a	5.60E-05	a	0.28	a
Naphthalene	91-20-3	3.10E+01	a	1.19E+03	a	128.00	8.50E-02	a	4.80E-04	a	3.40	a
Nitrobenzene	98-95-3	2.09E+03	a	1.19E+02	a	123.11	2.45E-01	a	2.40E-05	a	1.84	a
o-Nitrophenol	88-75-5	2.20E+03	a	6.50E+01	c	139.00	1.10E-01	a	9.50E-06	a	1.80	a

Table 3.1 Physical and Chemical Data for Selected Analytes Reported for Emissions from Hazardous Waste Incinerators (continued)

Chemical Name	CAS Registry No.	Solubility H_2O (mg/L)	Ref.	K_{oc} (L/kg)	Ref.	Molecular Weight (g/mol)	Vapor Pressure (mmHg)	Ref.	Henry's Law Constant (atm-m³/mole)	Ref.	Log K_{ow} (dimensionless)	Ref.
N-nitrosomethyl-ethylamine	10595-95-6	2.00E+04	a	3.90E+01	c	88.10	NF	—	NF	—	-0.12	a
N-nitroso-n-dibutylamine	924-16-3	1.30E+03	e	2.63E+02	c	158.24	3.00E-02(20°C)	a	5.10E-06	c	1.92	c
N-nitrosopyrrolidine	930-55-2	>1.0E+06	c	NF	—	100.12	9.20E-02	c	1.99E-07	c	-0.19	c
Pentachlorobenzene	608-93-5	2.40E-01	c	1.38E+04	c	250.34	2.16E-05	a	7.10E-04	c	5.17	c
Pentachloroethane	76-01-7	5.00E-02	c	2.44E+02	c	202.29	3.50E+00	a	1.90E-03	c	NF	—
Pentachlorophenol (PCP)	87-86-5	2.00E-03	a	2.00E+04	a	266.00	3.20E-05	a	2.40E-08	a	5.10	a
Phenanthrene	85-01-8	1.20E+00	b	1.40E+04	a	178.00	1.10E-04	a	2.30E-05	a	4.60	a
Phenol	108-95-2	8.30E+04	a	2.88E+01	a	94.10	2.80E-01	a	4.00E-07	a	1.50	a
Polychlorinated biphenyls (PCB)	1336-36-3	7.00E-02	a	3.09E+05	a	Varies	7.70E-05	a	2.60E-03	a	6.00	a
Pyrene	129-00-0	1.40E-01	a	6.80E+04	a	202.00	4.60E-06	a	1.10E-05	a	5.10	a
Styrene	100-42-5	3.10E+02	a	9.12E+02	a	104.15	6.12E+00	a	2.75E-03	a	2.94	a
1,2,4,5-Tetrachlorobenzene	95-94-3	5.90E-01	a	7.58E+03	c	216.00	5.40E-03	a	2.60E-03	a	4.60	a
1,1,1,2-Tetrachloroethane	630-20-6	1.10E+03	a	1.53E+02	c	168.00	1.20E+01	a	2.40E-03	a	2.60	a
1,1,2,2-Tetrachloroethane	79-34-5	3.00E+03	a	7.90E+01	a	168.00	4.60E+00	a	3.50E-04	a	2.40	a
Tetrachloroethene	127-18-4	2.00E+02	a	2.65E+02	a	165.83	1.85E+01	a	1.84E-02	a	2.67	a
Toluene	108-88-3	5.30E+02	a	1.40E+02	a	92.10	2.80E+01	a	6.60E-03	a	2.80	a
2,6-Toluenediisocyanate	91-08-7	NF	—	NF	—	174.16	8.00E-03	a	NF	—	NF	—
1,2,4-Trichlorobenzene	120-82-1	3.50E+01	a	1.66E+03	a	181.00	4.30E-01	a	1.40E-03	a	4.00	a
1,1,1-Trichloroethane	71-55-6	1.30E-03	a	1.35E+02	a	133.00	1.20E+02	a	1.70E-02	a	2.50	a
1,1,2-Trichloroethane	79-00-5	4.40E+03	a	7.50E+01	a	133.00	2.30E+01	a	9.10E-04	a	2.10	a
Trichloroethene (TCE)	79-01-6	1.10E+03	a	9.40E+01	a	131.00	7.30E+01	a	1.00E-02	a	2.70	a
Trichlorofluoromethane	75-69-4	1.10E+03	a	NF	—	137.00	8.00E+02	a	9.70E-02	a	2.50	a
2,4,5-Trichlorophenol	95-95-4	1.20E+03	a	2.38E+03	a	197.45	2.20E-02	a	4.33E-06	a	3.90	a
2,4,6-Trichlorophenol	88-06-2	8.00E+02	a	1.07E+03	a	197.45	1.00E+00(76.5°C)	a	7.79E-06	a	3.70	a
1,2,3-Trichloropropane	96-18-4	1.80E+03	a	7.20E+01	c	147.00	3.70E+00	a	4.10E-04	a	2.30	a
Vinyl chloride	75-01-4	2.80E+03	a	1.86E+01	a	62.50	3.00E+03	a	2.70E-02	a	1.50	a
m-Xylene	108-38-3	1.60E+02	a	4.07E+02	a	106.00	8.50E+00	a	7.30E-03	a	3.20	a
o-Xylene	95-47-6	1.80E+02	a	3.63E+02	a	106.00	6.60E+00	a	5.20E-03	a	3.10	a
p-Xylene	106-42-3	1.90E+02	a	3.89E+02	a	106.00	8.90E+00	a	7.70E-03	a	3.20	a
Xylenes (total)	1330-20-7	1.30E+02	t	2.49E+02	t	318.50	7.99E+00	a	7.00E-03	c	3.12	t

Inorganics	CAS											
Antimony	7440-36-0	NA		—	1.00E+00(886°C)	121.76	n	NA		—	NA	—
Arsenic, inorganic	7440-38-2	NA		—	1.00E+00(372°C)	74.92	p	NA		—	NA	—
Barium	7440-39-3	NA		—	1.00E+01(1049°C)	137.33	r	NA		—	NA	—
Beryllium	7440-41-7	NA		—	1.00E+00(1,520°C)	9.01	g	NA		—	NA	—
Cadmium	7440-43-9	NA		—	1.00E+00(394°C)	112.41	f	NA		—	NA	—
Carbon monoxide	630-08-0	NA		—	9.68E-01	28.01	c	NA		—	NA	—
Chlorine	7782-50-5	NA		—	5.83E-03	70.91	c	NA		—	NA	—
Chromium (III)	16065-83-1	NA		—	NA	52.00	—	NA		—	NA	—
Chromium (VI)	18540-29-9	NA		—	NA	52.00		NA		—	NA	—
Cobalt	7440-48-4	NA		—	1.00E+00(1,910°C)	58.93	cc	NA		—	NA	—
Copper	7440-50-8	NA		—	1.00E+01(1870°C)	63.55	s	NA		—	NA	—
Hydrogen chloride (hydrochloric acid)	7647-01-0	>1.0E+06	c	—	3.54E+04	36.46	c	NA		—	NA	—
Hydrogen fluoride	7664-39-3	>1.0E+06	mm	—	9.17E+02	20.01	c	NA		—	NA	—
Iron	7439-89-6	NA		—	1.00E+00(1787°C)	55.85	c	NA		—	NA	—
Lead	7439-92-1	NA		—	1.77E+00(1000°C)	207.20	o	NA		—	NA	—
Magnesium	7439-95-4	NA		—	1.00E+00(621°C)	24.31	o	NA		—	NA	—
Manganese	7439-96-5	Decomposes	nn	—	1.00E+00(1,292°C)	54.94	c	1.14E-02		—	NA	—
Mercury	7439-97-6	7.70E-08	e	—	1.20E-03	200.59	bb	NA	e	—	NA	—
Mercuric chloride	7487-94-7	NA		—	1.00E+00(136.2°C)	271.50	e	NA		—	NA	—
Molybdenum	7439-98-7	NA		—	NA	95.94	c	NA		—	NA	—
Nickel	7440-02-0	NA		—	1.00E+00(1,810°C)	58.69	—	NA		—	NA	—
Nitrogen dioxide	10102-44-0	NA		—	9.08E+01	46.01	z	NA		—	NA	—
Nitrogen monoxide	10102-43-9	NA		—	4.56E+04(-94.8°C)	30.01	c	NA		—	NA	—
Particulate matter	NF	NA		—	NA	NF	c	NA		—	NA	—
Scandium	7440-20-2	NA		—	NA	45.00	—	NA		—	NA	—
Selenium (and Compounds)	7782-49-2	NA		—	1.00E+00(356°C)	78.96	x	NA		—	NA	—
Silicon	7440-21-3	NA		—	1.33E-05	28.09	c	NA		—	NA	—
Silver	7440-22-4	NA		—	1.00E+02(1,865°C)	107.87	y	NA		—	NA	—
Sulfur dioxide	7446-09-5	NA		—	2.47E+03(20°C)	64.06	w	NA		—	NA	—
Sulfur trioxide	7446-11-9	NA		—	7.30E+01(25°C)	80.06	v	NA		—	NA	—
Tellurium	13494-80-9	NA		—	NF	127.60	—	NA		—	NA	—
Thallium	7440-28-0	NA		—	1.00E+01(1000°C)	204.38	u	NA		—	NA	—
Tin	7440-31-5	NA		—	1.0E+00(1492°C)	118.69	c	NA		—	NA	—
Titanium	7440-32-6	NA		—	NF	47.87	—	NA		—	NA	—
Vanadium	7440-62-2	NA		—	NF	50.94		NA		—	NA	—
Zinc	7440-66-6	NA		—	1.00E+00(487°C)	65.39	q	NA		—	NA	—

(continues)

Table 3.1 Physical and Chemical Data for Selected Analytes Reported for Emissions from Hazardous Waste Incinerators (continued)

a) U.S. EPA. 1996. Soil Screening Guidance: Technical Background Document.

b) Electronic Handbook of Risk Assessment Values (EHRAV).

c) Hazardous Substance Data Bank (HSDB). 1998.

d) Handbook of Chemical Property Estimation Methods. 1990.

e) Superfund Chemical Data Matrix (SCDM). 1996.

f) ATSDR Profile for Cadmium, Draft Update. September, 1997.

g) ATSDR Profile for Beryllium, Update. April, 1993.

h) ATSDR Profile for Acrylonitrile, Final. December, 1990.

i) ATSDR Profile for Benzidine, Update. August, 1995.

j) ATSDR Profile for Polycyclic Aromatic Hydrocarbons (PAHs), Update. August, 1995.

k) ATSDR Profile for Bis[chloromethyl]ether, Final. December, 1989.

l) ATSDR Profile for Formaldehyde, Draft. September, 1998.

m) ATSDR Profile for Methyl mercaptan. September, 1992.

n) ATSDR Profile for Antimony, Draft. February, 1991.

o) ATSDR Profile for Lead, Draft Update. September, 1997.

p) ATSDR Profile for Arsenic, Update Draft. October, 1992.

q) ATSDR Profile for Zinc, Update. May, 1994.

r) ATSDR Profile for Barium, Final. July, 1992.

s) ATSDR Profile for Copper, Final. December, 1990.

t) ATSDR Profile for Xylenes, Update. August, 1995.

u) ATSDR Profile for Thallium., Final. July, 1992.

v) ATSDR Profile for Sulfur Trioxide & Sulfuric Acid, Draft. September, 1998.

w) ATSDR Profile for Sulfur Dioxide, Draft. September, 1998.

x) ATSDR Profile for Selenium, Update. August, 1996.

y) ATSDR Profile for Silver, Final. December, 1990.

z) ATSDR Profile for Nickel, Update. September, 1997.

(continues)

aa) ATSDR Profile for Mercury, Update. May, 1994.
bb) ATSDR Profile for Manganese, Draft Update. September, 1997.
cc) ATSDR Profile for Cobalt, Final. July, 1992.
dd) ATSDR Profile for 1,3-Butadiene, Final. July, 1992.
ee) Value represents the midrange of the reported log K,
ff) Howard. 1991. Handbook of Environmental Fate and Exposure Data. Vol. 111. Pesticides
gg) Howard. 1991. Handbook of Environmental Fate and Exposure Data. Vol. II. Solvents
hh) Data for similar isomeric forms.
ii) PAH database. Website www.jaseo.co.jp/JINNO/DATABASE
jj) ASTDR Profile for Chlorinated Dibenzo-p-Dioxins, Update. September, 1997.
kk) ATSDR Profile for Chlorodibenzofurans. May, 1994.

Table 3.1 References

ATSDR (Agency for Toxic Substances and Disease Registry). 1989. Toxicological Profile for Bis[chloromethyl]ether, Final. December, 1989.
ATSDR (Agency for Toxic Substances and Disease Registry). 1990. Toxicological Profile for Acrylonitrile, Final. December, 1990.
ATSDR (Agency for Toxic Substances and Disease Registry). 1990. Toxicological Profile for Copper, Final. December, 1990.
ATSDR (Agency for Toxic Substances and Disease Registry). 1990. Toxicological Profile for Silver, Final. December, 1990.
ATSDR (Agency for Toxic Substances and Disease Registry). 1991. Toxicological Profile for Antimony, Draft. February, 1991.
ATSDR (Agency for Toxic Substances and Disease Registry). 1992. Toxicological Profile for Arsenic, Update Draft. October, 1992.
ATSDR (Agency for Toxic Substances and Disease Registry). 1992. Toxicological Profile for Barium, Final. July, 1992.
ATSDR (Agency for Toxic Substances and Disease Registry). 1992. Profile for 1,3-Butadiene, Final. July, 1992.
ASTDR (Agency for Toxic Substances and Disease Registry). 1997. Profile for Chlorinated Dibenzo-p-Dioxins, Update. September, 1997.
ATSDR (Agency for Toxic Substances and Disease Registry). 1994. Profile for Chlorodibenzofurans. May, 1994.
ATSDR (Agency for Toxic Substances and Disease Registry). 1992. Toxicological Profile for Cobalt, Final. July, 1992.
ATSDR (Agency for Toxic Substances and Disease Registry). 1992. Toxicological Profile for Methyl mercaptan. September, 1992.
ATSDR (Agency for Toxic Substances and Disease Registry). 1992. Toxicological Profile for Thallium, Final. July, 1992.

Table 3.1 Physical and Chemical Data for Selected Analytes Reported for Emissions from Hazardous Waste Incinerators (continued)

ATSDR (Agency for Toxic Substances and Disease Registry). 1993. Toxicological Profile for Beryllium, Update. April, 1993.

ATSDR (Agency for Toxic Substances and Disease Registry). 1994. Toxicological Profile for Mercury, Update. May, 1994.

ATSDR (Agency for Toxic Substances and Disease Registry). 1994. Toxicological Profile for Zinc, Update. May, 1994.

ATSDR (Agency for Toxic Substances and Disease Registry). 1995. Toxicological Profile for Benzidine, Update. August, 1995.

ATSDR (Agency for Toxic Substances and Disease Registry). 1995. Toxicological Profile for Polycyclic Aromatic Hydrocarbons (PAHs), Update. August, 1995.

ATSDR (Agency for Toxic Substances and Disease Registry). 1995. Toxicological Profile for Xylenes, Update. August, 1995.

ATSDR (Agency for Toxic Substances and Disease Registry). 1996. Toxicological Profile for Selenium, Update. August, 1996.

ATSDR (Agency for Toxic Substances and Disease Registry). 1997. Toxicological Profile for Cadmium, Draft Update. September, 1997.

ATSDR (Agency for Toxic Substances and Disease Registry). 1997. Toxicological Profile for Lead, Draft Update. September, 1997.

ATSDR (Agency for Toxic Substances and Disease Registry). 1997. Toxicological Profile for Manganese, Draft Update. September, 1997.

ATSDR (Agency for Toxic Substances and Disease Registry). 1997. Toxicological Profile for Nickel, Update. September, 1997.

ATSDR (Agency for Toxic Substances and Disease Registry). 1998. Toxicological Profile for Formaldehyde, Draft. September, 1998.

ATSDR (Agency for Toxic Substances and Disease Registry). 1998. Toxicological Profile for Sulfur Dioxide, Draft. September, 1998.

ATSDR (Agency for Toxic Substances and Disease Registry). 1998. Toxicological Profile for Sulfur Trioxide & Sulfuric Acid, Draft. September, 1998.

Electronic Handbook of Risk Assessment Values (EHRAV).

Hazardous Substance Data Base (HSDB). 1998. On-line computer database.

Howard. 1991. Handbook of Environmental Fate and Exposure Data. Vol. 11. Solvents

Howard. 1991. Handbook of Environmental Fate and Exposure Data. Vol. 111. Pesticides

Lyman, W.J., W.F. Reehl et al. (eds.). 1990. In: *Handbook of Chemical Property Estimation Methods*. American Chemical Society, Washington, DC.

PAH database. Website www.jaseo.co.jp/JINNO/DATABASE

U.S. EPA. 1996. Soil Screening Guidance: Technical Background Document. EPA/540/R-95/128. May, 1996.

U.S. EPA. 1996. Superfund Chemical Data Matrix (SCDM).

As described in greater detail in Chapter 2, the analytes presented in Tables 2.8 and 2.9 were compiled from a number of sources. These sources, listed in alphabetical order, include

Becker, R. A. 1994. Personal communication; memorandum from Air Board of the California Environmental Protection Agency.

Canadian Council of Ministers of the Environment (CCME). 1992. *National Guidelines for Hazardous Waste Incineration Facilities*. March, 1992.

Cooper, C. D. 1994. Personal communication, August 1994.

Dempsey, C. R. and E. T. Oppelt. 1993. Incineration of Hazardous Waste: A Critical Review Update. *Air Waste* 43: 25–73.

Travis, C. C. et al. 1984. *Inhalation Pathway Risk Assessment of Hazardous Waste Incineration Facilities*. Oak Ridge National Laboratory, U.S. Environmental Protection Agency. DE85-001694.

Travis, C. C. and S. C. Cook. 1989. *Hazardous Waste Incineration and Human Health*, CRC Press, Boca Raton, FL.

Trenholm, A. et al. 1984. *Performance Evaluation of Full-Scale Hazardous Waste Incinerators*. Volumes I–V, Industrial Environmental Research Laboratory, U.S. Environmental Protection Agency. EPA-600/2-84/181a-e, NTIS PB85-129500.

Trenholm, A. et al. 1987. *Total Mass Emissions from a Hazardous Waste Incinerator*. Midwest Research Institute, U.S. Environmental Protection Agency. EPA-600/2-87/064, NTIS PB87-228508.

U.S. Environmental Protection Agency. 1993. *Addendum to Methodology for Assessing Health Risks Associated with Indirect Exposure to Combustor Emissions*. (External Review Draft) EPA/600/AP-93/003.

U.S. Environmental Protection Agency. 1993. *Risk Assessment for the Waste Technologies Industries (WTI) Hazardous Waste Incinerator Facility (East Liverpool, Ohio)*. (External Review Draft) EPA 905-D95-002c.

Van Buren, D. et al. 1987. *Characterization of Hazardous Waste Incinerator Residuals*. Acurex Corporation. EPA-600/2-87/017, NTIS PB87-168159.

The following sections provide a discussion of the physical and chemical properties for the identified analytes or analyte groups as well as an indication of the relevance of each of these properties to the health risk assessment calculations or investigations pertaining to environmental fate and transport. For example, the organic carbon partition coefficient (expressed as K_{oc}, or as log K_{oc}) represents a measure of the affinity of substances for organic matter in soils, sediments, or tissues and hence the potential for leachability from soils or for bioconcentration in animals or humans. Similarly, water solubility (S) is important in evaluating the potential for movement in groundwater.

Table 3.2   Toxicological Data for Selected Analytes Reported in Emissions from Hazardous Waste Incinerators

Chemical Name	CAS Registry Number	Oral		Inhalation		Carcin. Classif.
		RfD (mg/kg·day)	Target Organ/Effect	RfC (mg/m^3)	Target Organ/Effect	
Organics						
Acenaphthene	83-32-9	6.0E–02^I	Lv	1.10E–01^I	NF	NF
Acenaphthylene	208-96-8	3.0–E–02^G	NF	1.10E–01^I	NF	D
Acetonitrile	75-05-8	6.0E–03^I	Bl,Lv	5.0E–02^A	Lv,Bl^H	NF
Acrylonitrile	107-13-1	1.0E–03^H	Re	2.0E–03^I	R	B1
Aniline	62-53-3	NF	NF	1.0E–03^I	Spl	B2
Anthracene	120-12-7	3.0E–01^I	NF	NF	NF	D
Azobenzene	103-33-3	NF	NF	NF	NF	B2
Azobenzene (cis)	1080-16-6	NF	NF	NF	NF	NF
Benzaldehyde	100-52-7	1.0E–01^I	K, Fore	NF	NF	NF
Benz[a]anthracene^c	56-55-3	NF	NF	NF	NF	B2
Benz[e]anthracene	NF	NF	NF	NF	NF	NF
Benzene	71-43-2	UR	NF	6.0E–03^E	NF	A
Benzidine	92-87-5	3.0E–03^I	B,Lv	NF	NF	A
Benzo[b]fluoranthene^c	205-99-2	NF	NF	NF	NF	B2
Benzo[j]fluoranthene	205-82-3	NF	NF	NF	NF	NF
Benzo[k]fluoranthene^c	207-08-9	NF	NF	NF	NF	B2
Benzo[g,h,i]perylene	191-24-2	NF	NF	NF	NF	D
Benzo[a]pyrene	50-32-8	NF	NF	NF	NF	B2
Benzo[e]pyrene	192-97-2	NF	NF	NF	NF	NF
Benzyl chloride	100-44-7	NF	NF	NF	NF	B2
1,1-Biphenyl	92-52-4	5.0E–02^I	K	NF	NF	D
Bis[2-chloroethoxy]methane	111-91-1	NF	NF	NF	NF	D
Bis[2-chloroethyl]ether (BCEE)	111-44-4	NF	NF	NF	NF	B2
Bis[chloromethyl]ether (BCME)	542-88-1	NF	NF	NF	NF	A
Bis[2-ethylhexyl]phthalate (DEHP)	117-81-7	2.0E–02^I	Lv	NF	NF	B2
Bromochloromethane	74-97-5	1.30E–02^H	NF	NF	NF	D
Bromoform	75-25-2	2.0E–02^I	Lv	NF	NF	B2
1,3-Butadiene	106-99-0	NF	NF	NF	NF	B2
Butyl benzyl phthalate	85-68-7	2.0E–01^I	Lv	NF	NF	C
Captan	133-06-2	1.3E–01^I	Wt Ls	UR	NF	B2^H
Carbon disulfide	75-15-0	1.0E–01^I	Fe	7.0E–01^I	Per Nerv	NF
Carbon tetrachloride	56-23-5	7.0E–04^I	Lv	2.0E–03^E	NF	B2
Chlorobenzene	108-90-7	2.0E–02^I	Lv	2.0E–02^A	NF	D
Chlorocyclopentadiene	41851-50-7	NF	NF	NF	NF	D
Chloroform	67-66-3	1.0E–02^I	Lv	NF	NF	B2
o-Chlorophenol	95-57-8	5.0E–03^I	Re	NF	NF	NF
Chrysene^c	218-01-9	NF	NF	NF	NF	B2
Coronene (hexabenzobenzene)	197-07-1	NF	NF	NF	NF	NF
Cumene	98-82-8	1.0E–01^I	K	4.0E–01^I	K, Ad	D
Dibenzo[a,e]fluoranthene	5385-75-1	NF	NF	NF	NF	NF
Dibenzo[a,h]fluoranthene	NF	NF	NF	NF	NF	NF
Dibenzodioxins, chlorinated						
2,3,7,8-TCDD (dioxin)^b	1746-01-6	NF	NF	NF	NF	B2^H
1,2,3,4,8-PCDD	40321-76-4	NF	NF	NF	NF	NF
1,2,3,4,6,7,8-HCDD	34465-46-8	NF	NF	NF	NF	NF
1,2,3,4,7,8,9-HCDD	37871-00-4	NF	NF	NF	NF	NF
1,2,3,4,6,7,8,9-OCDD	3268-87-9	NF	NF	NF	NF	NF
Dibenzofurans, chlorinated^b						
1,3,7,8-TCDF	57117-35-8	NF	NF	NF	NF	NF
1,2,3,4,8-PCDF	67517-48-0	NF	NF	NF	NF	NF
1,2,3,4,7,8-HCDF	70648-26-9	NF	NF	NF	NF	NF
1,2,3,4,6,7,8-HCDF	67562-39-4	NF	NF	NF	NF	NF
1,2,3,4,6,7,8,9-OCDF	39001-02-0	NF	NF	NF	NF	NF

Table 3.2 (continued)

Relevant Route	Oral CSF (mg/kg·day)$^{-1}$	Target Organ/Effect	Oral RsD 1.00E–06 (mg/kg·day)	Oral RsD 1.00E–05 (mg/kg·day)	Inhalation Unit Risk (ug/m^3)$^{-1}$	Inhalation Target Organ/Effect	Derm.Perm Const.(meas) (cm/hr)[a]	Derm.Perm Const.(est) (cm/hr)[a]
NF	NF	NF	NA	NA	NF	NF	NF	NF
NF	NF	NF	NA	NA	NF	NF	NF	NF
NF	NF	NF	NA	NA	NF	NF	NF	NF
O,I	5.4E–01[I]	Bl,CNS,Sto	1.9E–06	1.9E–05	6.8E–05[I]	R	NF	1.4E–03
O	5.7E–03[I]	Spl	1.8E–04	1.8E–03	NF	NF	4.1E–02	2.2E–03
NF	NF	NF	NA	NA	NF	NF	NF	NF
O,I	1.1E–01[I]	Abd Cav	9.1E–06	9.1E–05	3.1E–05[I]	Abd Cav[e]	NF	NF
NF	NF	NF	NF	NF	NF	NF	NF	NF
NF	NF	NF	NA	NA	NF	NF	NF	NF
O,I	7.3E–01[E]	NF	1.4E–06	1.4E–05	8.9E–04[E]	NF	NF	8.1E–01
NF	NF	NF	NF	NF	NF	NF	NF	NF
O,I	2.9E–02[I]	Leuk	3.5E–05	3.5E–04	8.3E–06[I]	Leuk	1.1E–01	2.1E–02
O,I	2.3E+02[I]	Blad[f]	4.4E–09	4.4E–08	6.7E–02[I]	Blad	NF	1.3E–03
NF	7.3E–01[E]	NF	1.4E–06	1.4E–05	8.9E–05[E]	NF	NF	1.2E+00
NF	NF	NF	NA	NA	NF	NF	NF	NF
NF	7.3E–02[E]	NF	NF	NA	8.9E–06[E]	NF	NF	NF
NF	NF	NF	NA	NA	NF	NF	NF	NF
O	7.3E+00[I]	Fore	1.4E–07	1.4E–06	8.9E–04[E]	NF	NF	1.2E+00
NF	NF	NF	NA	NA	NF	NF	NF	NF
O	1.7E–01[I]	Th	5.9E–06	5.9E–05	NF	NF	NF	1.4E–02
NF	NF	NF	NA	NA	NF	NF	NF	NF
NF	NF	NF	NA	NA	NF	NF	NF	NF
O,I	1.1E+00[I]	Lv	9.1E–07	9.1E–06	3.3E–04[I]	Lv[e]	NF	2.1E–03
O,I	2.2E+02[I]	R	4.6E–09	4.6E–08	6.2E–02[I]	R	NF	NF
O	1.4E–02[I]	Lv	7.1E–05	7.1E–04	NF	NF	NF	NF
NF	NF	NF	NA	NA	NF	NF	NF	NF
O,I	7.9E–03[I]	GI	1.3E–04	1.3E–03	1.1E–06[I]	GI	NF	2.6E–03
I	NF	NF	NA	NA	2.8E–04[I]	NF	NF	2.3E–02
O	NF	NF	NA	NA	NF	NF	NF	NF
O	3.5E–03[H]	NF	2.9E–04	2.9E–03	NF	NF	NF	1.3E–03
NF	NF	NF	NA	NA	NF	NF	5.0E–01	2.4E–02
O,I	1.3E–01[I]	Lv	7.7E–06	7.7E–05	1.5E–05[I]	Lv[e]	NF	2.2E–02
NF	NF	NF	NA	NA	NF	NF	NF	4.1E–02
NF	NF	NF	NA	NA	NF	NF	NF	NF
O,I	6.1E–03[I]	K	1.6E–04	1.6E–03	2.3E–05[I]	Lv	1.3E–01	8.9E–03
NF	NF	NF	NA	NA	NF	NF	3.3E–02	1.1E–02
O,I	7.3E–03[E]	NF	1.4E–05	1.4E–04	8.9E–07[E]	NF	NF	8.1E–01
NF	NF	NF	NA	NA	NF	NF	NF	NF
NF	NF	NF	NA	NA	NF	NF	NF	NF
NF	NF	NF	NA	NA	NF	NF	NF	NF
NF	NF	NF	NA	NA	NF	NF	NF	NF
O,I	1.5E+05[H]	R,Lv	6.7E–12	6.7E–11	4.3E+01[H]	NF	NF	1.4E+00
NF	7.5E+04	NF	1.3E–11	1.3E–10	2.2E+00	NF	NF	NF
NF	1.5E+04	NF	6.7E–11	6.7E–10	4.3E+00	NF	NF	NF
NF	1.5E+03	NF	6.7E–10	6.7E–09	4.3E–01	NF	NF	NF
NF	1.5E+02	NF	6.7E–09	6.7E–08	4.3E–02	NF	NF	NF
NF	1.5E+04	NF	NA	NA	4.3E+00	NF	NF	NF
NF	NF	NF	NA	NA	NF	NF	NF	NF
NF	1.5E+04	NF	6.7E–11	6.7E–10	4.3E+00	NF	NF	NF
NF	1.5E+03	NF	6.7E–10	6.7E–09	4.3E–01	NF	NF	NF
NF	1.5E+02	NF	6.7E–09	6.7E–08	4.3E–02	NF	NF	NF

(continues)

Table 3.2 Toxicological Data for Selected Analytes Reported in Emissions from Hazardous Waste Incinerators (continued)

Chemical Name	CAS Registry Number	Oral RfD (mg/kg·day)	Oral Target Organ/Effect	Inhalation RfC (mg/m³)	Inhalation Target Organ/Effect	Carcin. Classif.
Dibenzofuran	132-64-9	4.0E–03[E]	NF	NF	NF	D
Dibromochloromethane	124-48-1	2.0E–02[I]	Lv	NF	NF	C
1,2-Dibromo-3-chloropropane (DBCP)	96-12-8	NF	NF	2.0E–04[I]	Re	B2
Dibutyl phthalate	84-74-2	1.0E–01[I]	Mor	NF	NF	D
1,2-Dichlorobenzene	95-50-1	9.0E–02[I]	NOAEL	3.2E–02[A]	NF	D
1,3-Dichlorobenzene	541-73-1	8.9E–02[O]	NF	NF	NF	D
1,4-Dichlorobenzene	106-46-7	NF	NF	8.0E–01[I]	Lv	C[H]
3,3′-Dichlorobenzidine	91-94-1	NF	NF	NF	NF	B2
Dichlorobromomethane	75-27-4	2.0E–02[I]	Wt Ls	NF	NF	B2
3,4-Dichlorobutene	760-23-6	NF	NF	NF	NF	NF
Dichlorodifluoromethane	75-71-8	2.0E–01[I]	Wt Ls	2.0E–01[A]	NF	NF
1,1-Dichloroethane	75-34-3	1.0E–01[H]	NF	5.0E–01[A]	NF	C
1,2-Dichloroethane	107-06-2	NF	NF	4.9E–03[E]	NF	B2
1,1-Dichloroethene	75-35-4	9.0E–03[I]	Lv	NF	NF	C
2,4-Dichlorophenol	120-83-2	3.0E–03[I]	Hys	NF	NF	NF
1,3-Dichloropropene (cis)	10061-01-5	NF	NF	NF	NF	NF
Diethyl phthalate	84-66-2	8.0E–01[I]	Grw, Org Wt	NF	NF	D
2,4-Dimethylphenol	105-67-9	2.0E–02[I]	Cli	NF	NF	NF
Dimethyl phthalate	131-11-3	1.0E+01[H]	NF	NF	NF	D
Dimethyl sulfide	75-18-3	NF	NF	NF	NF	NF
1,4-Dioxane	123-91-1	NF	NF	NF	NF	B2
Epichlorohydrin	106-89-8	2.0E–03[H]	K	1.0E–03[I]	N	B2
Ethylbenzene	100-41-4	1.0E–01[I]	Lv, K	1.0E+00[I]	De	D
Fluoranthene	206-44-0	4.0E–02[I]	Lv,Ren,Cli,Bl	NF	NF	D
Fluorene	86-73-7	4.0E–02[I]	Bl	NF	NF	D
Formaldehyde	50-00-0	2.0E–01[I]	Wt Ls	NF	NF	B1
Hexachlorobenzene	118-74-1	8.0E–04[I]	Lv	NF	NF	B2
Hexachlorobutadiene	87-68-3	2.0E–04[I]	K	NF	NF	C
Hexachlorocyclopentadiene	77-47-4	7.0E–03[I]	Sto	7.0E–05[H]	N	D
Hexachloroethane	67-72-1	1.0E–03[I]	K	NF	NF	C
Hydrogen sulfide	7783-06-4	3.0E–03[I]	GI	1.0E–03[I]	N	NF
Methyl bromide (bromomethane)	74-83-9	1.4E–03[I]	Fore	5.0E–03[I]	N	D
Methyl chloride (chloromethane)	74-87-3	NF	NF	NF	NF	C[H]
Methyl mercaptan (methanethiol)	74-93-1	NF	NF	NF	NF	NF
Methyl mercury	22967-92-6	1.0E–04[I]	De	NF	NF	C
Methylene bromide	74-95-3	1.0E–02[A]	NF	NF	NF	NF
Methylene chloride (dichloromethane)	75-09-2	6.0E–02[I]	Lv	3.0E+00[H]	Lv	B2
Methyl ethyl ketone (MEK)	78-93-3	6.0E–01[I]	De	1.0E+00[I]	De	D
Naphthalene	91-20-3	4.0E–02[W]	NF	NF	NF	D
Nitrobenzene	98-95-3	5.0E–04[I]	Ad,Ren,Lv,Bl	NF	NF	D
o-Nitrophenol	88-75-5	NF	NF	NF	NF	NF
N-nitrosomethyl-ethylamine	10595-95-6	NF	NF	NF	NF	B2
N-nitroso-n-dibutylamine	924-16-3	NF	NF	NF	NF	B2
N-nitrosopyrrolidine	930-55-2	NF	NF	NF	NF	B2
Pentachlorobenzene	608-93-5	8.0E–04[I]	Lv,K	NF	NF	D
Pentachloroethane	76-01-7	NF	NF	NF	NF	NF
Pentachlorophenol (PCP)	87-86-5	3.0E–02[I]	Lv,K	NF	NF	B2
Phenanthrene	85-01-8	3.0E–02[H]	NF	NF	NF	D
Phenol	108-95-2	6.0E–01[I]	De	NF	NF	D
Polychlorinated biphenyls (PCB)	1336-36-3	NF	NF	NF	NF	B2
Aroclor 1016	12674-11-2	7.0E–05[I]	De	NF	NF	NF
Pyrene	129-00-0	3.0E–02[I]	K	NF	NF	D
Styrene	100-42-5	2.0E–01[I]	Bl	1.0E+00[I]	CNS	NF
1,2,4,5-Tetrachlorobenzene	95-94-3	3.0E–04[I]	K	NF	NF	NF
1,1,1,2-Tetrachloroethane	630-20-6	3.0E–02[I]	K,Lv	NF	NF	C
1,1,2,2-Tetrachloroethane	79-34-5	NF	NF	NF	NF	C

Table 3.2 (continued)

Relevant Route	Oral CSF (mg/kg·day)$^{-1}$	Target Organ/Effect	Oral RsD 1.00E–06 (mg/kg·day)	Oral RsD 1.00E–05 (mg/kg·day)	Inhalation Unit Risk (ug/m^3)$^{-1}$	Inhalation Target Organ/Effect	Derm.Perm Const.(meas) (cm/hr)^a	Derm.Perm Const.(est) (cm/hr)^a
NF	NF	NF	NA	NA	NF	NF	NF	NF
O	8.4E–02^I	Lv	1.2E–05	1.2E–04	NF	NF	NF	3.9E–03
O,I	1.4E+00^H	Sto,K,Lv	7.1E–07	7.1E–06	2.4E–03^I	Re	NF	NF
NF	NF	NF	NA	NA	NF	NF	NF	3.3E–02
NF	NF	NF	NA	NA	NF	NF	NF	6.1E–02
NF	NF	NF	NA	NA	NF	NF	NF	8.7E–02
O	2.4E–02^H	Lv	4.2E–05	4.2E–04	NF	NF	NF	6.2E–02
O	4.5E–01^I	Mm	2.2E–06	2.2E–05	NF	NF	NF	1.7E–02
O	6.2E–02^I	K	1.6E–05	1.6E–04	NF	NF	NF	5.8E–03
NF	NF	NF	NA	NA	NF	NF	NF	NF
NF	NF	NF	NA	NA	NF	NF	NF	1.2E–02
NF	NF	NF	NA	NA	NF	NF	NF	8.9E–03
O,I	9.1E–02^I	He Sarc, Ci	1.1E–05	1.1E–04	2.6E–05^I	He Sarc,Ci^e	NF	5.3E–03
O,I	6.0E–01^I	Ad	1.7E–06	1.7E–05	5.0E–05^I	K	NF	1.6E–02
NF	NF	NF	NA	NA	NF	NF	6.0E–02	2.3E–02
NF	NF	NF	NF	NF	NF	NF	NF	NF
NF	NF	NF	NA	NA	NF	NF	NF	4.8E–03
NF	NF	NF	NA	NA	NF	NF	1.1E–01	1.5E–02
NF	NF	NF	NA	NA	NF	NF	NF	1.6E–03
NF	NF	NF	NA	NA	NF	NF	NF	NF
O	1.1E–02^I	N	9.1E–05	9.1E–04	NF	NF	4.0E–04	3.6E–04
O,I	9.9E–03^I	Fore	1.0E–04	1.0E–03	1.2E–06^I	N	NF	3.7E–04
NF	NF	NF	NA	NA	NF	NF	1.0E+00	7.4E–02
NF	NF	NF	NA	NA	NF	NF	NF	3.6E–01
NF	NF	NF	NA	NA	NF	NF	NF	NF
I	NF	NF	NA	NA	1.3E–05^I	N	NF	2.2E–03
O,I	1.6E+00	Lv	6.3E–07	6.3E–06	4.6E–04^I	Lv^e	NF	2.1E–01
O,I	7.8E–02^I	K	1.3E–05	1.3E–04	2.2E–05^I	K^e	NF	1.2E–01
NF	NF	NF	NA	NA	NF	NF	NF	NF
O,I	1.4E–02^I	Lv	7.1E–05	7.1E–04	4.0E–06^I	Lv^e	NF	4.2E–02
NF	NF	NF	NA	NA	NF	NF	NF	NF
NF	NF	NF	NA	NA	NF	NF	NF	3.5E–03
O,I	1.3E–02^H	NF	7.7E–05	7.7E–04	1.8E–06^H	K	NF	4.2E–03
NF	NF	NF	NA	NA	NF	NF	NF	NF
NF	NF	NF	NA	NA	NF	NF	NF	NF
NF	NF	NF	NA	NA	NF	NF	NF	NF
O,I	7.5E–03^I	Lv	1.3E–04	1.3E–03	4.7E–07^I	Lv,R	NF	4.5E–03
NF	NF	NF	NA	NA	NF	NF	5.0E–03	1.1E–03
NF	NF	NF	NA	NA	NF	NF	NF	6.9E–02
NF	NF	NF	NA	NA	NF	NF	NF	NF
NF	NF	NF	NA	NA	NF	NF	1.0E–01	5.0E–03
O	2.2E+01^I	Lv	4.6E–08	4.6E–07	NF	NF	NF	NF
O,I	5.4E+00^I	Blad, Eso	1.9E–07	1.9E–06	1.6E–03^I	Blad, Eso^e	NF	4.8E–03
O,I	2.1E+00^I	Lv	4.8E–07	4.8E–06	6.1E–04^I	Lv^e	NF	NF
NF	NF	NF	NA	NA	NF	NF	NF	NF
NF	NF	NF	NA	NA	NF	NF	NF	NF
O	1.2E–01^I	Lv, Ad, Bl	8.3E–06	8.3E–05	NF	NF	NF	6.5E–01
NF	NF	NF	NA	NA	NF	NF	NF	2.3E–01
NF	NF	NF	NA	NA	NF	NF	8.2E–03	5.5E–03
O	2.0E+00^I	Lv	5.0E–07	1.3E–06	1.4E+03^I	Lv^e	NF	NF
NF	NF	NF	NA	NA	NF	NF	NF	NF
NF	NF	NF	NA	NA	NF	NF	NF	NF
NF	NF	NF	NA	NA	NF	NF	6.7E–01	5.5E–02
NF	NF	NF	NA	NA	NF	NF	NF	NF
O,I	2.6E–02^I	Lv	3.9E–05	3.9E–04	7.4E–06^I	Lv^e	NF	NF
O,I	2.0E–01^I	Lv	5.0E–06	5.0E–05	5.8E–05^I	Lv^e	NF	9.0E–03

(continues)

Table 3.2 Toxicological Data for Selected Analytes Reported in Emissions from Hazardous Waste Incinerators (continued)

Chemical Name	CAS Registry Number	Oral RfD (mg/kg·day)	Oral Target Organ/Effect	Inhalation RfC (mg/m³)	Inhalation Target Organ/Effect	Carcin. Classif.
Tetrachloroethene (Perc)	127-18-4	1.0E–02^I	Lv, Wt Gn	NF	NF	NF
Toluene	108-88-3	2.0E–01^I	Lv,K	4.0E–01^I	N	D
2,6-Toluenediisocyanate	91-08-7	NF	NF	UR	NF	NF
1,2,4-Trichlorobenzene	120-82-1	1.0E–02^I	Ad	2.0E–01^H	Lv	D
1,1,1-Trichloroethane	71-55-6	2.0E–02^E	NF	1.0E+00^W	NF	D
1,1,2-Trichloroethane	79-00-5	4.0E–03^I	Clin	NF	NF	C
Trichloroethene (TCE)	79-01-6	6.0E–03^I	NF	NF	NF	NF
Trichlorofluoromethane	75-69-4	3.0E–01^I	Su	7.0E–01^A	NF	NF
2,4,5-Trichlorophenol	95-95-4	1.0E–01^I	Lv,K	NF	NF	NF
2,4,6-Trichlorophenol	88-06-2	NF	NF	NF	NF	B2
1,2,3-Trichloropropane	96-18-4	6.0E–03^I	Bl, Clin	NF	NF	B2
Vinyl chloride	75-01-4	NF	NF	NF	NF	A^H
m-Xylene (see xylenes, total)	108-38-3	2.0E+00^H	CNS,Hyp,Wt Ls	7.0E–01^W	NF	NF
o-Xylene (see xylenes, total)	95-47-6	2.0E+00^H	CNS, Wt Ls	7.0E–01^W	NF	NF
p-Xylene (see xylenes, total)	106-42-3	NF	NF	3.0E–01^W	NF	NF
Xylenes, total	1330-20-7	2.0E+00^I	CNS, Bl, Mor	NF	NF	D
Inorganics						
Antimony	7440-36-0	4.0E–04^I	Gluc, Chol	NF	NF	NF
Arsenic, inorganic	7440-38-2	3.0E–04^I	Ci, Sk	NF	NF	A
Barium	7440-39-3	7.0E–02^I	Bl Press	5.0E–04^A	NF	NF
Beryllium	7440-41-7	2.0E–03^I	NOAEL	2.0E–05^I	CBD	B2
Cadmium	7440-43-9	5.0E–04^I	Pro	2.0E–04^W	NF	B1
Carbon monoxide (CO)	630-08-0	NF	NF	NF	NF	NF
Chlorine	7782-50-5	1.0E–01^I	NOAEL	NF	NF	NF
Chromium (III)	16065-83-1	1.0E+01^I	NOAEL	2.0E–06^W	NF	NF
Chromium (VI)	18540-29-9	5.0E–03^I	NOAEL	UR	NF	A
Cobalt	7440-48-4	6.0E–02^E	NF	NF	NF	NF
Copper	7440-50-8	1.8E+02^H	NF	NF	NF	D
Hydrogen chloride (HCl)	7647-01-0	NF	NF	2.0E–03^I	Rsp	NF
Hydrogen fluoride	7664-39-3	NF	NF	NF	NF	NF
Iron	7439-89-6	3.0E–01^E	NF	NF	NF	NF
Lead	7439-92-1	NF	NF	NF	NF	B2
Magnesium	7439-95-4	NF	NF	NF	NF	NF
Manganese	7439-96-5	1.4E–01^I	CNS	5.0E–05^I	CNS	D
Mercury	7439-97-6	NF	NF	3.0E–04^I	CNS	D
Mercuric chloride	7487-94-7	3.0E–04^I	Im	NF	NF	C
Molybdenum	7439-98-7	5.0E–03^I	Cli, K	NF	NF	NF
Nickel (soluble salts)	7440-02-0	2.0E–02^I	Wt Ls, OrgWt	NF	NF	NF
Nickel (refinery dust)	NF	NF	NF	NF	NF	A
Nitrogen dioxide	10102-44-0	1.0E+00^W	NF	NF	NF	NF
Nitrogen monoxide (NO)	10102-43-9	1.0E–01^W	NF	NF	NF	NF
Particulate matter	NF	NF	NF	NF	NF	NF
Scandium	7440-20-2	NF	NF	NF	NF	NF
Selenium (and Compounds)	7782-49-2	5.0E–03^I	Sel	NF	NF	D
Silicon	7440-21-3	NF	NF	NF	NF	NF
Silver	7440-22-4	5.0E–03^I	Sk	NF	NF	D
Sulfur dioxide	9/5/46	NF	NF	NF	NF	NF
Sulfur trioxide	11/9/46	NF	NF	NF	NF	NF
Tellurium	13494-80-9	NF	NF	NF	NF	NF
Thallium	7440-28-0	NF	NF	NF	NF	NF
Tin	7440-31-5	6.0E–01^H	Lv,K	NF	NF	NF
Titanium	7440-32-6	NF	NF	NF	NF	NF
Vanadium	7440-62-2	7.0E–03^H	Lifespan	NF	NF	NF
Zinc	7440-66-6	3.0E–01^I	Bl	NF	NF	D

Table 3.2   (continued)

Relevant Route	Oral CSF (mg/kg·day)⁻¹	Target Organ/Effect	Oral RsD 1.00E-06 (mg/kg·day)	Oral RsD 1.00E-05 (mg/kg·day)	Inhalation Unit Risk (ug/m³)⁻¹	Inhalation Target Organ/Effect	Derm.Perm Const.(meas) (cm/hr)[a]	Derm.Perm Const.(est) (cm/hr)[a]
O,I	5.2E–02[E]	NF	1.9E–05	1.9E–04	5.8E–07[E]	NF	3.7E–01	4.8E–02
NF	NF	NF	NA	NA	NF	NF	1.0E+00	4.5E–02
NF	NF	NF	NA	NA	NF	NF	NF	NF
NF	NF	NF	NA	NA	NF	NF	NF	1.0E–01
NF	NF	NF	NA	NA	NF	NF	NF	1.7E–02
O,I	5.7E–02[I]	Lv	1.8E–05	1.8E–04	1.6E–05[I]	Lv[e]	NF	8.4E–03
O,I	1.1E–02[W]	NF	NA	NA	1.7E–06[E]	NF	2.3E–01	1.6E–02
NF	NF	NF	NA	NA	NF	NF	NF	1.7E–02
NF	NF	NF	NA	NA	NF	NF	NF	NF
O,I	1.1E–02[I]	Leuk	9.1E–05	9.1E–04	3.1E–06[I]	Leuk[e]	5.9E–02	5.0E–02
O	7.0E+00[H]	Mult	NA	NA	NF	NF	NF	NF
O,I	1.9E+00[H]	Lv	5.3E–07	5.3E–06	8.6E–05[H]	Lvr	NF	7.3E–03
NF	NF	NF	NA	NA	NF	NF	NF	8.0E–02
NF	NF	NF	NA	NA	NF	NF	NF	NF
NF	NF	NF	NA	NA	NF	NF	NF	NF
NF	NF	NF	NA	NA	NF	NF	NF	NF
NF	NF	NF	NA	NA	NF	NF	NF	NF
I,O	1.8E+00[I]	NF	5.7E–07	5.7E–06	4.3E–03[I]	Lg	NF	NF
NF	NF	NF	NA	NA	NF	NF	NF	NF
O,I	NF[I]	NF	NF	NF	UR	NF	NF	NF
I	NF	NF	NA	NA	1.8E–03[I]	Lg, Trach[I]	NF	NF
NF	NF	NF	NA	NA	NF	NF	NF	NF
NF	NF	NF	NA	NA	NF	NF	NF	NF
NF	NF	NF	NA	NA	NF	NF	NF	NF
I	NF	NF	NA	NA	1.2E–02[I]	Lg	NF	NF
NF	NF	NF	NA	NA	NF	NF	NF	NF
NF	NF	NF	NA	NA	NF	NF	NF	NF
NF	NF	NF	NA	NA	NF	NF	NF	NF
NF	NF	NF	NA	NA	NF	NF	NF	NF
NF	NF	NF	NA	NA	NF	NF	NF	NF
NF	NF	NF	NA	NA	NF	NF	NF	NF
NF	NF	NF	NA	NA	NF	NF	NF	NF
NF	NF	NF	NA	NA	NF	NF	NF	NF
NF	NF	NF	NA	NA	NF	NF	NF	NF
NF	NF	NF	NA	NA	NF	NF	NF	NF
NF	NF	NF	NA	NA	NF	NF	NF	NF
I	NF	NF	NA	NA	2.4E–04[I]	Lg	NF	NF
NF	NF	NF	NA	NA	NF	NF	NF	NF
NF	NF	NF	NA	NA	NF	NF	NF	NF
NF	NF	NF	NA	NA	NF	NF	NF	NF
NF	NF	NF	NA	NA	NF	NF	NF	NF
NF	NF	NF	NA	NA	NF	NF	NF	NF
NF	NF	NF	NA	NA	NF	NF	NF	NF
NF	NF	NF	NA	NA	NF	NF	NF	NF
NF	NF	NF	NA	NA	NF	NF	NF	NF
NF	NF	NF	NA	NA	NF	NF	NF	NF
NF	NF	NF	NA	NA	NF	NF	NF	NF
NF	NF	NF	NA	NA	NF	NF	NF	NF
NF	NF	NF	NA	NA	NF	NF	NF	NF
NF	NF	NF	NA	NA	NF	NF	NF	NF
NF	NF	NF	NA	NA	NF	NF	NF	NF

(continues)

LEGEND FOR TABLE 3.2

Target Organ/Effect Abbreviations:

Abd Cav	– Abdominal Cavity	Lg	– Lung
Ad	– Adrenal	Lv	– Liver
B	– Brain	Meth	– Methemoglobinemia
Bl	– Blood	Mor	– Mortality
Bl Press	– Blood Pressure	Mm	– Mammary
Blad	– Bladder	Mult	– Whole Body, Liver, Kidney,
Chol	– Cholesterol		Erythrocytes, Blood
Cli	– Clinical Biochemistry Changes	N	– Nasal
Ci	– Circulatory	NOAEL	– No Observed Adverse Effects Level
Chol	– Cholesterol	Per Nerv	– Peripheral Nervous System
CNS	– Central Nervous System	Pro	– Proteinuria
De	– Development	Org Wt	– Organ Weight Change
Eso	– Esophagus	R	– Respiratory
Fe	– Fetal	Re	– Reproductive
Feto	– Fetotoxicity	Ren	– Renal
FI	– From Inhalation	Sel	– Selenosis
Fore	– Forestomach	Sk	– Skin
GI	– Gastrointestinal	Sm Int	– Small Intestines
Gluc	– Glucose	Spl	– Spleen
Grw	– Growth Rate Changes	Sto	– Stomach
He Sarc	– Hemangiosarcomas	Su	– Survival
Hyp	– Hyperactivity	Th	– Thyroid
Hys	– Hypersensitivity	Trach	– Trachea
Im	– Immune	UR	– Under Review
K	– Kidney	Wt Gn	– Body Weight Gain
Leuk	– Leukemia	Wt Ls	– Body Weight Loss

Footnotes:

I — IRIS

H — HEAST

E — EPA-NCEA Regional Support provisional value (obtained through Region III) (U.S. EPA, 1998b).

W — Withdrawn from IRIS or HEAST (obtained through Region III) (U.S. EPA, 1998b).

O — Other EPA documents as reported in the Region III RBC table (U.S. EPA, 1998b).

A — HEAST alternate value as reported by U.S. EPA Region III (U.S. EPA, 1998b).

a) Dermal values are found in Dermal Exposure Assessment: Principles and Applications, Interim Report (U.S. EPA, 1992).

b) Toxicity equivalent values (TEFs) for CDDs and CDFs are based on 2,3,7,8-TCDD (U.S. EPA, 1995).

c) Toxicity equivalent values (TEFs) are based on reported toxicity values for benzo(a)pyrene (U.S. EPA, 1995).

d) Cancer Classifications reported on IRIS unless otherwise noted (U.S. EPA, 1998a).

e) Value was derived from the oral RfD, per U.S. EPA.

f) Value was derived from the inhalation RfC, per U.S. EPA.

g) Withdrawn value from U.S. EPA.

h) Reported in Brownfields Rule Florida Department of Environmental Protection (FDEP, 1998). Some RfC values may be calculated based on RfD_o values.

Table 3.2 References

U.S. EPA. 1992. Dermal Exposure Assessment: Principles and Applications. EPA/600/8-91/011B.

U.S. EPA. 1995. Supplemental Guidance to RAGS: Region 4 Bulletins. November, 1995.

U.S. EPA. 1997. Health Effects Assessment Summary Tables. FY 1997 Update. EPA-540-R-97-036. July, 1997.

U.S. EPA. 1998a. IRIS (Integrated Risk Information System). On-line computer database.

U.S. EPA. 1998b. Risk-Based Concentration Table: Region III. May, 1998.

Table 3.3 U.S. EPA Water Quality Data for Selected Analytes Reported for Emissions from Hazardous Waste Incinerators

Chemical Name	CAS Registry Number	U.S. EPA water quality (acute, freshwater) (mg/l)	U.S. EPA water quality (chronic, freshwater) (mg/l)	U.S. EPA water quality (acute, marine) (mg/l)
Organics				
Acenaphthene	83-32-9	NF	NF	3.00E–01
Acenaphthylene	208-96-8	NF	NF	3.00E–01
Acetonitrile	75-05-8	NF	NF	NF
Acrylonitrile	107-13-1	7.55E+00	2.60E+00	NF
Aniline	62-53-3	NF	NF	NF
Anthracene	120-12-7	NF	NF	NF
Azobenzene (cis)	103-33-3	NF	NF	NF
Benzaldehyde	100-52-7	NF	NF	NF
Benz[a]anthracene	56-55-3	NF	NF	NF
Benz[e]anthracene	NF	NF	NF	NF
Benzene	71-43-2	5.30E+00	NF	5.10E+00
Benzidine	92-87-5	2.50E+00	NF	NF
Benzo[b]fluoranthene	205-99-2	NF	NF	NF
Benzo[j]fluoranthene	205-82-3	NF	NF	NF
Benzo[k]fluoranthene	207-08-9	NF	NF	NF
Benzo[g,h,i]perylene	191-24-2	NF	NF	NF
Benzo[a]pyrene	50-32-8	NF	NF	NF
Benzo[e]pyrene	192-97-2	NF	NF	NF
Benzyl chloride	100-44-7	NF	NF	NF
1,1-Biphenyl	92-52-4	NF	NF	NF
Bis[2-chloroethoxy]methane	111-91-1	NF	NF	NF
Bis[2-chloroethyl]ether	111-44-4	2.38E+02	NF	NF
Bis[chloromethyl]ether	542-88-1	2.38E+02	NF	NF
Bis[2-ethylhexyl]phthalate-(DEHP)	117-81-7	9.40E–01	3.00E–03	2.94E+00
Bromochloromethane	74-97-5	1.10E+01	NF	1.20E+01
Bromoform	75-25-2	NF	NF	NF
1,3-Butadiene	106-99-0	NF	NF	NF
Butyl benzyl phthalate	85-68-7	9.40E–01	3.00E–03	2.94E+00
Captan	133-06-2	NF	NF	NF
Carbon disulfide	75-15-0	NF	NF	NF
Carbon tetrachloride	56-23-5	3.52E+01	NF	5.00E+01
Chlorobenzene	108-90-7	2.50E–01	NF	1.60E–01
Chlorocyclopentadiene	41851-50-7	NF	NF	NF
Chloroform	67-66-3	2.89E+01	1.24E+00	NF
o-Chlorophenol	95-57-8	4.38E+00	NF	NF
Chrysene	218-01-9	NF	NF	NF
Coronene (hexabenzobenzene)	197-07-1	NF	NF	NF
Cumene	98-82-8	NF	NF	NF

(continues)

Table 3.3 U.S. EPA Water Quality Data for Selected Analytes Reported for Emissions from Hazardous Waste Incinerators (continued)

Chemical Name	CAS Registry Number	U.S. EPA water quality (acute, freshwater) (mg/l)	U.S. EPA water quality (chronic, freshwater) (mg/l)	U.S. EPA water quality (acute, marine) (mg/l)
Organics (continued)				
Dibenzo[a,e]fluoranthene	5385-75-1	NF	NF	NF
Dibenzo[a,h]fluoranthene	NF	NF	NF	NF
Dibenzodioxins, chlorinated				
2,3,7,8-TCDD	1746-01-6	NF	NF	NF
1,2,3,4,8-PCDD	40321-76-4	NF	NF	NF
1,2,3,4,6,7,8-HCDD	34465-46-8	NF	NF	NF
1,2,3,4,7,8,9-HCDD	37871-00-4	NF	NF	NF
1,2,3,4,6,7,8,9-OCDD	3268-87-9	NF	NF	NF
Dibenzofurans, chlorinated				
1,3,7,8-TCDF	57117-35-8	NF	NF	NF
1,2,3,4,8-PCDF	67517-48-0	NF	NF	NF
1,2,3,4,7,8-HCDF	70648-26-9	NF	NF	NF
1,2,3,4,6,7,8-HCDF	67562-39-4	NF	NF	NF
1,2,3,4,6,7,8,9-OCDF	39001-02-0	NF	NF	NF
Dibenzofuran	132-64-9	NF	NF	NF
Dibromochloromethane	124-48-1	1.10E+01	NF	1.20E+01
1,2-Dibromo-3-chloropropane (DBCP)	96-12-8	NF	NF	NF
Dibutyl phthalate	84-74-2	NF	NF	NF
1,2-Dichlorobenzene	95-50-1	1.12E+00	7.63E–01	1.97E+00
1,3-Dichlorobenzene	541-73-1	1.12E+00	7.63E–01	1.97E+00
1,4-Dichlorobenzene	106-46-7	1.12E+00	7.63E–01	1.97E+00
3,3'-Dichlorobenzidine	91-94-1	NF	NF	NF
Bromodichloromethane	75-27-4	1.10E+01	NF	1.20E+01
3,4-Dichlorobutene	760-23-6	NF	NF	NF
Dichlorodifluoromethane	75-71-8	1.10E+01	NF	1.20E+01
1,1-Dichloroethane	75-34-3	NF	NF	NF
1,2-Dichloroethane	107-06-2	1.18E+02	NF	1.13E+02
1,1-Dichloroethene	75-35-4	1.16E+01	NF	2.24E+02
2,4-Dichlorophenol	120-83-2	2.02E+00	3.65E–01	NF
1,3-Dichloropropene (cis)	542-75-6	6.06E+00	2.44E–01	7.90E–01
Diethyl phthalate	84-66-2	NF	NF	NF
2,4-Dimethylphenol	105-67-9	2.12E+00	NF	NF
Dimethyl phthalate	131-11-3	NF	NF	NF
Dimethyl sulfide	75-18-3	NF	NF	NF
1,4-Dioxane	123-91-1	NF	NF	NF
Epichlorohydrin	106-89-8	NF	NF	NF
Ethylbenzene	100-41-4	3.20E+01	NF	4.30E–01
Fluoranthene	206-44-0	3.98E+00	NF	4.00E–02
Fluorene	86-73-7	NF	NF	3.00E–01

Table 3.3 U.S. EPA Water Quality Data for Selected Analytes Reported for Emissions from Hazardous Waste Incinerators (continued)

Chemical Name	CAS Registry Number	U.S. EPA water quality (acute, freshwater) (mg/l)	U.S. EPA water quality (chronic, freshwater) (mg/l)	U.S. EPA water quality (acute, marine) (mg/l)
Organics (continued)				
Formaldehyde	50-00-0	NF	NF	NF
Hexachlorobenzene	118-74-1	NF	NF	NF
Hexachlorobutadiene	87-68-3	9.00E–02	9.30E–03	3.20E–02
Hexachlorocyclopentadiene	77-47-4	7.00E–03	5.20E–03	7.00E–03
Hexachloroethane	67-72-1	5.40E–01	NF	9.40E–01
Hydrogen sulfide	7783-06-4	NF	2.00E–03	NF
Methyl bromide (bromomethane)	74-83-9	1.10E+01	NF	1.20E+01
Methyl chloride (chloromethane)	74-87-3	1.10E+01	NF	1.20E+01
Methyl mercaptan	74-93-1	NF	NF	NF
Methyl mercury	22967-92-6	NF	NF	NF
Methylene bromide	74-95-3	NF	NF	NF
Methylene chloride (dichloromethane)	75-09-2	1.10E+01	NF	1.20E+01
Methyl ethyl ketone (MEK)	78-93-3	NF	NF	NF
Naphthalene	91-20-3	2.30E+00	6.20E–01	2.35E+00
Nitrobenzene	98-95-3	2.70E+01	NF	6.68E+00
o-Nitrophenol	88-75-5	2.30E–01	NF	4.85E+00
N-Nitrosomethylethyl-amine	10595-95-6	5.85E+00	NF	3.30E+03
N-Nitroso-n-dibutylamine	924-16-3	5.85E+00	NF	3.30E+03
N-Nitrosopyrrolidine	930-55-2	NF	NF	NF
Pentachlorobenzene	608-93-5	NF	NF	NF
Pentachloroethane	76-01-7	1.10E+00	NF	3.90E–01
Pentachlorophenol (PCP)	87-86-5	5.50E–02	3.20E–03	5.30E–02
Phenanthrene	85-01-8	NF	NF	NF
Phenol	108-95-2	1.02E+01	2.56E+00	5.80E+00
Polychlorinated biphenyls (PCB)	1336-36-3	1.40E–05	NF	3.00E–05
Aroclor 1016	12674-11-2	NF	NF	NF
Pyrene	129-00-0	NF	NF	NF
Styrene	100-42-5	NF	NF	NF
1,2,4,5-Tetrachlorobenzene	95-94-3	NF	NF	NF
1,1,1,2-Tetrachloroethane	630-20-6	9.32E+00	NF	NF
1,1,2,2-Tetrachloroethane	79-34-5	9.32E+00	NF	9.02E+00
Tetrachloroethene (Perc)	127-18-4	5.28E+00	8.40E–01	1.02E+01
Toluene	108-88-3	1.75E+01	NF	6.30E+00
2,6-Toluenediisocyanate	584-84-9	NF	NF	NF
1,2,4-Trichlorobenzene	120-82-1	NF	NF	NF

(continues)

Table 3.3 U.S. EPA Water Quality Data for Selected Analytes Reported
for Emissions from Hazardous Waste Incinerators (continued)

Chemical Name	CAS Registry Number	U.S. EPA water quality (acute, freshwater) (mg/l)	U.S. EPA water quality (chronic, freshwater) (mg/l)	U.S. EPA water quality (acute, marine) (mg/l)
Organics (continued)				
1,1,1-Trichloroethane	71-55-6	1.80E+01	NF	3.12E+01
1,1,2-Trichloroethane	79-00-5	1.80E+01	NF	NF
Trichloroethene	79-01-6	4.50E+01	NF	2.00E+00
Trichlorofluoromethane	75-69-4	1.10E+01	NF	1.20E+01
2,4,5-Trichlorophenol	95-95-4	NF	NF	NF
2,4,6-Trichlorophenol	88-06-2	NF	9.70E–01	NF
1,2,3-Trichloropropane	96-18-4	NF	NF	NF
Vinyl chloride	75-01-4	NF	NF	NF
m-Xylene (see xylenes, total)	108-38-3	NF	NF	NF
o-Xylene (see xylenes, total)	95-47-6	NF	NF	NF
p-Xylene (see xylenes, total)	106-42-3	NF	NF	NF
Xylenes, total	1330-20-7	NF	NF	NF
Inorganics				
Antimony	7440-36-0	9.00E+00	1.60E+00	NF
Arsenic, inorganic	7440-38-2	NF	NF	NF
Barium	7440-39-3	NF	NF	NF
Beryllium	7440-41-7	1.30E–01	5.30E–03	NF
Cadmium	7440-43-9	NF	NF	NF
Carbon monoxide (CO)	630-08-0	NF	NF	NF
Chlorine	7782-50-5	NF	NF	NF
Chromium (III)	16065-83-1	NF	NF	NF
Chromium (VI)	18540-29-9	NF	NF	NF
Cobalt	7440-48-4	NF	NF	NF
Copper	7440-50-8	NF	NF	NF
Hydrogen chloride (HCl)	7647-01-0	NF	NF	NF
Hydrogen fluoride	7664-39-3	NF	NF	NF
Iron	7439-89-6	1.00E+00	1.00E+00	NF
Lead	7439-92-1	NF	NF	NF
Magnesium	7439-95-4	NF	NF	NF
Manganese	7439-96-5	NF	NF	NF
Mercury	7439-97-6	NF	NF	NF
Mercuric chloride	7487-94-7	NF	NF	NF
Molybdenum	7439-98-7	NF	NF	NF
Nickel (soluble salts)	7440-02-0	NF	NF	NF
Nickel (refinery dust)	NF	NF	NF	NF
Nitrogen dioxide	10102-44-0	NF	NF	NF
Nitrogen monoxide (NO)	10102-43-9	NF	NF	NF
Particulate matter	NF	NF	NF	NF
Scandium	7440-20-2	NF	NF	NF

Table 3.3   U.S. EPA Water Quality Data for Selected Analytes Reported for
Emissions from Hazardous Waste Incinerators (continued)

Chemical Name	CAS Registry Number	U.S. EPA water quality (acute, freshwater) (mg/l)	U.S. EPA water quality (chronic, freshwater) (mg/l)	U.S. EPA water quality (acute, marine) (mg/l)
Inorganics (continued)				
Selenium (and Compounds)	7782-49-2	7.60E–01	NF	NF
Silicon	7440-21-3	NF	NF	NF
Silver	7440-22-4	4.10E–03	1.20E–04	2.30E–03
Sulfur dioxide	7446-09-5	NF	NF	NF
Sulfur trioxide	7446-11-9	NF	NF	NF
Tellurium	13494-80-9	NF	NF	NF
Thallium	7440-28-0	1.40E+00	4.00E–02	2.13E+00
Tin	7440-31-5	NF	NF	NF
Titanium	7440-32-6	NF	NF	NF
Vanadium	7440-62-2	NF	NF	NF
Zinc	7440-66-6	NF	NF	NF

Water solubility

Water solubility (S) represents the maximum concentration of a chemical
that will dissolve in pure water at a specific temperature and pH. It is a
critical property that typically exerts a significant effect on the environmental
fate and transport of a contaminant. As such, water solubility data are
required for calculation of the Henry's law constant and for calculating
certain partition coefficients (e.g., K_{oc} or K_{ow}). The values that are reported
for water solubility in Table 3.1, expressed in units of mg/L, are provided
for conditions of neutral pH (approximate pH = 7) and a temperature range
of 20 to 30°C, unless otherwise specified. Chemicals listed in the literature
as being "infinitely soluble" or "miscible" were assigned a solubility value
of 1,000,000 mg/L for purposes of consistency.

Vapor pressure

The vapor pressure (VP) of an organic chemical is a measure of the volatility
or evaporative potential of the chemical in its pure state. Vapor pressure is
an important determinant of the rate of vaporization of liquids. Values for
VP, expressed in units of mm Hg, are given for a temperature range of 20
to 30°C, unless otherwise noted in Table 3.1. Although some inorganic analytes may exhibit measurable vapor pressures (e.g., approximately 0.00001
mm Hg for mercury), the volatility of most inorganic analytes is not of
environmental significance under typical conditions.

Henry's law constant

The Henry's law constant (H) is a comparative measure of the net tendency for a dissolved material to leave aqueous solution and go to the gas phase, based on the balance between its solubility and its vapor pressure. This property, or constant, is important in evaluating potential air exposure pathways or in the determination of half-life values for chemicals in water. Tabulated empirical Henry's law constants exist for many substances, and these may differ substantially from calculated values under optimal conditions, depending on the experimental conditions. The ideal, defined values for Henry's law constants, expressed in units of $(atm \bullet m^3/mole)$ were calculated using the following equation and the values that previously were reported for solubility, VP, and for molecular weight:

$$H = [\text{vapor pressure (atm)}] \times [\text{MW (g/mole)}] / [\text{water solubility (g/m}^3)]$$

Organic carbon partition coefficient (K_{oc})

As noted previously, the organic carbon partition coefficient (K_{oc}) is a measure of the tendency for a substance to be adsorbed to organic materials in soil, in sediment, or in biological tissues and is expressed as shown in the following equation:

$$K_{oc} = (\text{mg chemical adsorbed/kg organic carbon})/(\text{mg solute/L of solution})$$

The degree of adsorption to organics may affect not only the environmental mobility of a chemical but also may be an important parameter in evaluating fate processes such as volatilization, photolysis, hydrolysis, biodegradation, and bioaccumulation. The K_{oc} is chemical specific and is largely independent of other soil properties. It may be expressed as K_{oc} or as log K_{oc}, depending on the literature source, but conversion is readily accomplished as shown in the following example:

$$K_{oc} \text{ for pyrene} = 3.8 \text{ E}+04 \text{ or } 38,000 \text{ (see Table 3.1)}$$

$\log_{10}$ of 38,000 is expressed as $10^{4.58}$; thus the log K_{oc} for pyrene is 4.58

Many of the K_{oc} values that are presented in Table 3.1 were recorded directly from the cited literature sources, but some were estimated using methods specified by Lyman et al. (1982). In such instances, the log K_{oc} was estimated by the following expression:

$$\log K_{oc} = (-0.55 \bullet \log S) + 3.64$$
(Note: water solubility (S) in units of mg/L)

A number of other expressions for estimation of the K_{oc} for specific classes of analytes have been proposed and tested by empirical observations

(Lyman et al. 1982). However, the expression shown above is commonly applicable to a wide variety of analytes.

The overall distribution coefficient for an analyte (typically expressed as K_d) is dominated by the contribution of the organic carbon fraction in the medium (defined as f_{oc}). This relationship is expressed by the relationship $K_d = K_{oc} \bullet f_{oc}$. This is true for most environmentally relevant situations. However, at very low organic carbon concentrations in soil (e.g., less than 0.01% to 0.1%), other aspects of adsorption beyond adsorption to organics (e.g., inorganic binding) may become increasingly important.

Octanol–water partition coefficient (K_{ow})

The octanol–water partition coefficient (K_{ow}) is a measure of how a chemical is distributed at equilibrium between an eight-carbon alcohol (octanol) and water. K_{ow} is defined as the ratio of a chemical's concentration in the octanol phase to its concentration in the aqueous phase of a two-phase octanol/water experimental system. Values of K_{ow} are thus unitless, as illustrated in the following expression.

$$K_{ow} = \text{(concentration in octanol phase) / (concentration in aqueous phase)}$$

This parameter is measured using low solute concentrations, where K_{ow} is a very weak function of solute concentration. Values of K_{ow} typically are measured at room temperature (20 to 30°C). K_{ow} has been used in medical and environmental science as a reasonable measure of the hydrophobicity and hydrophilicity of chemicals and as a reasonable indicator of lipid solubility.

As such, K_{ow} often is a key variable that is used in the estimation of other physical/chemical characteristics of organic analytes.

Bioaccumulation factor (K_B)

Bioaccumulation refers to the uptake and concentration of a xenobiotic (a foreign compound) by an organism following exposure. The bioaccumulation factor (K_B) expressed in terms of the equilibrium ratio of the chemical concentration in the organism to the concentration in the surrounding medium (i.e., the food or water ingested by the organism) is an accurate measure of the chemical's bioaccumulation potential. A chemical's lipophilicity (K_{ow}), the rate of metabolism, and the biological half-life, are the primary determinants of the rate and degree of bioaccumulation. Since K_{ow} can be measured easily, and K_B is reliably calculated using the appropriate K_{ow}, the following equation can be used for calculation of K_B (Esser and Moser, 1982).

$$\log K_B = n \log K_{ow} + b$$

where n and b are regression coefficients.

This equation assumes that *n*-octanol has similar solubility properties to the lipids of the cell walls. Reliable correlations between K_B and K_{ow} have been demonstrated and occur when K_{ow} is between 10^2 and 10^6, which corresponds to K_B values between 10^1 and 10^4 (Shane, 1994). For practical purposes, K_B values of less than 100 to 1,000 do not indicate significance from a toxicological perspective.

Toxicological properties and reference values

Introduction

This section discusses the toxicology of the principal classes of emissions that are attributed to hazardous waste incinerators and related facilities. The organic analytes may be subdivided into three groups. First, there are the principal organic hazardous constituents (POHCs), represented by those RCRA Appendix VIII constituents that are present in emissions and that are detectable in the incoming waste stream at concentrations greater than 100 mg/kg (100 ppm). Second, there are products of incomplete combustion (PICs), which, as defined previously, represent regulated organic compounds (i.e., those listed in Appendix VIII of CFR 40 Part 261 under RCRA) that are detected in stack emissions but not detected at a concentration greater than 100 mg/kg in the initial waste stream. Third, there are those organic constituents that may be present in the incoming waste stream but that are *not* regulated as Appendix VIII parameters. These categories are discussed in greater detail in the following sections with regard to their toxicological importance. The organic and inorganic analytes presented in Table 3.2 and Table 3.3 include selected members of the inorganics and members of all three organic groups that have been reported or projected in hazardous waste incinerator emissions and that are considered to be a reasonable representation upon which to base evaluations of potential risk.

Advantages and limitations of reference values

Where agency-derived consensus toxicological values exist, they typically provide an adequate basis to start an evaluation. In those instances where toxicological guidance is not available, it may be necessary to derive guidance concentrations or proposed acceptable exposure levels on the basis of the published toxicological literature. This approach, while time consuming, is preferable to ignoring the health-based evaluation of an analyte simply because there is not guidance readily available. In such instances, it is necessary to submit the appropriate documentation and perhaps the appropriate toxicological literature to the reviewing agency to achieve concurrence on the final target concentration or guidance value.

Toxicological guidance for selected organic analytes

In keeping with the intent of a hazardous waste incinerator facility, most of the POHCs are thermally destroyed during incineration of hazardous wastes.

A smaller proportion is removed from the waste stream as bottom ash or trapped as fly ash in air pollution control devices (APCDs). Although destruction and removal efficiencies (DREs) at adequately designed and operated hazardous waste incinerators range from 99.99% to 99.9999%, it still is possible for a very small proportion of the POHCs (e.g., less than 0.01% to 0.0001%) to be released from the incinerator stack into the atmosphere with the theoretical potential to affect human health or environmental receptors.

The wide variability in structure and properties for the organic chemicals makes it extremely difficult to make generalizations regarding potential adverse effects. At sufficient dosage, there is some organic substance that may affect virtually any organ system. Table 3.2 summarizes selected relevant toxicity end points for many organic analytes of interest to the extent that such data are involved in the establishment of regulatory toxicological guidance. The most significant distinction relates to the classification of a substance as a carcinogen or as a noncarcinogen. This classification may be route independent (e.g., benzene) or may be route specific (e.g., some inorganics as discussed in the next section).

A wealth of other health-based information is available for many of these analytes, either by class of compound or by individual analyte. One major source for such information is the Agency for Toxic Substances and Disease Registry (ATDSR, 1989–1997), which has produced an extensive set of toxicological profiles for well over 100 organic (and some inorganic) analytes. In addition, several sources of U.S. EPA information and information from other agencies are excellent compilations of existing studies. These sources include

- Hazardous Substances Data Bank (HSDB)
- U.S. EPA Health Assessment documents
- Health Effects Assessment Summary Tables (HEAST)
- World Health Organization (WHO) Environmental Criteria documents

Given the large number of analytes selected for potential evaluation or monitoring application, detailed toxicological summaries are beyond the scope of this document. A brief summary discussion concerning the dioxins/furans is presented in Chapter 2.

Toxicological guidance for selected metals and other inorganic parameters

Although the focus of hazardous waste incineration is on the destruction of organic compounds in the waste feed, metals that may be present in hazardous waste streams are not thermally destroyed but can be partially or completely vaporized during incineration, may be associated with particulates, and may accumulate in ash. Vaporized metals may condense to form particles that are then removed by APCDs. However, some proportion of these metals may be released to the atmosphere and, therefore, has the potential to affect human health.

As noted in the previous section regarding organics, metals may be classified as carcinogens or as noncarcinogens, and such classifications may be route specific. Carcinogenic potential is estimated in terms of lifetime excess cancer risk, while noncarcinogenic toxicity is reflected in terms of an acceptable daily intake (ADI) or reference dose (RfD). The range of toxicological end points is described in Table 3.2 for the approximately 35 metals and other inorganic analytes of interest. This list of inorganics was compiled from the available sources that address inorganic emissions from hazardous waste incinerators. It, therefore, is a broader list than that which strictly represents the "hazardous" inorganics. Given the number of analytes selected for potential evaluation or monitoring application, detailed toxicological summaries are beyond the scope of this document. However, the following general discussion highlights the toxicology of some of the inorganics that are likely to be components of emissions from hazardous waste incinerators.

Metals represent the oldest recognized group of toxicants, having been used for nearly 4,000 years in the case of lead. Arsenic and mercury have been used for cosmetic, medicinal, and manufacturing purposes for over 2,000 years. Many other metals, however, are of much more recent concern; for example, cadmium was only identified as a component of zinc ore in the early 1800s. Although the well-recognized, historically described adverse effects associated with exposure to metals were often of an acute nature, such as gastrointestinal problems from lead or mercury poisoning, the chronic effects of low-level exposures have become the focus of more recent concerns about the metals. This includes neurological and neurobehavioral effects that have been attributed to low blood lead concentrations in children and renal toxicity from accumulative exposure to cadmium. Often the site of action and the progression of the disease state can be important diagnostic aids in the evaluation of metal intoxication. The following page presents a list of some important, common metals with a description of the principal observed effects from acute or chronic exposure.

Specific interest, in the context of incinerator facilities, often is focused on mercury. The sources, the fate and transport, and the environmental impacts of the various forms of mercury are not well understood, despite the efforts conducted to date. One observation that seems to have remained constant, however, is that incineration sources, whether they be power plants, municipal refuse combustors, or hazardous waste incinerators, are likely to be regional rather than local contributors to mercury loading; that is, it has been difficult to demonstrate that mercury emissions from these combustion facilities are deposited locally. A better understanding of the sources and movement of mercury will provide a sounder foundation for policies concerning mercury emissions and related topics such as health advisories for accumulation in edible fish species.

Metal	Sites of acute effects	Sites of chronic effects
Arsenic	Respiratory	Nervous system
	Liver	Respiratory
	Nervous	Circulatory
	Circulatory	Cancer
Beryllium	Respiratory	Respiratory
	Dermal	Cancer
Cadmium	Gastrointestinal	Respiratory
	Respiratory	Kidney
		Skeletal
		Cardiovascular
		Cancer
Chromium	Renal	Respiratory
		Cancer
Lead	Gastrointestinal	Nervous
		Cardiovascular
		Cancer
Mercury	Respiratory	Renal
		Nervous
		Reproductive

Regulatory guideline values

Toxicological information and regulatory guideline values for selected analytes that have been reported in emissions from hazardous waste incinerators are presented in Table 3.2. The table includes noncarcinogenic toxicity data, organs of effect, carcinogenic toxicity data, dermal permeability data, and the regulatory classification where appropriate.

Noncarcinogenic effects data

The oral reference dose (RfD; expressed in units of mg/kg•day) and the inhalation reference concentration (RfC; expressed in units of mg/m^3) indicate the potential noncarcinogenic benchmark risk. The oral RfD and the inhalation RfC are route-specific estimates of the daily exposure to the human population that is likely to be without an appreciable risk of detrimental effects during a lifetime. Table 3.2 provides quantitative estimates of the RfD and RfC for the given analytes and specifies the route-specific target organ or the effect that may be caused by an analyte at sufficient exposures.

The critical dose (RfD) or air concentration (RfC) of an analyte usually is based upon the no-observed-adverse-effect level (NOAEL) or a lowest-observed-adverse-effect level (LOAEL). The RfC or RfD typically is derived by dividing the NOAEL or the LOAEL by an uncertainty factor (UF) and, at times, a modifying factor (MF) as illustrated in the following expression:

$$RfC \text{ or } RfD = (NOAEL \text{ or } LOAEL)/(UF * MF)$$

The uncertainty factor that is used in calculating the RfC or RfD reflects the prevailing toxicological and scientific judgment regarding the available data that were used to estimate the RfC or RfD values. An uncertainty factor of 10 usually is used to account for variation in human sensitivity among populations. An additional 10-fold uncertainty factor normally is used to account for each of the uncertainties that may exist when extrapolating from animal data to humans, when extrapolating from a LOAEL to a NOAEL, and when extrapolating from subchronic to chronic exposures.

The current methodology for the derivation of inhalation RfCs is explained in the document "Interim Methods for Development of Inhalation Reference Doses" (U.S. EPA, 1990). These methods are different from those used for oral RfDs because parameters such as deposition, clearance mechanisms, and differences in the physical/chemical properties of the analyte must be considered (such as the size and shape of a particle). RfC and RfD values that have been derived for carcinogens are based on noncancer end points only and historically have not been used in cases that are designed to be protective against carcinogenic effects.

Carcinogenic effects data

In assessing the carcinogenic potential of a chemical, the Human Health Assessment Group (HHAG) of U.S. EPA classifies the chemical into one of the following groups, according to the weight of evidence from epidemiologic and animal studies:

Group A: Human carcinogen (sufficient evidence of carcinogenicity in humans)

Group B: Probable human carcinogen (Group B1 — limited evidence of carcinogenicity in humans; Group B2 — sufficient evidence of carcinogenicity in animals with inadequate or lack of evidence in humans)

Group C: Possible human carcinogen (limited evidence of carcinogenicity in animals and inadequate or lack of human data)

Group D: Not classifiable as to human carcinogenicity (inadequate data or no evidence available)

Group E: Not a human carcinogen

Each analyte in Table 3.2 is assigned to a carcinogenicity group, and the relevant route (either oral or inhalation) is specified as appropriate. Quantitative carcinogenic risk assessments typically are performed for chemicals in group A and group B, and on a case-by-case basis for chemicals from group C.

It should be noted that the Carcinogenic Assessment Guidelines are currently undergoing extensive revision, and the proposed changes are likely to dramatically alter the way in which carcinogens are regulated (U.S. EPA,

1996). Greater emphasis will be placed on mechanistic considerations and modes of action rather than on discrete classifications.

Oral cancer slope factor

Cancer slope factors (CSF) also known as cancer potency factors (CPF in the *Superfund Public Health Evaluation Manual* and other agency documents) are estimated through the use of mathematical extrapolation models applied to lifetime or chronic animal bioassay data. The most common model employed by regulatory agencies is the linearized multistage model. This statistical model is used to estimate the slope of the cancer dose-response curve as developed from the chronic animal studies, adjusted to reflect the slope associated with the 95% upper confidence limit from the data. The implicit conditions of this approach include the assumptions that low extrapolated doses will be proportional in linear fashion to much higher doses and that animal studies are accurately reflective of human responses. Both of these assumptions have numerous exceptions, and these assumptions have been used as criticisms of the present modeling approach.

$$\text{Slope Factor} = \text{risk per unit dose}$$
$$= \text{risk per mg/kg} \bullet \text{day or } (\text{mg/kg} \bullet \text{day})^{-1}$$

The unit risk for oral exposure conditions (risk per concentration unit in water, e.g., per $\mu g/L$) can be determined by dividing the appropriate slope factor by 70 kg (assumed human weight) and multiplying by the standard, if conservative, assumption regarding water consumption rate (2 L/day).

Oral risk-specific dose

Oral risk-specific doses, expressed in mg/kg•day, may be calculated for doses that correspond to upper bound increased lifetime cancer risks of both 1E-05 (1/100,000) and 1E-06 (1/1,000,000). These values, known as the RsD (also expressed in mg/kg•day), are obtained as follows:

$$\text{Oral RsD (1E-06)} = (1E-06)/(\text{oral slope factor})$$

or

$$\text{Oral RsD (1E-05)} = (1E-05)/(\text{oral slope factor})$$

Inhalation unit risk

The inhalation unit risk is an estimate of the inhalation exposure risk per concentration unit in air (per $\mu g/m^3$). To estimate the risk-specific concentrations that would be permissible in air from the unit risk in air, the specific level of risk (e.g., one in 100,000) is divided by the unit risk for air, expressed in $(\mu g/m^3)^{-1}$. For instance, the air concentration (in $\mu g/m^3$) that corresponds to an upper bound increased lifetime cancer risk of one in 100,000 (also expressed as 1×10^{-5} or 1E-05) is calculated as follows:

$$\mu g/m^3 \text{ in air} = (1E-05)/[\text{unit risk, in } (\mu g/m^3)^{-1}]$$

Dermal permeability data

Table 3.2 lists the estimated and, if available, the measured dermal permeability constants (K_p) for the selected analytes. The dermal permeability coefficient can be a key parameter in estimating dermally absorbed dose in site- or scenario-specific exposure evaluations and risk assessments. Its application typically has been in terms of absorption from aqueous solutions rather than gaseous exposures. However, some limited information is available regarding the dermal permeability of vapors and, in some situations, such exposures and subsequent absorption may be of toxicological significance (U.S. EPA, 1992). Vapor K_p values are available for few compounds, and the reported vapor values often differ dramatically from those that have been estimated or empirically derived for aqueous K_p values. For purposes of consistency, the estimated or empirical K_p values for aqueous conditions are presented in Table 3.2, in cases where they were available. As a practical matter, the dermal absorption of vapor phase materials typically is on the order of 0.1% to 1% of (i.e., 100- to 1,000-fold less than) the inhalation intake of the same substance under the same exposure conditions.

The aqueous dermal permeability constant (K_p), generally expressed in units of cm/hour or $(cm \bullet hr)^{-1}$, is a flux value that represents the rate at which the chemical penetrates the skin, normalized for concentration. The permeability coefficient is derived from Fick's first law of diffusion, which is used to relate the steady-state flux (J_{ss}) of a compound through the membrane to the concentration differential (ΔC) across the membrane. The resulting proportionality constant (K_p), shown in the following expression, also is known as the dermal permeability constant or coefficient.

$$K_p = (\Delta C)/(J_{ss})$$

The dermal permeability coefficient is a function of several factors, including the path length of chemical diffusion, the membrane/vehicle partition coefficient of the chemical, and the diffusion coefficient of the chemical in the membrane. Experimental values of permeability coefficients can be measured directly under in vitro conditions or may be evaluated indirectly from in vivo data by the use of appropriate pharmacokinetic models. Where data are inadequate or are lacking, the dermal permeability constant can be estimated from appropriate physical/chemical property relationships. A discussion of the methodology for evaluation of permeability constants and their applicability is provided in U.S. EPA (1992).

Inhalation considerations

The most common route of exposure to toxicants from hazardous waste incinerator emissions typically is considered to be through direct inhalation. Hazardous waste incinerator facilities operate at varied degrees of efficiency, and accurate knowledge of fugitive emission rates is critical to an accurate characterization of inhalation exposures and hence risks. Calculation of potential doses or inhalation exposure concentrations (IECs) can be achieved using the following formula (Foster, 1994). IECs are required to properly characterize the carcinogenic and noncarcinogenic risk associated with exposure to a particular chemical. It is important that the IEC units match those of the available risk specific dose-response data.

$$IEC = C_a \cdot ET/24 \cdot EF/365 \cdot ED/70 \cdot BIO$$

where C_a = chemical concentration in air (mg/m^3), ET = exposure time (hours/day), EF = frequency of exposure (days/year), ED = duration of exposure (days/year), and BIO = relative inhalation bioavailability factor.

Bioavailablity is the quantity of a chemical that is available to be taken into systemic circulation; it consists of absorption of the chemical across a membrane and desorption from its matrix. Toxicants absorbed by the lungs include gases, vapors, aerosols, and other particulate matter. Of these toxicants, vapors and gases are the most easily absorbed into the bloodstream. Once inside the lungs, gases and vapors are available for absorption within the alveolar space. Upon absorption, gases and vapors will continue to dissolve until an equilibrium is reached with the chemical concentration in the alveolar space (the ratio is constant and not the concentrations). This chemical-specific solubility ratio is known as the blood-to-gas partition coefficient. This partition coefficient determines how quickly a gas or vapor can be absorbed into the bloodstream. For aerosols and other particles, the size of the particle and its water solubility largely determine how quickly it will be absorbed.

Evaluation of analytes for which regulatory guidelines are not available

With organic and inorganic analytes, there are not always peer-reviewed benchmarks or consensus toxicological guidance or reference values available (e.g., RfD or CSF values through IRIS or HEAST). In some instances then, it may be necessary to review alternate sources of information to identify or derive a preliminary benchmark value. For example, there may be preliminary or draft RfD-equivalent or CSF-equivalent values available through the U.S. EPA Environmental Criteria and Assessment Office (ECAO). A number of these have been included in Table 3.2, referenced as such.

If preliminary, or draft, agency reference values are not available, it may be possible to identify potential "third tier" guidance values such as acceptable daily intake values from the World Health Organization. These values may be usable as developed, or they may be converted to RfD-equivalent values by the application of appropriate modifying factors to account for uncertainties in the database. However, these ADI values often are derived by the application of 100- or 1,000-fold "safety factors" and thus may already be similar to RfD equivalents. Only by reviewing the supporting documentation and references is it possible to determine the applicability and relevance of these values for risk assessment purposes.

In those cases where no formal, toxicologically based guidance values have been developed, it may be appropriate or necessary to conduct this process de novo for specific analytes. Depending on the magnitude, the accessibility, and the quality of the published or unpublished toxicological literature, this process may be straightforward or it may be quite time consuming and uncertain. Nevertheless, for proper risk assessment, an evaluation of this kind may serve a useful purpose. It is likely that the benchmark values thus derived will be subject to rigorous agency scrutiny, and, to the extent possible, procedures should be consistent with those used to derive official RfD or CSF values. This includes the selection of studies by which to document the NOAEL or the LOAEL, studies from which to derive alternative RfD or CSF values, and a measure of the uncertainty as reflected in the modifying factor.

In addition to the RfD equivalent or CSF equivalent, air concentrations for several hundred substances are available from the Occupational Safety and Health Administration (OSHA) or the American Conference of Governmental Industrial Hygienists (ACGIH). These guidance concentrations are developed on the basis of industrial exposure circumstances but in some instances are used, in modified form, for comparison with modeled air concentrations (e.g., in the derivation of the Florida no-threat levels). This approach, while not sanctioned by the ACGIH or by OSHA, provides a preliminary point of departure for evaluation of potential airborne hazards.

Table 3.4 Estimated Average Daily Intake of
Metals in Tests of Five Hazardous Waste
Incinerators

Metal	Average daily intake ADI (mg/d)	Estimated intake as fraction of acceptable intake
Antimony	0.292	1.5E-07 to 1.5E-05
Arsenic	0.07	9.5E-07
Barium	0.179	6.2E-07 to 2.5E-06
Beryllium	0.00071	5.2E-07 to 1.0E-05
Cadmium	0.0179	6.9E-07 to 1.1E-04
Chromium	0.175	1.4E-07 to 1.0E-05
Lead	0.0045	1.3E-05 to 5.9E-03
Mercury	0.02	8.0E-07
Nickel	1.46	1.0E-08 to 1.5E-06
Selenium	0.7	2.5E-07 to 2.4E-05
Silver	0.016	2.4E-07 to 1.5E-06
Thallium	0.0373	4.0E-07 to 1.2E-06

Source: From Travis, C. C. and Cook, S. C. *Hazardous Waste Incineration and Human Health*, CRC Press, Boca Raton, FL, 1989, 1. With permission.

References

Agency for Toxic Substances and Disease Registry (ATSDR), *Toxicological Profiles for Chemicals of Concern.* U.S. Department of Health & Human Services, Atlanta, GA. Series of reports regarding individual chemicals from 1989–1997.

Becker, R. A., personal communication, 1994.

Canadian Council of Ministers of the Environment (CCME), *National Guidelines for Hazardous Waste Incineration Facilities,* March, 1992.

Cooper, C. D., personal communication, 1994.

Dempsey, C. R. and Oppelt, E. T., Incineration of hazardous waste: a critical review update, *Air Waste* 43, 25, 1993.

Esser, H. O. and Moser, P., An appraisal of problems related to the measurement and evaluation of bioaccumulation, *Ecotoxicol. Environ. Safety,* 6, 131, 1982.

Foster, S. A., Human intake, in *Toxic Air Pollution Handbook,* Patrick, D. R., ed., Van Nostrand Reinhold, New York, 1994, 249.

Lyman, W. J. et al., *Handbook of Chemical Property Estimation Methods,* McGraw-Hill, New York, 1982, 1.

Shane, B. S., Principles of ecotoxicology, in *Basic Environmental Toxicology,* Cockerman, L. G. and Shane, B. S., eds., CRC Press, Boca Raton, FL, 1994, 11.

Travis, C. C., et al, *Inhalation Pathway Risk Assessment of Hazardous Waste Incineration Facilities,* Oak Ridge National Laboratory, U.S. Environmental Protection Agency, DE85-001694, 1984.

Travis, C. C. and Cook, S. C., *Hazardous Waste Incineration and Human Health,* CRC Press, Boca Raton, FL, 1989, 1.

Trenholm, A. et al., *Performance Evaluation of Full-Scale Hazardous Waste Incinerators,* Vol. I–V, Industrial Environmental Research Laboratory, U.S. Environmental Protection Agency, EPA-600/2-84/181a-e, NTIS PB85-129500, 1984.

Trenholm, A. et al., *Total Mass Emissions from a Hazardous Waste Incinerator,* Midwest Research Institute, U.S. Environmental Protection Agency, EPA-600/2-87/064, NTIS PB87-228508, 1987.

U.S. Environmental Protection Agency, *Interim Methods for Development of Inhalation Reference Doses,* U.S. EPA/600/8-88/066F, January 1990.

U.S. Environmental Protection Agency, *Dermal Exposure Assessment: Principles and Applications,* Interim Report, EPA/600/8-91/011B, 1992.

U.S. Environmental Protection Agency, *Addendum to Methodology for Assessing Health Risks Associated with Indirect Exposure to Combustor Emissions,* (external review draft) EPA/600/AP-93/003, 1993.

U.S. Environmental Protection Agency, *Risk Assessment for the Waste Technologies (WTI) Hazardous Waste Incinerator Facility (East Liverpool, Ohio),* EPA/905/D95/002c, 1993.

U.S. Environmental Protection Agency, *Carcinogen Assessment Guidelines,* 1996.

Van Buren, D., *Characterization of Hazardous Waste Incinerator Residuals.* Acurex Corporation, EPA-600.2-87/017, NTIS PB87-168159, 1987

chapter four

Modeling the atmospheric dispersion of hazardous waste incinerator emissions

C. David Cooper

Contents

Although the atmosphere is not the only medium to which pollutants from a hazardous waste incinerator (HWI) are emitted, air emissions are often considered to be the most serious threat. Air emissions can be classified as stack or fugitive emissions. *Stack* emissions are those contained in the gases

1-56670-250-X/99/$0.00+$.50

that exhaust through the stack or chimney of the facility. These exhaust gases form in the combustor, travel through the air pollution control (APC) equipment, and are emitted from a tall stack. *Fugitive* emissions come from leaking valves, pumps, or instruments, from evaporating spills or liquids in tanks, or wind-blown dust. After pollutants have been emitted from an HWI, they are transported downwind and dispersed to relatively low concentrations in the atmosphere. As the pollutants spread out in the atmosphere, they may react and be transformed, they may be deposited onto the earth's surface (either dry, as particles, or wet, as contaminated rain, snow, and so forth), or they may be carried in the air down to the ground level where they may be inhaled. The health effects of exposure to individual pollutants (or combinations of them) are proportional to the concentrations or deposition rates of the pollutants. The prediction of air concentrations and surface deposition rates is the task of dispersion modeling.

What is dispersion modeling, and why is it used?

Simply put, dispersion modeling is the mathematical representation of what happens to the air emissions from an HWI, resulting in predictions of pollutant concentrations at various distances from the HWI. These models range in sophistication, but even the simpler ones require a computer.

Modeling can always be criticized on a number of counts. First, any model of reality must necessarily make a number of simplifying assumptions. Next, modeling is expensive and requires much time and effort to gather input data as well as significant computing power and people with specific technical expertise. Modeling is definitely not 100% accurate; the many pieces of data required for input to the model cannot be known with absolute certainty. Also, the predictions from a dispersion model are subject to the uncertainties of the algorithms within the model and may in fact be accurate only to within a factor of two! Thus, to critics, modeling may seem a waste of time and money. With all these problems, why not forego modeling and use "high-tech" instruments to measure the desired information?

Despite its shortcomings, modeling *must* be conducted for several reasons. First, it is impossible to measure the actual impact from a facility that is not yet in existence. Still, it is crucial to have an estimate of what those impacts may be *before* allowing the facility to be constructed. Second, for those HWIs that are in existence, comprehensive measurement programs could be a factor of 1,000 times more expensive than comparable modeling studies. Third, modeling helps identify critical parameters that can be changed or sensitive areas that can be protected before a facility is permitted to be built. Fourth, models give the capability to analyze and differentiate many sources at once. Finally, modeling may not be 100% accurate, but it is precise (reproducible). Thus, modeling provides an impartial and reproducible yardstick for assessing different proposed projects. For all these reasons, models are good tools to help regulators make decisions. However, sound judgment must always be applied when interpreting the results of any modeling study.

The Gaussian plume model

The Gaussian plume model has long been the key element of air dispersion modeling. To better explain the features of a Gaussian model, graphical representations of a plume of pollution being released from a stack and being dispersed in the atmosphere are shown in Figure 4.1a and 1b. The plume rises from the stack, bends over in the wind, and spreads out. How fast and how far it rises, the direction of travel, and the extent of the horizontal and vertical spreading depend on a number of variables, including but not lim-

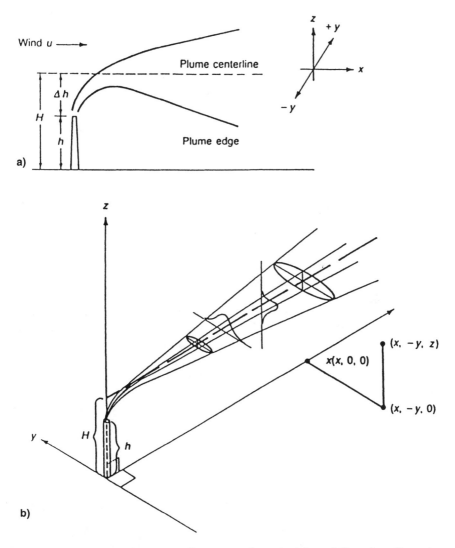

Figure 4.1 Schematic diagrams illustrating the principles of Gaussian dispersion models. (a) The spreading of a bent-over plume. (b) The two-dimensional spreading of a Gaussian plume.

ited to the height of the stack, the temperature and velocity of the exhaust gases, the wind speed and direction, the atmospheric stability, the terrain, and the distance downwind from the stack. The 1-hour average predicted concentration of a pollutant is a complex function of all those variables. The general shape of that function plotted with distance is shown in Figure 4.2. Over long periods of time (months or years) the long-time average concentrations (or depositions) surrounding an isolated source tend to be much lower than the peak 1-hour values, and the gradient diagram tends to be as shown in Figure 4.3.

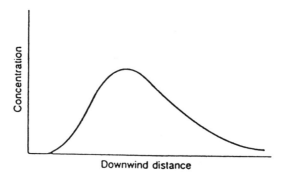

Downwind distance

Figure 4.2   Variation of ground-level 1-hour concentration with distance downwind from an elevated source.

Current modeling approaches

Modeling has long been accepted and required by the U.S. EPA for regulatory purposes. A recent EPA policy directive affects hazardous waste combustors directly in that (1) site-specific, comprehensive multipathway risk assessments are now required and (2) facility-specific permit emission limits will be set for dioxins/furans and for particulate matter to control the risks from trace organics and from metals, respectively (U.S. EPA, 1993). The use of maximum expected emissions for both the permit and the risk assessment would greatly overestimate risk because permit limits are written on a "not to exceed" basis (Rothstein et al., 1994). However, maximum emission rates are typically used in the dispersion modeling for risk assessments.

There are three broad categories of models currently used for regulatory purposes — Gaussian steady-state plume models, Gaussian puff models, and numerical models. The original modeling efforts of the 1970s were, by today's standards, very simplistic. The approach was to calculate a worst-case concentration for nonreactive, nondepositing gases based on a continuous release at a constant rate of emissions into a steady uniform wind field. The released plume of pollutants rose and bent over in the wind, then dispersed smoothly in the vertical and horizontal directions with rates of spread that were statistical functions of the atmospheric stability class and

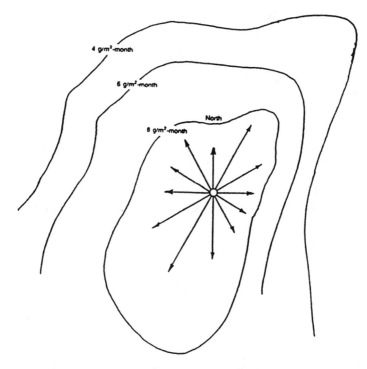

Figure 4.3   Wind rose and corresponding particulate fallout pattern.

the distance away from the source. Simple as they were, these approaches met a number of needs and were used for many years. Some recent models are discussed below.

TSCREEN

Gaussian steady-state models (with refinements) are still used extensively today. For initial screening, the EPA program TSCREEN (EPA-454/B-94-023, *User's Guide to TSCREEN,* July, 1994) is often useful in the flat or gently rolling terrain found throughout Florida and many parts of the nation. Many of the assumptions required for steady-state Gaussian models are still required, but better science has been incorporated to account for plume downwash owing to the aerodynamic effects of nearby buildings or even the stack itself. Plume rise formulas have improved considerably, and there is a choice of urban or rural dispersion parameters. TSCREEN gives a "worst-case" estimate of the 1-hour average concentration. Estimates for longer averaging times are obtained by multiplying the 1-hour value by a persis-tence factor, which can be done internally by the model if the user so chooses.

TSCREEN is very easy to use and requires no meteorological data (a number of meteorological conditions are built into the model). It is interac-tive and user friendly and can be learned in minutes or hours by most people with any air quality modeling background. From this author's limited experience, the worst-case 1-hour concentration predicted from TSCREEN is about the same as the highest 1-hour concentration obtained from ISCST3 when run for one year's worth of hourly meteorology (see discussion of ISCST3). However, the use of TSCREEN to obtain longer time averages is too conservative; that is, the longer time averages predicted from TSCREEN are significantly higher than those obtained using actual meteorology with ISCST3.

The philosophy of screening models is that one assumes very conservative values for emissions, meteorology, etc., and a maximum expected concentration (MEC) is calculated. If this MEC is below some maximum acceptable concentration (MAC) for each pollutant, then one feels reasonably well assured that all real concentrations will also be below the MAC. Sometimes, though, the public tends to interpret this result as an "implied guarantee" that the MACs will never be exceeded.

One problem is that no one really knows how conservative the assumptions are. Is the risk of violating the implied guarantee 1 in 10,000, or 1 in a million, or 1 in 100 million? Another problem is that once the worst-case values are published, the public "believes" them. Modelers know that if a worst-case model predicts a violation of a standard, it only means that we should do a more detailed model (which in all likelihood will show more realistic, lower concentrations). The public might view this process as "cheating"; if you first get answers that are too high, then change the model and try again. From this perspective, it might be wise not to do screening models on HWIs. Rather, the process should just begin with a refined or detailed model.

ISCST3

The refined Gaussian plume model most frequently used today (and for the past few years) for regulatory purposes is ISCST [the most recent version of which is ISCST3: EPA-454/B-95-003a, *User's Guide for the Industrial Source Complex (ISC3) Dispersion Models*, September, 1995]. Because ISCST is so widely used, it will be discussed in some detail in the following few paragraphs.

The Industrial Source Complex (ISC) models were initially developed in 1979. By mid-1980, ISC had become very popular and had displaced CRSTER, MPTER, and CDM (older EPA Gaussian plume models). The EPA refers to ISCST3 as a refined, regulatory model, and it has been used in literally thousands of studies for permitting PSD analysis and for other purposes.

There are two ISC models: ISCST3 (short term) and ISCLT3 (long term), depending on whether you need concentrations for short averaging times (hourly to 24 hour) or for long times (annual averages). ISCST3 uses *actual historical hourly* meteorological data for 1 to 5 years from a weather station close to the proposed HWI site. Based on user input for source data, receptor geometry, and output formats, ISCST3 calculates hourly average concentrations for every hour of every year at every receptor. It then can sort that morass of data for, say, the 50 highest concentrations and second-highest concentrations that have been modeled to occur anywhere during the whole time frame. It can also combine the hourly concentrations to yield 3-hour, 8-hour, 24-hour, and annual averages at each receptor and then sort those as well.

ISCLT3 employs STAR data, a frequency distribution of wind speeds, wind directions, and stability classes to calculate long-term averages (monthly, quarterly, and annual). It is not useful for pollutants where 1-hour to 24-hour concentrations are needed.

ISCST3 can handle literally dozens of point sources, hundreds of receptors, and tens of thousands of hours of meteorological data. There are about 4,000 lines of code, but the real core algorithms are contained in less than 200 lines — the rest is input, output, and averaging routines.

Dispersion is based on the classic Gaussian plume model with plume reflections at the ground and at an elevated inversion layer. Plume rise is based on Briggs models. Stack tip downwash is used if $u > 0.667\ v_s$, and building downwash is included if $H_s < H_{GEP}$. ISC can handle line, area, or volume sources but mostly finds use for elevated point sources. Chemical transformations can be handled in a very simplified way through a "pollutant half-life"; the maximum range of predictions is 50 km. It can handle PM fallout with the "tilted plume" approach (tilt angle is a function of wind speed and deposition velocity). The dry deposition algorithm of ISCST2 was criticized because it applied only to large dry particles and has been replaced in ISCST3.

The model assumes steady state — both for emissions and for meteorology. This is an acceptable assumption for nearby receptors (note that if the wind is blowing at 1 m/s, a parcel of air takes 1 hour to travel 3.6 km). However, these assumptions are very questionable for travel times of 5 hours or more (receptors farther than about 18 km); it is very unlikely that the wind will be constant in speed and direction for 5 hours. It is possible to adjust emissions by the hour on a daily basis. For a single source, the computed concentrations for each different pollutant are scaled by making one run with a unit emissions rate of, say, 1.0 g/s and then scaling by the actual emissions rate for each pollutant. Terrain is limited to flat or gently rolling conditions, where all receptors are less than H_s (stack height), but this is acceptable in Florida. ISC allows for flagpole receptors (at elevations above the ground plane).

Meteorological data must have been preprocessed (using the same preprocessor as for CRSTER or RAM). The inputs required include wind speed, direction, ambient temperature, stability class, and height of the mixed layer. Calms are handled in the same manner as other EPA programs (u is reset to 1.000 m/s, and wind angles are interpolated between noncalm hours). The concentrations are calculated for these calm hours but not used in the calculation of 3-, 8-, and 24-hour averages.

Since the wind can blow from any direction during the year, the receptors must be placed all around the source. Various distances from the source must be used to ensure that the maximums will be captured. Also, the receptor spacing must be dense enough to find the maximums. The model can "place" receptors either using a polar coordinate system or a Cartesian coordinate system (or a combination).

The model uses Pasquill–Gifford (P–G) dispersion coefficients in rural areas and McElroy–Pooler (Briggs) coefficients in urban areas. Keep in mind that the P–G coefficients were developed from measurements made with a 3- to 10-minute averaging time. Thus they are somewhat conservative because you would expect more variability over 1 hour. However, they seem to work well, and EPA continues to use them as-is in the models. The Briggs coefficients were measured over 30- to 60-minute averages. The real question is when to use rural or urban classification. The answer is not always obvious.

There are two algorithms in the model for calculating wind-direction-specific building wake effects: the Schulman–Scire algorithm for shorter stacks and the Huber–Snyder method for stacks up to GEP height. These are not perfect and can make a huge difference in your answers if the stack is calculated to be just within the zone of influence or just outside of it. Nevertheless, the model is much improved to have building downwash considered automatically than to ignore it.

Interpretation of the thousands of predicted concentrations that are generated by the model often reduces to looking for the highest or the second-highest concentration anywhere in the modeling domain. If that one number is acceptable, then the modeling has shown that the HWI will not produce violations of the ambient standards. For health-based risk assessments, the highest concentrations or deposition amounts predicted anywhere within the domain are the usual data for starting the exposure and risk calculations.

ISCST3 runs on a 386 or higher personal computer with 1 MB RAM, MS-DOS 3.2 or higher, and large hard drive. A math coprocessor is not required but is highly recommended. The time for a run is *greatly* reduced with a 486-DX and is extremely quick with a Pentium chip. The input format for the latest version of ISC uses a keyword approach that is significantly different from the older standard rigid-format input versions with numeric codes. Output options are many and varied. For more detailed information, refer to *User's Guide for the Industrial Source Complex (ISC3) Dispersion Models*, EPA-454/B-95-003a, U.S. EPA, Research Triangle Park, NC, September, 1995. Please note that the entire user's manual can be downloaded from the EPA's SCRAM Bulletin Board System in Word Perfect format.

MESOPUFF II

An alternative to the steady-state plume models is the class of models that deal with individual "puffs" of pollutants, including the EPA models

INPUFF and MESOPUFF. The models take a Lagrangian approach in that they follow the puff along a wind vector line and allow it to disperse in all three coordinate axes directions. By tracking thousands of puffs emitted sequentially, these models can simulate a continuous plume of pollutants. The puff approach was developed to circumvent the need to calculate concentrations at each and every receptor surrounding a facility. The computational effort thus could be focused on tracking the pollutants to where they were having their impacts.

Recent improvements to the puff modeling approach are incorporated into MESOPUFF II and include the adjustment of the friction velocity to account for the differences in surface roughness between a particular grid cell and the location of the meteorological station and the ability to output and store wet and dry flux predictions. MESOPUFF II is "a Gaussian, variable-trajectory, puff superposition model designed to account for the spatial and temporal variations in transport, diffusion, chemical transformation, and removal mechanisms encountered on regional scales." Each puff is transported independently of other puffs and is subject to growth by dispersion and pollutant removal by chemical transformations, precipitation, and dry deposition. In addition to the wind field module for transport, it has a boundary layer module that computes (from routinely available data) micrometeorological variables that describe the structure of the boundary layer (i.e., surface friction velocity, convective velocity scale, Monin–Obukhov length, and boundary layer height).

Unfortunately, at this time MESOPUFF II only accommodates five pollutants: sulfur dioxide, sulfate ions, nitrogen oxides (as the sum of NO_2 and NO), nitric acid, and nitrate ions. So it is not useful at present for HWI impact assessment. Also, it requires powerful PCs with much hard drive space available for each monthly simulation. Lastly, this model was intended for longer range transport studies; it will not give adequate results for distances shorter than about 20 km.

Urban airshed model

The urban airshed model (UAM) is the best example of a numerical (non-Gaussian) model. UAM is a very large, complex, and user-unfriendly model and was intended for regional applications to model the photochemical reactions that produce ozone. It starts with the fundamental unsteady-state mass balance equation that accounts for bulk transport, diffusion, chemical transformation, and deposition. UAM uses numerical integration techniques to solve hundreds of equations for hundreds of grid squares for thousands of hours of meteorology, taking into account time and space variable emissions rates of many pollutants from point, area, mobile, and biogenic sources. This large numerical model requires many hours of running time on a mainframe or a workstation and huge amounts of gridded input data. Although it may be possible to reformulate the UAM for a more local assessment of an HWI, that has not been done. However, as personal computers become more powerful, and as meteorological data become more readily available, it seems likely that a smaller-scale, unsteady state, numerical grid model will be developed for such assessments.

Models for hazardous/toxic releases

In the toxics modeling business, the term hazardous/toxic air release usually refers specifically to the accidental release of a concentrated gas, a liquid, or a gas/liquid mixture (such as from a ruptured tank car of ammonia) that forms a toxic or hazardous cloud or plume that disperses slowly as it moves downwind. Often the plume or cloud is denser and/or colder than air, and this affects its transport and dispersion behavior.

EPA has published a detailed review of the science and engineering behind these computer models as well as the models themselves (*Guidance on the Application of Refined Dispersion Models for Hazardous/Toxic Air Releases,* U.S. EPA, 1993). The models listed in that review are *ADAM, ALOHA, DEGADIS, HGSYSTEM,* and *SLAB.* The transport and dispersion of these materials are complicated by the fact that the release rate is dependent on the molecular weight, viscosity, volatility, and other properties of the material released, the size of the spill or rupture in the tank, the ambient temperature, and heat transfer characteristics of the surface onto which the spill occurs (and many other variables). Many of these parameters must be specified as inputs to the model, and the release rate, the rate of vaporization or condensation, and the transport rate must be calculated within the model.

Each of the above-listed models has some advantages and disadvantages, and each was designed for modeling certain types of releases. None of these releases is typical of a hazardous waste incinerator, although one or more could be applicable to an accident involving a tanker truck that was transporting certain hazardous substances to the incineration facility.

New and emerging methods and models

ISC-COMPDEP

In the 1990 EPA document *Methodology for Assessing Health Risk Associated with Indirect Exposure to Combustor Emissions* (U.S. EPA, 1990), the model COMPDEP was recommended for the air modeling. COMPDEP predicts air concentrations and surface deposition fluxes, incorporating dry and *wet* deposition. It was especially made for predictions in complex terrain and included building wake effects also. However, after detailed study, all those aforementioned areas were found to need improvement. The EPA undertook to combine the best of ISC and COMPDEP and chose to use the modeling platform of ISCST3. The differences between COMPDEP and ISC-COMPDEP that are likely to cause the greatest impacts on concentration and deposition flux are in the algorithms that handle building effects, dry deposition, wet deposition, and complex terrain. All are important in Florida except for the complex terrain portion.

The details of these differences were reviewed by Schwede (Schwede and Scire, 1994) recently. The new dry deposition algorithms are well described in detail elsewhere [*Development and Testing of a Dry Deposition Algorithm (revised)*, U.S., EPA, 1994], and the new wet deposition approach in ISC-COMPDEP is handled by a scavenging coefficient. The new approaches require estimation of the friction velocity and the Monin–Obukhov length, both of which can be estimated from routine meteorological observations, although the procedures are not trivial.

It was concluded that ISC-COMPDEP was an improvement in the arsenal of models available for indirect exposure assessment but that future improvements should include the capability to model the dry deposition of gases in addition to particles. In a private conversation with Schwede, it was learned that ISC-COMPDEP was further along in the regulatory approval process than another important modeling improvement, AERMOD.

AERMOD

Perhaps the most exciting development in dispersion modeling to occur in 20 years is a result of the combining of forces of EPA and the American Meteorological Society (AMS) to develop a new and fundamentally better dispersion model. This model (currently dubbed AERMOD) is still in the development and testing stage, but current plans are for it to be submitted for regulatory approval soon. The objective is for it to replace ISC as the workhorse of regulatory modeling (Perry et al., 1994).

AERMOD currently contains new or improved algorithms for dispersion in both the convective and stable boundary layers; plume rise; penetration of elevated inversions; computing vertical profiles of the wind; handling turbulence and temperature profiles; and treatment of receptors on all terrain. High priority is being given to improvements in plume downwash theory. AERMOD will incorporate ISC input format, but the improved theoretical basis for AERMOD should greatly increase confidence in its applications, particularly in situations where the models have yet to be fully evaluated.

AERMOD represents a fundamental change from Gaussian (Pasquill–Gifford) approaches to more current planetary boundary layer theory. It includes nonsymmetric updrafts and downdrafts, nonhomogeneity in the planetary boundary layer (PBL), non-Gaussian probability density functions (pdf's), and better plume rise theory. AERMOD uses both surface and mixed layer scaling to characterize the dispersion in the PBL. Dispersion in the convective boundary layer is based on Gifford's meandering plume concept and extends it to account for continuous plume rise. The paper on AERMOD is not easy reading but clearly represents the latest and most sound scientific thinking about atmospheric dispersion of pollutants.

AERMOD is still a steady-state, plume model and so is still bound by the limitations of a steady source of emissions and a steady "average" wind over a period of 1 hour. However, using similarity scaling and non-Gaussian pdf's, AERMOD has significantly changed (and improved) the theory of atmospheric dispersion. Promulgations of a new regulatory model (gaining "Appendix A" status) is a long process. It seems likely that ISC will be with us for some time to come, but changes are under way and improvements will continue to be added.

TOXST

In a very good article presented at a recent international meeting, the different approaches needed for air quality modeling of long-term (chronic) exposures, short-term peak (acute) exposures, or accidental exposures were discussed (Sullivan and Guinnup, 1994). Whereas ISCST is a very good application for estimating chronic exposures, it may not be so good for peak exposures and cannot be used at all for accidental exposures or to model batch operations. There is a clear need to represent the case of peak exposures resulting from randomly variable and intermittent emissions combined with randomly variable meteorology. Using maximum emissions with historical hourly meteorology is too conservative, and using average emissions would typically underestimate peak exposures. The third alternative is a Monte Carlo technique, which may offer the best chance to most accurately assess peak exposures (Sullivan and Guinnup, 1994).

The TOXST (Sullivan and Guinnup, 1994) was designed to execute the ISCST model and postprocess the output through the TOXX program to account for emissions variability. TOXX uses a Monte Carlo routine and simulates a user-specified number of years (between 100 and 1,999 years). The output is an expected number of exceedances of various threshold values (up to six different ones) specified by the user.

TOXST requires additional work to apportion emissions into emission distributions. Therefore, those authors suggested screening using the old "maximum ISCST" approach and only if that showed potential exceedances would TOXST be run. However, then TOXST could be used "to avoid (1) expending control resources solely because of excessive (and reducible) conservatism in the modeling analysis, or (2) establishing future operating permits that are unnecessarily restrictive" (Sullivan and Guinnup, 1994).

CAirTOX

CAirTOX (McKone, 1994) is an integrated multimedia model to help conduct a quantitative risk assessment for toxic chemical emissions into the air. In any risk assessment, five steps are included: (1) determine source emissions,

(2) apply transport and dispersion models to determine air concentrations, (3) calculate exposures, (4) assess toxicity, and (5) characterize the risk. CAir-TOX has been developed as a spreadsheet model to address how pollutants released to an air basin can lead to contamination of soil, food, surface water, and sediments. It is compatible with add-on Monte Carlo programs to conduct an uncertainty analysis.

CAirTOX is a seven-compartment, regional, dynamic, and multimedia-fugacity model (McKone, 1994). Multipathway exposure models are used to estimate average daily doses within a human population in a region. The seven compartments are (1) air, (2) ground-surface soil, (3) root-zone soil, (4) plant leaves, (5) plant roots, (6) surface water, and (7) sediments. CAirTOX is different from many existing models for assessing environmental fate in that CAirTOX imposes mass balance on each contaminant and systematically accounts for gains or losses among compartments.

As expected, the model requires a considerable amount of input data. For each contaminant assessed, one needs physical–chemical properties (such as molecular weight, octanol–water partition coefficient, solubility, Henry's law constant, etc.), as well as media specific transformation rates or rate constants. Meteorological data include average annual wind speed, deposition velocities, and air temperatures; hydrologic data include annual rainfall, runoff, and soil infiltration; and soil properties include bulk density, porosity, erosion rate, and root zone depth. Human exposure input data include exposure duration, anatomical and dietary properties, activity patterns, food consumption patterns, soil ingestion, and breast milk intake among others.

The model at present has several capabilities and limitations (McKone, 1994). It often starts with a "screened" worst-case or average air concentration in every cell to avoid air dispersion modeling. It is best applied to nonionic organic chemicals, radionuclides, fully dissociating organic and inorganic chemicals, and solid-phase metal species. It should not be used for surfactants or volatile metals. It is intended for long time scales, several months to years, and should not be applied to landscapes with more than 10% of the surface area taken up by water. Thus, it probably is not a good model at this time for Florida, with its many lakes and long coastline.

Case study

The purpose of this section is to give an example of what is being done at present to assess health risk uncertainty for a toxic waste incinerator as reported at a recent international meeting (DiCristofaro et al., 1994). An air quality analysis and risk assessment of an existing waste oils and solvents incinerator in Massachusetts was conducted to determine the facility's continued ability to meet the requirements to burn PCBs. Trial burns were conducted in October and December of 1990. The authors of the paper presented the results of the dispersion modeling and public health risk assessment and quantified the uncertainty in the risk estimates via three different methods: (1) standard EPA exposure assumptions and toxicity values, (2) alternate exposure assumptions and toxicity values (for PCBs and dioxin), and (3) a probabilistic analysis.

The air quality modeling used the five most recent years of NWS surface meteorological data from a nearby airport. The ISCST2 model was used for dispersion calculations. Due to a short stack, terrain exceeded stack height in many locations, so the VALLEY model was used. (The ISCST2 concentrations were much larger, so no further complex terrain modeling was done.) They used 625 receptors in a square array spaced at 100-m intervals and ran ISCST2 with the regulatory default option set for the urban setting (DiCristofaro et al., 1994).

Exposure could occur through multiple pathways — direct inhalation of gases and particles, and contact with particles that are deposited on the ground. Indirect exposure could occur through the food chain or other routes. The particle size distribution and chemical composition was needed to perform detailed modeling but did not exist. However, it was postulated that for this liquid waste incinerator, particle emissions would not be a major threat. Therefore, only the risks associated with inhalation of gases was evaluated. This illustrates what compromises can sometimes occur during an actual risk assessment.

Using the standard EPA risk assessment assumptions, the total incremental cancer risks for combined exposures of child and adult over 30 years were calculated to range from 1×10^{-5} for the maximum exposed individual to 8×10^{-7} for the average exposed individual, which are within the EPA target risk range of 1×10^{-6} to 1×10^{-4}.

To provide a more realistic risk estimate, the emission and exposure assumptions were altered. Specifically, the chromium emissions were assumed to be 1% as hexavalent chromium instead of 100% (measurements at two incinerators have shown that Cr^{+6} is actually about 0.1% of the total). Also, the exposed individuals were assumed to be exposed only 16 hours per day in this scenario instead of 24 hours per day in the standard EPA scenario (DiCristofaro et al., 1994). Under this scenario, the total incremental cancer risks were calculated to range from 9×10^{-7} for the maximum exposed individual to 7×10^{-8} for the average exposed individual, about one order of magnitude less than those obtained with the standard EPA assumptions.

A probabilistic risk assessment was conducted using a Monte Carlo analysis. This technique uses the full distribution of pertinent data (as probability density functions — pdfs), instead of point estimates, to characterize a number of variables. When applied to a risk assessment model, exposure variables (expressed as pdfs) are randomly sampled thousands of times with risk also calculated each time. The calculated risks are then presented as a cumulative probability distribution. This method is a more accurate portrayal of risk assessment and has been endorsed by EPA (*Guidance for Exposure Assessment*, EPA 1992).

Those authors used the software "@Risk" by Palisades, Inc., with Microsoft Excel. They simulated the parameters ground-level concentrations within the receptor grid, variation in duration and frequency of exposure, and physical characteristics of body weight and ventilation rate. No attempt was made to vary waste composition, plant operations, or emission rates. To avoid the codependence of body weight and ventilation rate, the inhalation rate was input as a function of body weight. Both 1,000 and 5,000 simulations were run. The results indicated that there was a 90% probability that the total cancer risks are less than 7×10^{-8} and a 50% probability that they were less than 1×10^{-8}. The 98th percentile risk was 3×10^{-7}.

Near-term trends in modeling

Emissions modeling

After reviewing numerous articles on emissions, it was concluded that a definitive model to predict emission rates of all pollutants does not (indeed, cannot) exist. A large database does exist from numerous plant tests including some trial burn tests. In the past, applications for new HWIs have been submitted that included estimated stack emission rates for numerous pollutants (after considering estimated APC equipment efficiencies). Some of these estimates were based on data from the open literature but some were based on proprietary information. In some cases it was extremely difficult to substantiate the emissions estimates. Also, in many cases these were maximum limits, and long-term average actual emissions were probably significantly less (although short-term excursions could certainly occur during upset conditions).

Essentially there are four methods of estimating emission rates: (1) emission factors, (2) data from the open literature for similar facilities, (3) proprietary data, and (4) trial burn test results. Each has its advantages and disadvantages.

More and more data are now being published in the open literature, and future applications should become more accurate and defensible with regard to emissions estimates. The U.S. EPA has recently developed an extensive computerized database called FIRE (Pope and Blackley, 1994) that eventually should allow quick and easy determination of "best" emission factors. It should also enable one to put upper and lower accuracy bounds on the data and indicate the reliability of the data source for each emission factor. Perhaps one day in the not-too-distant future, engineers will be able to assign probabilities to estimated emission rates from a proposed HWI based on the FIRE database and other site-specific data. This probabilistic approach to emission rates might fit very well with the trend toward probabilistic risk assessments.

Dispersion modeling

The recent and near-term trends in dispersion modeling indicate that current EPA dispersion and deposition models will continue to be used, although they are constantly being updated as new methods that reflect better under-standing of the physical and chemical processes are incorporated. The pro-cedure for gaining regulatory status is long and drawn out, so it seems unlikely that any new approaches will be introduced within the next few years. The new efforts with AERMOD will be incorporated into an ISC-type model because of the wide use of that modeling platform and because of the "satisfaction" an air quality modeler has in using actual meteorology over a long period of time.

As indicated by the efforts in California, the trend is to more complete models, in which the air modeling, the exposure modeling, and the risk calculations are all incorporated into one large model. This requires more and more computing power, but the ongoing "PC power wars" show no sign of abating, so that should not be a limitation. The advantages to all-inclusive models include completeness and standardization; the disadvan-tages often are the requirement for more input data and a reduced level of understanding by the analysts.

The recent case study shows the power of probabilistic risk assessment. Combined with faster and more powerful computers, future risk assessments will be probabilistic in nature. This will give a more realistic appraisal of the consequences of hazardous waste incineration, but require a better means of communicating and explaining the results of the assessment to the public, who tend to have little tolerance for externally imposed risks, no matter how small.

Summary

Emissions from an HWI are transported by the wind and dispersed into nearby areas where they may have negative impacts. It is important to be able to predict these impacts *before* an HWI is built. Models, with all their shortcomings, help analysts do that impartially and reproducibly. They are useful tools to help regulators make decisions. Models are constantly being updated as scientific understanding of atmospheric phenomena advances. The latest EPA models (either just released or in a late stage of development) appear to offer further improvements in the art of atmospheric dispersion modeling.

References

"Comments on the Report Development and Testing of a Dry Deposition Algorithm (Revised)," a memorandum issued by Touma, J. S., Pleim, J., and Binkowski, F. S., Office of Air Quality Planning and Standards, U.S. Environmental Protection Agency, Research Triangle Park, NC, May 6, 1994.

Development and Testing of a Dry Deposition Algorithm (Revised), EPA-454/R-94-015, published by the Office of Air Quality Planning and Standards, U.S. Environ-mental Protection Agency, Research Triangle Park, NC, April, 1994.

D. C. DiCristofaro, D. Pedersen, and S. McCarthy. *Assessing Health Risk Uncertainty at a PCB Thermal Incinerator,* presented at the 87th Annual Meeting and Exhibition of the Air and Waste Management Association, Cincinnati, 1994.

T. G. Grosch, D. A. Sullivan, and D. Hlinka. *Toxic Modeling System Short-Term (TOXST) User's Guide,* Volume I (draft), Alexandria, VA, 1993.

U.S. EPA Guidance on the Application of Refined Dispersion Models for Hazardous/Toxic Air Releases, EPA-454/R-93-002, Office of Air Quality Planning and Standards, U.S. Environmental Protection Agency, Research Triangle Park, NC, May 1993.

Guidance for Exposure Assessment, U.S. Environmental Protection Agency, Fed. Reg., 57 (104) 22888-22938, May 1992.

T. E. McKone. CAirTOX, *An Inter-Media Transfer Model for Assessing Indirect Exposures to Hazardous Air Contaminants,* presented at the 87th Annual Meeting and Exhibition of the Air and Waste Management Association, Cincinnati, 1994.

T. E. McKone. *CalTOX, A Multimedia Total-Exposure Model for Hazardous-Wastes Sites, Part I: Executive Summary,* UCRL-CR-111456PtI, prepared for the State of California, Department Toxic Substances Control, Lawrence Livermore National Laboratory, Livermore, CA, 1993.

T. E. McKone. *CalTOX, A Multimedia Total-Exposure Model for Hazardous-Wastes Sites, Part II: The Dynamic Multimedia Transport and Transformation Model,* UCRL-CR-111456PtII, prepared for the State of California, Department of Toxic Substances Control, Lawrence Livermore National Laboratory, Livermore, CA, 1993.

T. E. McKone. *CalTOX, A Multimedia Total-Exposure Model for Hazardous-Wastes Sites, Part III: The Multiple-Pathway Exposure Model,* UCRL-CR-111456PtIII, prepared for the California Department Toxic Substances Control, Lawrence Livermore National Laboratory, Livermore, CA, 1993.

Methodology for Assessing Health Risks Associated with Indirect Exposure to Combustor Emissions, EPA/600/6-90/003, Office of Health and Environmental Assessment, Washington, DC, 1990.

S. G. Perry, A. J. Cimorelli, R. F. Lee, R. J. Paine, A. Venkatram, J. C. Weil, and R. B. Wilson. *AERMOD: A Dispersion Model for Industrial Source Applications,* presented at the 87th Annual Meeting and Exhibition of the Air and Waste Management Association, Cincinnati, 1994.

A. A. Pope and C. R. Blackley. EPA's FIRE (Factor Information Retrieval) — The System and its Contents, presented at the 87th Annual Meeting and Exhibition of the Air and Waste Management Association, Cincinnati, 1994.

R. A. Rothstein, P. Billig, and G. W. Siple. Implications of EPA's Draft Combustion Strategy on Permitting and Retrofit of Hazardous Waste Incinerators and BIFs, presented at the 87th Annual Meeting and Exhibition of the Air and Waste Management Association, Cincinnati, 1994.

D. B. Schwede and J. S. Scire. *Improvements in Indirect Exposure Assessment Modeling: A Model for Estimating Air Concentrations and Deposition,* presented at the 87th Annual Meeting and Exhibition of the Air and Waste Management Association, Cincinnati, 1994.

J. S. Scire, F. Lurmann, A. Bass, and S. Hanna. *Development of the MESOPUFF II Dispersion Model,* EPA-600/3-84-057, U.S. Environmental Protection Agency, Research Triangle Park, NC, 1984.

J. S. Scire, F. Lurmann, A. Bass, and S. Hanna. *User's Guide to the MESOPUFF II Model and Related Processor Programs,* EPA-600/8-84-013, U.S. Environmental Protection Agency, Research Triangle Park, NC, 1984.

W. M. Smith, F. O. Weyman, and M. T. Alberts. *Pros and Cons of Including Indirect Exposure Assessment in the Human Health Risk Assessments for Permitting of Boilers and Industrial Furnaces*, presented at the 87th Annual Meeting and Exhibition of the Air and Waste Management Association, Cincinnati, 1994.

D. A. Sullivan, T. G. Grosch, and D. Hlinka. *The Evaluation of Peak Exposures to Meet State and Clean Air Act Requirements: Benefits of Establishing a Proactive Position*, presented at the 86th AWMA Annual Meeting, Denver, 1993.

D. A. Sullivan and D. Guinnup. *Enhanced Version of TOXST*, presented at the 87th Annual Meeting and Exhibition of the Air and Waste Management Association, Cincinnati, 1993.

Toxic Modeling System Short-Term (TOXST) Modeling Guide, Office of Air Quality Planning and Standards, U.S. EPA, Research Triangle Park, NC, 1992.

User's Guide for the Industrial Source Complex (ISC2) Dispersion Models, EPA-450/4-92-008a, U.S. EPA, Research Triangle Park, NC, March, 1992.

G. K. Whitmyre, S. R. Baker, and D. A. Sullivan. *Evaluation of State Regulatory Initiatives for Assessing Peak Exposures to Air Toxics*, presented at the 85th AWMA Annual Meeting, Kansas City, 1992.

chapter five

Methods for health risk assessment of combustion mixtures*

Richard C. Hertzberg, Glenn Rice, and Linda K. Teuschler

Contents

---

* The views expressed in this paper are those of the authors and do not necessarily reflect the views and policies of the U.S. Environmental Protection Agency. Mention of trade names or commercial products does not constitute endorsement or recommendation for use.

Introduction

The estimation of health risk from exposure to emissions from municipal incinerators should consider at least two complicating factors. First, emissions alone are not necessarily good estimates of population exposure; second, health risk is caused by the simultaneous exposure to all the chemicals in the emission, yet most environmental health risk estimates are for single chemical exposures. Many procedures and models have been published dealing with toxicity and risk of multichemical exposures, called here "chemical mixtures," ranging from purely conceptual approaches to official guidelines for regulatory policy.

The usual methods and protocols for investigations of mixtures in the field or laboratory and the common techniques for analysis of multidimensional data do not work very well in the case of health risk estimation for chemical mixtures. Such methods are best applied to simple defined mixtures consisting of only a few chemicals, usually five or fewer. Actual mixture risk assessments usually involve exposures to tens or hundreds of chemicals, many of which may be unknown. For most of the identified chemicals, toxicologic mechanisms are not known. The regulatory assessment methods for mixtures are then developed from plausible generalizations and extensive extrapolation. As a consequence, accurate estimates of exposure and health risk for the situation being assessed are not feasible. The risk methods and risk characterizations for chemical mixtures exposures are then rough descriptions based on what is known and what can be reasonably assumed. The results are most often presented as qualitative criteria or quantitative indicators that assist regulatory decisions. The procedures described herein are then not prescriptive but are intended to guide the assessor in evaluating the available information. As more studies and data become available, the procedures are expected to change accordingly.

The discussion of mixtures approaches in this chapter is oriented toward the practice of the U.S. Environmental Protection Agency (EPA) and is organized by the type of available data. Next is the discussion of multipathway exposures and the implications of assessing chemical mixtures for incineration risk assessment. The biological details and evidence for toxicologic interactions and other cumulative effects from combined chemical exposures will not be presented, as they are covered well in several publications and are beyond the intended scope of this chapter. Some of the recent work on mixtures risk assessment and mixtures toxicology include the NAS/NRC book on testing,[1] the EPA's Technical Support Document that supplemented the 1986 guidelines,[2] two books on toxicology of mixtures,[3,4] the review on human concordance for mixtures,[5] and proceedings of two international symposia.[6,7]

Of the several guidelines written by the EPA on risk assessment, only a few will be discussed here, primarily the Guidelines for Health Risk Assessment of Chemical Mixtures and the Methodology for Assessing Health Risks

Associated with Indirect Exposure to Combustor Emissions. Several other risk assessment guidelines have been published by EPA and should be consulted for additional details on exposure assessment, both modeling and monitoring, and specific health end points, such as cancer and neurotoxicity. For mixtures assessment, many conceptual approaches have recently been developed. A few are included in the proposed revision of the EPA mixtures guidelines, but others will also be discussed because of their potential for marked improvement over current approaches. Much of the discussion will be on suggestions for selecting a particular approach. Older methods will only be summarized and referenced, whereas newer methods will be described in more detail. Because most incinerator emissions are highly complex mixtures, most of the following discussion will concern the use of whole mixture data; procedures based on chemical components will only be summarized.

For this chapter, "mixture" will be defined as in the original EPA's mixtures guidelines:[8] a mixture consists of those chemicals, regardless of spatial or temporal proximity, that contribute to the actual or potential toxic effects. The intent of that definition was to avoid focusing only on commercially produced mixtures (e.g., pesticides) or on simultaneous exposure to multiple chemicals from a single source (e.g., a landfill). For EPA, the mixture of concern may also be further defined by the applicable regulatory statutes. For the following discussion, the definition will allow consideration of sequential (over a short period) as well as simultaneous exposures from multiple chemicals by several exposure routes, where the original source is incinerator emissions.

The next section will focus on procedures for risk assessment of chemical mixtures. The general principles in existing and proposed EPA guidelines will be discussed, and special considerations for incinerator emissions will be highlighted. The last topical section will summarize the recent draft guidance for multipathway exposures, which is based on the U.S. EPA guidance for assessment of indirect exposures to combustor emissions.

Risk assessment of chemical mixtures

The U.S. EPA's Guidelines for Health Risk Assessment of Chemical Mixtures[8] describe three basic approaches to the risk estimation, based mainly on the nature of the available data. The preferred method is based on data for the mixture of concern. Next is to base the assessment on a mixture that is toxicologically similar to the mixture of concern. Last is to base the assessment on data for the chemical components of the mixture. All risk assessment procedures we have located to date fall into one of these three groups. The following discussion will be structured accordingly. Such an organization matches well the actual risk assessment process, where the selection of a risk assessment procedure is determined by the nature of the available data.

The application of any of these risk assessment approaches to municipal incinerators has several complications. First is the definition of the mixture of concern. The mixture being emitted will change over distance and over time, so the mixture to which the public is actually exposed may be substantially different. The "mixture of concern" may be defined as the mixture being regulated. For incinerators, that mixture may be defined by the chemicals identified in the emission. In contrast, the health effects decision criteria would be developed for the actual mixture that contacts the public. Furthermore, the public's exposure will vary depending on the distance from the stack and the nature of the intervening terrain. A second complication is that incineration mixtures often include several different states of matter, including volatiles, semivolatiles, and particulates. Chemical and physical interactions between particulates and other chemicals may be manifested as changes in transport of the chemicals or changes in deposition of these chemicals in the airways and lungs of an exposed individual.[9,10] No solutions to these issues will be presented, but they should be kept in mind when judging the relevance and accuracy of any mixture assessment method.

Procedure for selecting a risk assessment method

The development of a risk assessment for a chemical mixture will generally involve the examination of complex exposures and toxicities and the application of specific methods as well as scientific judgment. Because of the uncertainties inherent in exposure assessment, environmental fate, uptake and pharmacokinetics, and the magnitude and nature of toxicity and toxicant interactions, the assessment of health risk from exposure to chemical mixtures must include a thorough discussion of all limitations and assumptions. The situations addressed by mixtures risk assessments are highly varied. For combustion products of incinerators, there can be considerable variations in the source (feed) material, operating characteristics (e.g., temperature), and the completeness of chemical identification in the emissions estimates. Consequently, consideration must be given to the use of several approaches depending on the nature and quality of the available data, the type of mixture, the known toxic effects of the mixture or of its components, the toxicologic or structural similarity of a class of mixtures or of mixture components, and the nature of the environmental exposure.

Proposed approach

Consistent and clear terminology is critical to the discussion of chemical mixtures risk assessment methodology. The selection of a mixtures risk assessment approach is dependent on the available toxicity and exposure data. It is important, then, to articulate differences among classification terms that group chemicals according to the assumptions and requirements of the methodologies. Table 5.1 presents definitions suggested for this purpose in terms of specific criteria including the complexity of the mixture, similarity

Table 5.1 Definitions of Chemical Mixtures

Chemical Mixture

Any set of two or more chemical substances, regardless of their sources, that may jointly contribute to toxicity in the target population; may also be referred to as a "whole mixture" or as the "mixture of concern."

Components

Single chemicals that make up a chemical mixture that may be further classified as systemic toxicants, carcinogens, or both.

Simple Mixture

A mixture containing two or more identifiable components, but few enough that the mixture toxicity can be adequately characterized by a combination of the components' toxicities and the components' interactions.

Complex Mixture

A mixture containing so many components that any estimation of its toxicity based on its components' toxicities contains too much uncertainty and error to be useful; the chemical composition may vary over time or with different conditions under which the mixture is produced; complex mixture components may be generated simultaneously as by-products from a single source or process, intentionally produced as a commercial product, or may coexist because of disposal practices; risk assessments of complex mixtures are preferably based on toxicity and exposure data on the complete mixture; gasoline is an example.

Similar Components

Single chemicals that cause the same biologic activity or are expected to cause a type of biologic activity based on chemical structure; evidence of similarity may include parallel log-probit dose-response curves and same mechanism of action or toxic end point; these components are expected to have comparable characteristics for fate, transport, physiologic processes, and toxicity.

Similar Mixtures

Mixtures that are slightly different, but are expected to have comparable characteristics for fate, transport, physiologic processes, and toxicity; these mixtures may have the same components but in slightly different proportions or have most components in nearly the same proportions with only a few different (more or fewer) components; similar mixtures cause the same biologic activity or are expected to cause the same type of biologic activity due to chemical composition; similar mixtures act by the same mechanism of action or affect the same toxic end point; diesel exhausts from different engines are an example.

Chemical Classes

Groups of components that are similar in chemical structure and biologic activity, and that frequently occur together in environmental samples, usually because they are generated by the same commercial process; the composition of these mixtures is often well controlled, so that the mixture can be treated as a single chemical; dibenzodioxins are an example.

of biologic activity, similarity of chemical structure or mixture composition, the environmental source of the mixture, toxic end point, etc. Table 5.1 can be used by the risk assessor to classify available toxicity and exposure data to choose from among the risk assessment methods for chemical mixtures.

The approach for the selection of a procedure for a mixture risk assessment can be as outlined in the flow chart shown in Figure 5.1. Figure 5.1 begins with an assessment of data quality and then asks a series of questions that lead the risk assessor to selection of a method. Figure 5.1 is deceptively simple, however, as many of the issues that are represented in the diagram require the use of extensive scientific judgment or of data that may not be readily available. Table 5.2 presents a classification scheme for assessing the quality and nature of the available mixtures data. Consideration of the factors presented in Table 5.2 can be used to guide the risk assessor through the steps in Figure 5.1. For example, when there is high quality information on exposure, health effects, and interactions, the risk assessor should consider the data quality to be adequate for a quantitative risk assessment with good data available for both exposure and toxicity on the mixture of concern. The risk assessor would then perform a risk assessment directly on the mixture of concern, and a component-based approach would be supplemental.

The available information is usually less than perfect. With intermediate quality for exposure information and health effects information and minimally acceptable information on interactions, the risk assessor would conclude that data quality was adequate to estimate both the exposure and toxicity of the components of the mixture. The available interactions data

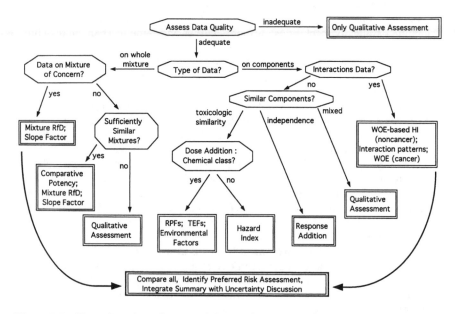

Figure 5.1   Flow chart for selection of chemical mixtures risk assessment procedures.

Table 5.2 Classification Scheme for the Quality of Available Mixtures Data

Exposure Information

A. Monitoring information either alone or in combination with modeling information is sufficient to accurately characterize human exposure to the mixture or its components

B. Modeling information is sufficient to reasonably characterize human exposure to the mixture or its components

C. Exposure estimates for some components are lacking, uncertain, or variable; information on health effects or environmental chemistry suggest that this limitation is not likely to substantially affect the risk assessment

D. Not all components in the mixture have been identified or levels of exposure are highly uncertain or variable; information on health effects or environmental chemistry is not sufficient to assess the effect of this limitation on the risk assessment

E. The available exposure information is insufficient for conducting a risk assessment

Health Effects Information

A. Full health effects data are available and relatively minor extrapolation is required

B. Full health effects data are available but extensive extrapolation is required for route or duration of exposure or for species differences, these extrapolations are supported by pharmacokinetic considerations, empirical observations, or other relevant information

C. Full health effects data are available, but extensive extrapolation is required for route or duration of exposure or for species differences; these extrapolations are not directly supported by the information available

D. Certain important health effects data are lacking, and extensive extrapolations are required for route or duration of exposure or for species differences

E. A lack of health effects information on the mixture and the components in the mixture precludes a quantitative risk assessment

Information on Interactions

A. Assessment is based on toxicologic data on the mixture of concern

B. Assessment is based on data on a sufficiently similar mixture

C. Quantitative interactions of all components are well characterized

D. The assumption of additivity is justified based on the nature of the health effects and on the number of component compounds

E. Interactions information is inadequate, an assumption of additivity cannot be justified, and no quantitative risk assessment can be conducted

Note: See text for discussion of sufficient similarity, adequacy of data, and justification for additivity assumptions.

might be inadequate for quantitative assessment but could be used to suggest at least the direction of bias in the assessment. Depending on whether the effect of concern is carcinogenicity or systemic toxicity, Figure 5.1 indicates the methodology to employ under these conditions. Failing to meet even the minimally acceptable classification for one or more of these categories would likely lead the risk assessor to decide that data quality was inadequate for quantification, so that only a *qualitative* risk assessment should be performed.

Tables 5.3 and 5.4 present additional tools for use in selection of a chemical mixtures risk assessment method for whole mixtures data or component data, respectively. The features of each method are compared in terms of type of available data, toxic end point of concern, limits on the complexity of the mixture, applicability of method, and uncertainties and assumptions.

Table 5.3   Methods for Whole Mixture Data

Procedure	Toxicity end point	Number of chemicals	Applicability	Uncertainty and assumptions
Mixture of concern	All	Any	Data rarely available	Composition of test mixture vs. field mixture; data cover all sensitive end points
Similar mixture	All	Any	Available data limited	Similarity judgment; plus above
Comparative potency	Cancer, genetic toxicity	Any	Data limited, short-term assay	Short-term assay vs. chronic *in vivo* assay; plus above

Data available on the mixture of concern

For predicting the effects of subchronic or chronic exposure to mixtures, the preferred approach usually will be to use subchronic or chronic health effects data on the mixture of concern and adopt procedures similar to those used for single compounds, either systemic toxicants or carcinogens.[11,12] Unfortunately, such information is rare. Exposure and toxicity data on the mixture of concern are most likely to be available on highly complex mixtures that are generated in large quantities and associated with or suspected of causing adverse health effects, such as coke oven emissions or diesel exhaust. For some mixtures with widespread exposure, the health risk has been attributed to a small number of chemical components (e.g., benzene for the cancer risk of inhaled gasoline vapor). For other mixtures, there is insufficient knowledge of the changes in composition over time (e.g., pesticides that degrade when present in soil for several years). When available, whole mixture data allow for a simpler and clearer evaluation of risk.

Table 5.4   Methods for Component Data

Procedure	Toxicity end point	Number of chemicals	Applicability	Uncertainty and assumptions
Hazard index	All	Limited by data quality and similarity	Good dose-response data, exposure data at low levels	Dose addition (toxicologic similarity) at doses near NOAELs; relative potency factor
Toxicity equivalence factor	All	Limited by data quality and similarity	Rare data; restricted by strong similarity, so few chemical classes will qualify; applied to all end points	Requires strong degree of toxicologic similarity based on dose addition; judgment of relative potency factor
Relative potency factor	All	Limited by data quality and similarity	Some data, restricted by similarity, restricted to specific conditions	Requires toxicologic similarity, but for specific conditions (end point, route, duration); based on dose addition; judgment of relative potency factor
WOE noncancer	Noncancer	Limited by data quality	Limited interactions data	Assumes binary interactions are most important, model with relative proportions untested; interaction magnitude not dose dependent

One of the main reasons for preferring data on the mixture of concern is that any interactions between component chemicals in the mixture, no matter how many or how complex, are represented by the whole mixture data. The availability of whole mixture data does not, however, guarantee an accurate risk assessment. For example, high dose to low dose extrapolation is much more complicated than with a single chemical. Some toxicologic interactions are associated with saturation phenomena, where the rate of a process has reached a ceiling. Mixture data at doses where those saturation conditions are present may then include interactions that would be negligible at much lower doses. The general issues needing to be considered then include stability of the mixture in the environment, variability of the mixture composition over time, variation across sources of the mixture, differences between mixtures tested in the laboratory and those in the environment, and the need for specialized dose-response models for mixtures data. These factors must be taken into account or the confidence in and applicability of the risk assessment will be diminished.

One example of a whole mixture approach is EPA's dose-response assessment for coke oven emissions, the results of which are on the Agency's on-line database IRIS.[13] Coke oven emissions were determined to be a human carcinogen, causing increased risk of mortality in coke oven workers from cancer of the lung, trachea, and bronchus; cancer of the kidney; cancer of the prostate; and cancer at all sites combined. The inhalation unit risk is defined by EPA[13] as the quantitative estimate of risk per $\mu g/cm^3$ air breathed. The value of 6.2×10^{-4} per ($\mu g/cm^3$) for coke oven emissions was based on respiratory cancer in males exposed in an occupational setting to coke oven emissions. This assessment is uniquely different from most cancer risk values found on IRIS because it is based on epidemiologic data and because the coke oven emissions mixture is evaluated as if it were a single chemical. An uncertainty analysis of this assessment would need to address the issues listed in the previous paragraph. Because the exposures are to humans and at the source, the main issues would be variability of the mixture composition over time and among different factories and the extrapolation of the risk estimate to low doses.

Of particular concern is when the toxicity data for the complete mixture is actually obtained for concentrates or extracts of the original mixture, i.e., the laboratory vs. environment issue. Because the risk assessment must extrapolate between different bioassays (different sensitivity of detecting various toxicity end points) as well as different mixtures (concentrates have higher concentrations and may not include all components), such data may not be predictive of human toxicity to the original mixture. Those data are more properly handled using procedures described below that were developed for toxicologically similar mixtures.

Data available on similar mixtures

Whole mixture data may be available, but for a mixture that is not exactly the same as the one of concern. An examination of the chemical, physical, and toxicologic properties of the mixture may lead to the judgment that the data are suitable, i.e., that the mixture represented by the data is "sufficiently similar" to the mixture of concern, and that the differences can be described. For example, the data may come from a single mixture that is known to be generated with varying compositions depending on time or different emission sources, such as municipal combustor emissions where the operating conditions are somewhat controlled but not constant. In such a case, the confidence in the applicability of the data to the risk assessment is diminished. This can be offset to some degree if data are available on several mixtures with the same components but with different component exposure levels (e.g., same incinerator over several months or different incinerators using similar source material), so that the likely range of compositional variation is covered. If such data are available, an attempt should be made to determine if significant and systematic differences exist among the chemical mixtures. If significant differences are noted, ranges of risk can be estimated based on the toxicologic data of the various mixtures. If no significant differences are noted, then a single risk assessment may be adequate, although the range of ratios of the components in the mixtures to which the risk assessment applies should also be given. Other examples of similar mixtures are discussed in the following sections.

Criteria for sufficient similarity
If no adequate data are available on the mixture of concern, but health effects data are available on a similar mixture, a decision must be made whether the mixture represented by the available data is or is not "sufficiently similar" to the mixture of concern to permit a risk assessment. The determination of sufficient similarity must be made on a case-by-case basis, considering not only the uncertainties associated with using data on a dissimilar mixture but also the uncertainties of using other approaches such as dose additivity of the component chemicals.

A mixture could be considered sufficiently similar to the mixture of concern in the following situations:

- The two mixtures have the same components but in slightly different ratios.
- The two mixtures have most components in common. The components that differ between the mixtures should be judged to be minor contributors to the mixture toxicity. The common components should also constitute similar proportions of the mixtures.
- The mixtures (or most of their components) display only minor differences in fate and transport in the environment, in uptake and pharmacokinetics, and in toxicologic effects.

If toxicity data for the candidate mixture are only available for a different exposure route than the environmental route being addressed, extreme care should be used to ensure that the results are applicable and that any effects restricted to the portal of entry to the body are appropriately discounted. For example, data for metals in drinking water may have little relevance to health effects from inhalation. Some differences concern the valence state of the metal and the physical state (dissolved or particulate).

Even if a risk assessment can be performed using data on the mixture of concern or a sufficiently similar mixture, it may be desirable to conduct a risk assessment based on toxicity data on the components in the mixture. In the case of a mixture containing carcinogens and toxicants, an approach based on the mixture data alone may not be sufficiently protective. For example, consider the case involving small dose group sizes and a two-chemical mixture of one carcinogen and one noncarcinogenic toxicant. The whole mixture approach would use toxicity data on the mixture of the two compounds. However, in a chronic study of such a mixture, the noncarcinogenic effects could mask the activity of the carcinogen; that is, at doses of the mixture sufficient to induce a carcinogenic effect, the doses could induce mortality so that at the maximum tolerated dose of the mixture, no carcinogenic effect could be observed. At lower doses, the small dose group size would reduce the statistical power so that there may be no significant carcinogenic responses. Because carcinogenicity may be a nonthreshold phenomenon, it may not be prudent to construe the lack of positive results of such a bioassay as indicating the absence of risk at lower doses. Consequently, the mixture approach should be modified to allow the risk assessor, on a case-by-case basis, to evaluate the mixture component data in terms of the potential for masking of one effect by another.

The comparative potency method

One of the few procedures for similar mixtures that has been developed and applied to data on environmental mixtures is the comparative potency method. In this procedure, a set of mixtures of highly similar composition is used to estimate a scaling factor that relates toxic potency between two different assays of the same toxic end point. The mixture of concern can then be tested in a simple assay (e.g., *in vitro* mutagenicity), and the resulting potency can then be scaled up to estimate the human cancer potency.

Comparative potency approaches were developed as a means of estimating the toxicity of a complex mixture in its entirety. Thus far, this method has been applied to data from the testing of mixtures of emissions released upon the combustion of organics.[14-16] In addition, the comparative potency procedure has only been applied to estimation of long-term cancer risk, using surrogate test information from short-term cancer bioassays and *in vitro* mutagenicity assays. Comparable efforts for noncancer effects are just beginning to be developed.[17]

The comparative potency method involves extrapolation across mixtures and across assays. It is restricted to a set of different assays that monitor the same, single type of health effect and to different mixtures that are considered toxicologically similar. The basic assumption is that the ratio of toxic potencies between any two mixtures is the same, regardless of which assay is used. This means that if mixture X is twice as potent as mixture Y in assay 1, then X is twice as potent as Y in assay 2. An equivalent assumption is that, for any mixture in the similarity group, the change in potency from assay 1 to assay 2 is the same. This constancy of potency ratios can then be used to estimate risk for one mixture in one assay by using data from other assays and on other similar mixtures.

The comparative potency approach is an example of a similar mixtures approach to risk assessment. It is assumed that the mixture of concern can be considered a member of a class of similar mixtures based on similarity of biologic activity or reasonable expectation of a type of biologic activity based on chemical composition. To use a comparative potency method, the risk assessor must test the consistency of dose-response for the class of mixtures in question and test the assumption of a uniform proportionality constant between assays for all mixtures in the similarity class and for the series of bioassays under consideration.

Theoretical Development. The major assumption in the comparative potency method is that there exists a simple linear relationship between the mixtures' potencies from two assays for all members of the group of similar mixtures. Consider the application to cancer risk estimation. A mixture with zero potency (i.e., it is not carcinogenic) must have zero potency in all bioassays for carcinogenicity, so the linear relationship across assays must pass through the origin (0,0) and is then a simple proportionality constant. This relationship is not chosen because it is simple but is used because the mixtures are deemed toxicologically similar and thus can serve as surrogates for one another. These mixtures must then change in potency from one assay to another in the same fashion.

In general, this assumption can be expressed as follows. Define:

$$\{ X_i \} = \text{group of m similar mixtures, where } i=1,...,m \tag{1}$$

$$\{ A_j \} = \text{the group of n bioassays, where } j=1,...,n \tag{2}$$

Let P represent the toxic potency. Then the above proportionality assumption can be written as

$$P_{A1}(X_i) = k * P_{A2}(X_i), \text{ for any } X_i \text{ in the similarity group} \tag{3}$$

where k is the proportionality constant that relates the potencies across the two assays.

When three or more assays are used to establish the necessary relationships, there will be several such proportionality constants. In general, for assays Ar and As (where r and s are different and each in the range 1,...,n), the constant is k_{sr}

$$P_{Ar}(X_i) = k_{sr} * P_{As}(X_i) \tag{4}$$

Example with two assays.   Suppose that we wish to estimate the human cancer potency for mixture X_2; thus X_2 is the mixture of concern. Although direct estimation of human cancer risk usually comes from epidemiological or occupational studies, not actual bioassays on humans, we will stay with that nomenclature for consistency with the preceding discussion. Suppose that the available information is the following:

- The group of similar mixtures contains four mixtures X_1 through X_4.
- Mixture X_1 is twice as potent for human cancer (assay A1) as it is for tumors from mouse skin painting (assay A2), and the cross-assay potency ratios for mixtures X_3 and X_4 are also roughly 2.
- The only potency estimate for X_2 is from mouse skin painting studies.

The human cancer potency for X_2 is then estimated as follows. First, k in Equation 3 can be estimated to be 2. Because X_2 is a member of the similarity class that includes mixtures X_1, X_3 and X_4, the same cross-assay ratio (k) holds for X_2 as for all the other similar mixtures. From Equation 3 and the estimate of $k = 2$, we then have the human potency estimate for X_2 as

$$P_{A1}(X_2) = 2 * P_{A2}(X_2) \tag{5}$$

Note that if a graph were created using the data for these mixtures with the potency for A1 on the y-axis and the potency for A2 on the x-axis, then the slope would be roughly 2. The decision to use this risk (potency) estimate from Equation 5 is stronger as the graph becomes more linear.

Example with three assays.   A slightly more complicated situation involves three assays, with incomplete data for each one. Suppose again that we wish to estimate the human cancer potency for mixture X_2, and that the available data are as follows:

- A potency estimate for mixture X_2 has only been measured with the *in vitro* study (assay A3).
- Three or more mixtures have been studied with both assays A3 and A2 (short term *in vivo* rodent study), and three or more mixtures (not the same group) have been studied with both assays A2 and A1.
- The two "cross-assay" constants k_{32} and k_{21} have been estimated separately using these two subsets of the class of similar mixtures.

The estimate of human risk (assay A1), using the notation in Equation 4, is then calculated by extrapolating from assay A3 to A2 and then from assay A2 to A1. The calculation is just the potency of X_2 from assay A3 multiplied ("scaled up") by the product of the two cross-assay constants:

$$P_{A1}(X_2) = k_{32} * k_{21} * P_{A3}(X_2) \tag{6}$$

Note that, because data for X_2 only exist with assay A3, the constants k_{32} and k_{21} are based only on data for the other mixtures and do not use data on mixture X_2 at all.

Table 5.5 Comparative Potency Method for Emission Extracts

Combustion product	Mouse skin tumor initiation[a]	Human lung cancer unit risk[b] $(\mu g/m^3)^{-1}$
Coke oven emissions	2.1	9.3×10^{-4}
Roofing tar	0.40	3.6×10^{-4}
CSC	0.0024	2.2×10^{-6}
Diesel	0.31	0.7×10^{-4}

[a] Given as number of papillomas/mouse at 1 mg organics.

[b] Direct estimates from human data. The diesel value was based on rat inhalation data[60] and was adjusted for the percentage of organics on the particulates.

Source: Nesnow, S. 1990. Mouse skin tumours and human lung cancer: Relationships with complex environmental emissions, in *Complex Mixtures and Cancer Risk*, IARC Scientific Publ. No. 104, p. 44–54. With permission.

Example with combustion emissions. The following application of this methodology is to the estimation of human cancer risk from exposure to polycyclic organic matter (POM) from mixtures such as cigarette smoke, coke oven emissions, internal combustion engine emissions, and coal burned for heat and cooking.[18] The data are given in Table 5.5. The linearity is illustrated in Figure 5.2. An example of the use of this method can be seen by pretending the diesel estimate for humans was not available. The

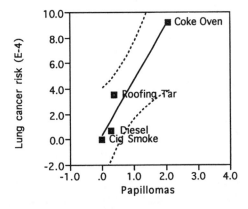

Figure 5.2 Linear regression of human lung cancer unit risk vs. mouse skin tumor initiation (as papillomas per mouse). (From Nesnow, S. *Complex Mixtures and Cancer Risk*. (ARC Scientific Publ. No. 104, p. 44–54. With permission.)

approach is then to use the remaining mixtures to estimate the human risk from diesel by scaling from the mouse skin potency for diesel. The linear regression performed on the data without diesel gives a cross-assay proportionality constant of $k = 4 \times 10^{-4}$. The intercept is not significant ($p = 0.57$) and the proportionality constant is only weakly significant ($p = 0.14$), though the adjusted model r-square is 0.91. This roughly consistent group of mixtures predicts a diesel lung cancer unit risk of 2×10^{-4}, which compares reasonably well with the actual estimate of 0.7×10^{-4}.

Presentation of data and results.   Previous publications on comparative potency[15,19] have performed the calculations using the "relative potency," i.e., the ratio of the potency of the mixture of concern to that of a "reference mixture" in the same assay, instead of using the actual mixture potencies. This scaling of the actual potencies does not add any information nor does it increase the flexibility of the approach. Consider a graph of P_{A2} vs. P_{A1} (i.e., the mixture potencies for assay A2 plotted vs. the mixture potencies for assay A1). Scaling a quantity by a constant (e.g., the reference mixture) only changes the numbers on the axes of the graph, but the shape of the curve through the data points remains unchanged. Thus, regardless of the reference mixture used for scaling the potencies, even if different in each assay, the only required relationship is that the same proportionality constant across assays holds for all the similar mixtures.

The use of "relative potency" for comparing assays has some advantages, however, because all potencies are then "standardized" to be numbers near one (1.0), and the differences are more easily visualized. The problem occurs when tables of these standardized values are used for calculations instead of statistical methods such as regression. The weakness with using *relative* potencies is that the relative potency for the reference mixture (relative to itself) is always viewed as exactly 1.0; it is no longer a measured random variable but is presumed to be exact, and the variation is all assumed to lie with the other mixtures' potencies. This is clearly wrong. Regression across all mixtures should be used instead. However, even if regression is used and the index mixture value is given with a confidence interval [e.g., 1.0 (0.5 – 2.8)], the visual comparison will still tend to focus on other values vs. 1.0. To avoid misinterpretation, it is better to give the analysis of the constant ratio assumption separately from the table of potency data.

Development of the comparative potency approach.   The implementation of the comparative potency method requires the gathering and analysis of data on several mixtures along with considerable judgment of toxicologic similarity. The approach should be limited to the assessment of mixtures for which whole mixture *in vivo* toxicity studies have not been done, and where the compositions of the mixtures are deemed too complex for the application of component-based assessment methods. The following main steps have been identified:

- Determine the class characteristics or other similarity criteria for the group of mixtures that are to be judged "toxicologically similar" to the mixture of concern.
- Compile the available toxicity data on the mixtures in the similarity class and evaluate them for general quality and applicability to the toxic end points of interest for the mixture of concern.
- Estimate the degree of consistency in the assay ratios and estimate values to support the constant potency ratio relationship.
- Describe the best estimates of the cross-assay ratios along with all uncertainties in their application to human risk assessment for the mixture of concern.

Given that this is a methodology based on the comparison of different mixtures and different types of data and not on an extrapolation from directly related human health data, it is expected that these estimates are accurate only within an order of magnitude. The comparative potency procedure should be considered when direct toxicity data on the whole mixture from *in vivo* studies are not available and when the mixture is deemed too complex for component-based methods to be plausible.

Data available on mixture components

Most of the experience of EPA in the risk assessment of mixtures has involved data on the chemical components, not the whole mixture. This unfortunate situation is not surprising. Historically, the federal legislation has emphasized single chemical evaluations (e.g., the water quality criteria of 1980 for the list of 129 priority pollutants), and the risk-related projects in EPA have been similarly constrained. For example, the EPA risk assessment database IRIS[13] is dominated by single chemical evaluations. In addition, the scientific literature is similarly proportioned. Except for research on tobacco smoke and a few other combustion emissions, the studies of whole mixture toxicity and exposure and their corresponding research on quantitative risk methods are extremely rare. For example, in a recent EPA symposium proceedings on mixtures,[6] of the 28 articles (not including the session summary papers), only three dealt with toxicology studies or methods development for whole mixtures.

There are many publications on component approaches, ranging from highly theoretical statistical models to simple testing of the interaction between a particular pair of chemicals.[20] Combustion mixtures often involve hundreds of component chemicals, so it is unlikely that a component approach would be feasible. The following discussion then only sketches the ideas in the component approaches being proposed by EPA. For details, the reader should consult the references.

Introduction to additivity and interaction effects

If data are not available on an identical or reasonably similar mixture, the risk assessment may be based on exposure and toxicity data for the components in the mixture. When quantitative information on toxicologic interaction exists, even if only on chemical pairs, it should be incorporated into the component-based approach. When there is no adequate interactions information, dose or risk additive models have been most commonly used for risk assessment and are recommended in the current EPA guidance.[8,11]

Several studies have demonstrated that dose (or concentration) addition often predicts reasonably well the toxicities of mixtures composed of a substantial variety of both similar and dissimilar compounds,[21-26] although exceptions have been noted. For example, Feron et al.[26] discusses studies where, even at the same target organ (the nose), differences in mode of action led to other than dose-additive response. Consequently, depending on the available information on modes of action and patterns of joint action, the most reasonable model should be used.

Criteria for dose addition vs. response addition

Toxicologic interactions are defined in this chapter only to facilitate the selection and application of specific risk assessment methods. Other definitions, including some with mechanistic interpretations, are also used in the scientific literature, but will not be discussed here. When adequate evidence for toxicologic interactions is not available, the no interaction approach (detailed below) using dose or response addition could be employed. Toxicologic "interactions" are then operationally defined by data showing significant deviations from the prediction made using dose or response addition. The assumed form of additivity (dose or response) must be clearly stated when deciding that an interaction has been observed.

Several differing definitions of "no interaction" are discussed in the scientific literature. Plaa and Vénzina[27] provide a nice historical overview of the differences in definitions, and Kodell and Pounds[28] discuss some of the implications of these differences. Muska and Weber[29] introduced the terms "concentration addition" and "response addition." Their definitions are based on ideas related to general toxicologic mechanisms, i.e., concentration addition applies when the components act on similar biological systems and elicit a common response, whereas response addition applies when components act on different systems that do not affect each other.

The term "no interaction" is defined here using the two common concepts of Muska and Weber:[29] dose addition and response addition. These definitions have been selected because the underlying concepts are extremely simple and because hypothesis tests exist to determine whether data are consistent with each of these concepts. These definitions are not intended to infer specific toxicologic mechanisms, although they should be somewhat consistent with the major examples and concepts of toxicologic interaction. The risk assessment using component data should then begin by the selection of one of these two concepts as most appropriate for the chemicals being considered. There will be many cases where the information does not support either dose or response addition. In those cases, the mixture should be further investigated, and consideration should be given to using methods that incorporate toxicologic interactions.

The primary criterion for choosing dose or response addition is the similarity between the chemicals in the mixture. This judgmental decision should be based on information about the toxicologic and physiological processes involved, the single chemical dose-response relationships, and the type of response data available. To facilitate understanding, the discussions that follow will initially consider only two-chemical mixtures. An illustration of the difference between these two approaches in terms of dose-response curves is given in Figure 5.3.

Dose addition.   In the simplest terms, two chemicals are dose additive if chemical 2 is functionally a clone of chemical 1. The chemicals are assumed to behave similarly in terms of the primary physiologic processes (uptake, metabolism, distribution, elimination) as well as the toxicologic processes. The mathematical definition of dose addition requires a constant proportionality between the effectiveness of the two chemicals. This means that, for equal effects, the dose of chemical 2 is a constant multiple of the dose of chemical 1. The dose-response functions are then congruent in shape. Let t be the proportionality constant that denotes the relative toxicity of chemical 2 to chemical 1, and let r_1 and r_2 be the responses of doses d_1 and d_2 for chemicals 1 and 2, respectively. Then

$$r_1 = f(d_1),$$
$$r_2 = g(d_2)$$
$$= f(t^*d_2)$$

The last equation shows this "functional clone" behavior: the dose (d_2) of chemical 2 is converted into an equivalent dose of chemical 1 and then the dose-response function f of chemical 1 is used to predict the response. For a mixture of these two chemicals, the mixture response r_m is then determined by converting the dose into an equivalent total dose of chemical 1, and then using the dose-response function for chemical 1:

$$r_m = f(d_1 + t^*d_2) \tag{7}$$

Among the many ways to decide dose addition, the isobole is one of the more common graphical methods. The isobole for a two-chemical mixture

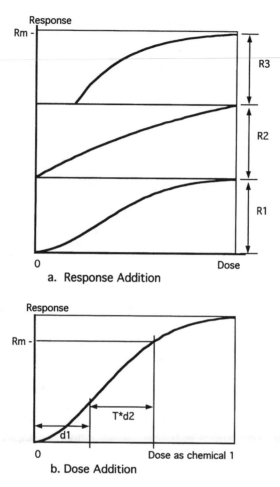

Figure 5.3 Comparison of dose and response addition using component dose-response curves.

is the graph of the various combinations of doses (d_1, d_2) at which a fixed response is observed.[30] This means that, for all points plotted on the graph, the same response occurs. When the set of equal-response points is a straight line, the two chemicals are said to be dose additive. Deciding whether the points are linear is often judgmental, but statistical methods also exist.[31] Note that in the simple "clone" definition of dose addition, all isoboles for different response rates will be parallel. Generalizations of dose addition have also been proposed,[32] primarily where the lines for different response rates are linear, but not parallel.[20] To ensure applicability to the low doses commonly used in environmental assessments, the isoboles should be evaluated for appropriately low doses, or at least for doses associated with low responses.

Response addition. Under response addition, the chemicals behave independently of one another, so that the body's response to the first chemical is the same whether or not the second chemical is present. In simplest terms, classical response addition is described by the statistical law of independent events. With the notation defined above,

$$r_m = 1 - (1 - r_1) * (1 - r_2) \tag{8}$$

In terms of risk, this equation says that the response to either chemical 1 or 2 is just one minus the probability of not responding to either chemical. Expanding the right-hand-side, one obtains

$$r_m = r_1 + r_2 - r_1 * r_2$$

which, for small single chemical risks, is well approximated by the simple summation of risks, i.e.,

$$r_m = r_1 + r_2$$

Several variations of response addition have been developed. Many of these variations require additional information and assumptions. When reviewing the literature for evidence supporting response addition, the risk assessor should understand the definitions and assumptions that are used. Conclusions that only state that the chemicals were "additive" clearly need further investigation and analysis to determine the kind of additivity.

Thresholds. The common assumption for noncancer (and some cancer) toxicity is that it exhibits a threshold exposure level, below which effects will not occur. For mixtures at low exposures, dose addition will combine subthreshold doses, possibly giving a mixture dose that could produce toxic effects. Under the assumptions for response addition, two subthreshold doses should not predict any toxicity because the body should be handling the two chemicals by independent processes, and so the two chemicals will not have any cumulative effect. For example, consider the case where thresholds exists at 10 dose units. Let $f(d)$ be the dose-response function for chemical 1 and t be the toxicity of chemical 2 relative to chemical 1. Then the two additive approaches give

D1	R1	D2	R2	Mixture dose	Dose add	Response add
8	0	8	0	16	$f(8 + t*8)$	$0 + 0 - 0*0 = 0$

Depending on the value of t, the dose additive formula may or may not predict a response. If $t < 0.125$, then the combined dose is less than the threshold dose of 10.

Note that the dose addition formula has no standard quantitative constraints, whereas the response addition formula has an upper bound. If "response" is incidence, then $r \leq 1$. If response is the measured effect, then there are physiological limits to that measurement. With dose addition, the same constraints on response are present but are automatically met because the combined dose follows the actual dose-response curve of one of the chemicals.

Evidence for dose or response additivity.   Several studies have been published that suggest dose or response additivity. In too many cases, however, a study was not designed properly for detecting departures from additivity. Some sense of the opinion of toxicologists, however, can be gained from some recent publications in which dose or response addition is recommended as a plausible default procedure.

Ikeda[25] surveyed the literature and found few cases, by his judgment, that showed "clear-cut cases of potentiation," and he concluded (p. 418), "Thus, the most practical approach in evaluating the combined effect of chemicals seems to be the assumption of additive effects." He also noted that assuming additivity of effects for chemicals with dissimilar modes of action is more protective than independence. Furthermore, except for their initial overview, Plaa and Vénzina[27] focus on concentration (dose) addition. Response addition, when it is mentioned, is usually just stated to be a possible no-interaction model, without any motivating discussion. For example, the recent NAS book[33] on complex mixtures notes that "no interaction" in chapter 1 was dose addition, while for ordinary linear statistical models, no interaction refers to response addition. The original U.S. EPA guidelines for mixtures risk assessment[8] recommend default approaches of dose addition for threshold toxicants acting by similar modes of action or affecting common organs and response addition for carcinogenic risk. The latter was not so much based on strong evidence of independence as on the prevalence of probabilistic risk estimates for carcinogens and on the numerical closeness between dose and response addition at low doses.

The evidence for either response addition or dose addition is then not strong and clearly is not comprehensive for the varying types of chemicals considered in environmental risk assessment. The decision to include only these two concepts as default "no interaction" definitions is then primarily based on clarity, simplicity, and ease of implementation. Whenever evidence exists that clearly disagrees with both dose and response addition, then alternative approaches should be considered, such as those developed to incorporate weight-of-evidence judgments on data for pairwise interactions.[34,35]

Definitions of toxicologic interactions

Several quantitative descriptions of interaction have been proposed during the past 50 years. One of the earliest quantitative characterizations of interactions is by Bliss,[36] who describes similar joint action, independent joint action, and synergistic and antagonistic joint action. Plaa and Vénzina[27] propose the terms *additive* (sum of individual effects, an admittedly vague definition), *infra-additive*, and *supra-additive* as having the advantage of not requiring consideration of mechanisms. Table 5.6 recommends a set of definitions for use in chemical mixtures risk assessment. That table clarifies the terminology related to additivity and interaction effects for both cancer and noncancer end points.

Risk assessment strategy

Approaches based on the mixture's chemical components are recommended for relatively simple, identified mixtures with approximately a dozen or so chemical constituents. For exposures at low doses with low component risks, the likelihood of significant interaction is usually considered to be low. For example, interaction arguments based on saturation of metabolic pathways or competition for cellular sites usually imply an increasing interaction effect with dose, so that the importance at low doses is probably small. The default component procedure is then to assume risk (response) addition when the component toxicological processes are assumed to act independently and dose (or concentration) addition when the component toxicological processes are similar. When adequate data on interactions suggests other than dose or response additivity at low doses, such information must be incorporated into the assessment.

Noncancer effects.   Noncancer effects are often assessed for single chemicals by evaluating a lifetime daily exposure that is associated with negligible risk of adverse effects, even in sensitive subgroups of the exposure population. In EPA procedures, this value for oral exposures is called the reference dose (RfD) and for inhalation exposures is called the reference concentration (RfC).[13] For mixtures, the equivalent is the hazard index (HI), which is based on dose addition but normalized so that the corresponding decision threshold value is at HI = 1. Dose addition requires some measure of relative toxicity to use as a weighting value in the addition (denoted as t in Equation 7). Although several such measures could be estimated, such as an ED_{10} for the effect of concern, few such values have been determined in the low-risk exposure range. EPA operational guidance for Superfund waste sites[11] then replaces the weighting factor by the inverse of the RfD (or RfC for inhalation exposures), i.e., the larger the RfD the less toxic and hence smaller is the scaled dose. The calculation of HI for a mixture of n chemicals is then:

$$HI = \sum_{i=1}^{n} \left(E_i / RfD_i \right) \tag{9}$$

where   n = the number of components in the mixture,

E_i = the estimated intake for the ith component, and

RfD_i = the Reference Dose (RfD) for the ith component, i.e., an estimated chronic daily oral intake believed to be below the threshold for human toxicity.

Table 5.6 Definitions of Toxicologic Interactions between Chemicals

---

Additivity

When the "effect" of the combination is estimated by the sum of the exposure levels or the effects of the individual chemicals; the terms "effect" and "sum" must be explicitly defined; effect may refer to the measured response or the incidence of adversely affected animals; the sum may be a weighted sum (see "dose addition") or a conditional sum (see "response addition")

Antagonism

When the effect of the combination is less than that suggested by the component toxic effects; antagonism must be defined by identifying the type of additivity (dose or response addition) from which the combination effect deviates

---

Chemical antagonism

When a reaction between the chemicals has occurred and a new chemical is formed; the toxic effect produced is less than that suggested by the component toxic effects

---

Chemical synergism

When a reaction between the chemicals has occurred and a different chemical is formed; the toxic effect produced is greater than that suggested by the component toxic effects, and may be different from effects produced by either chemical by itself

---

Complex interaction

When three or more compounds combined produce an interaction that cannot be assessed according to the other interaction definitions

---

Dose additivity

When the effect of the combination is the effect expected from the equivalent dose of an index chemical; the equivalent dose is the sum of component doses scaled by their potency relative to the index chemical

---

Index chemical

The chemical selected as the basis for standardization of toxicity of components in a mixture; the index chemical must have a clearly defined dose-response relationship

---

Inhibition

When one substance does not have a toxic effect on a certain organ system, but, when added to a toxic chemical, it makes the latter less toxic

---

Masking

When the compounds produce opposite or functionally competing effects at the same site or sites, so that the effects produced by the combination are less than suggested by the component toxic effects

---

Table 5.6 Definitions of Toxicologic Interactions between Chemicals (continued)

No apparent influence

When one substance does not have a toxic effect on a certain organ or system, and, when added to a toxic chemical, it has no influence, positive or negative, on the toxicity of the latter chemical

No observed interaction

When neither compound by itself produces an effect, and no effect is seen when they are administered together

Potentiation

When one substance does not have a toxic effect on a certain organ or system, but, when added to a toxic chemical, it makes the latter more toxic

Response additivity

When the response (rate, incidence, risk or probability) of effects from the combination is equal to the conditional sum of component responses as defined by the formula for the sum of independent event probabilities

Synergism

When the effect of the combination is greater than that suggested by the component toxic effects; synergism must be defined in the context of the definition of "no interaction," which is usually dose or response addition

Unable to assess

Effect cannot be placed in one of the above classifications; common reasons include lack of proper control groups, lack of statistical significance, and poor, inconsistent, or inconclusive data

Note: Based on definitions in U.S. EPA.[2] These definitions of interaction refer to the influence on observed toxicity, without regard to the actual mechanisms of interaction.

The mixture exposure is then usually judged acceptable when HI < 1. As with the RfD and RfC, the HI is not precise and not very accurate. When HI > 1, toxicity is not necessarily "expected," but is usually just considered "possible," and further investigation is indicated. Because data are usually only available for the mixture component chemicals, the HI method has become the de facto default method for mixtures risk assessment.[8,37] Dose addition requires toxicologic similarity of all the chemicals included in its calculation. The calculation then makes the most sense when an HI is specific to a single toxic effect. A consequence is that the EPA guidelines recommend determining a separate HI for each toxic effect of interest. An acceptable exposure is then indicated whenever *all* HIs are less than 1.

Interactions-based hazard index. Evidence of toxicologic interaction should be reflected in the mixture assessment. No standard methods are yet in place in regulatory agencies to incorporate interactions and no biologically motivated mathematical models have been developed that could serve as a default method. However, this is where risk assessment differs from traditional modeling: the procedures need only be toxicologically plausible and practical, not necessarily exact. One approach is to begin with the dose additive HI and then modify it to reflect the interactions results, using plausible assumptions to fill in the data gaps. EPA developed such a method previously[34] and is revising that approach to reflect current ideas about interactions.[35] The procedure begins with two evaluations of the weight of the evidence (WOE) for interaction for each pair of component chemicals in the mixture: one WOE for the influence of chemical A on the toxicity of chemical B and one for the reverse. This qualitative judgment is then changed into a numerical score. Other common assumptions and desirable properties could also be included:

1. The pairwise interactions capture most of the interaction effects in the mixture.
2. The interaction is highest when both chemicals in the interacting pair are at equally toxic doses (neither chemical is dominant).
3. The interaction HI must reduce to the additive HI as the interaction magnitudes decrease.
4. The main toxicologic effects from the mixture exposure are limited to those effects induced by the individual component chemicals.
5. The interaction magnitude is likely to decrease as mixture dose decreases.

Many formulas could be derived that reflect these ideas. Consider the following as a simple example. Assumptions 1 and 4 are simplifications in the data gathering stage. Let the terms in Equation 7 be defined as the hazard quotients (HQ) for the component chemicals. Assumption 2 can then be modeled by a simple convex function that is maximal when $HQ_i = HQ_j$. Assumption 5 has no empirical support we could find, and may be more reflective of the reduction in toxicity as dose decreases, making detection of an interaction more difficult. Consequently, assumption 5 will not be included here. Pairwise interaction studies usually show the influence of one chemical on the toxicity of the other chemical. If each HQ is used as the measure of that component chemical's toxicity, then we can modify the HI by multiplying each HQ in the formula by a function of the following quantities: the HQs of the other chemicals (to reflect the actual component exposure levels), the estimated magnitude of each pairwise interaction, and the two WOE scores. In this way, we are incorporating the interactions by modifying each HQ by the influences of all the other interacting chemicals. These modified HQs are then summed to get the interaction HI for the mixture.

Such a formula for the interactions HI might look like the following:

$$HI_{INT} = \sum_{i=1}^{n} \left(HQ_i \cdot \sum_{j \neq i}^{n} f_{ij} \, M_{ij}^{B_{ij}\theta_{ij}} \right) \quad (10)$$

where HQ is the hazard quotient and the other terms are identified in the following text.

The term M_{ij} in Equation 10 represents the magnitude of the interaction, preferably measured within the exposure range of concern. More specifically, M_{ij} is an estimate of the maximum effect, synergistic or antagonistic, that chemical j has on the threshold or risk-specific dose (e.g., ED_{10}) of chemical i. Thus, if the two chemicals are additive, then M = 1. If chemical j shifts the threshold either upward or downward by a factor of 5, M is five. The first exposure-related factor is f_{ij}, which is a rough measure of the relative toxicity of the jth component compared to all other potential interacting chemicals.

$$f_{ij} = \frac{HQ_j}{HI_{add} - HQ_i} \quad (11)$$

The denominator normalizes these weighting factors so that if no interactions were significant, i.e., if all $M_{ij} = 1$, then the interactions HI would reduce to the additive HI.

The binary weight-of-evidence factor B_{ij} reflects the strength of evidence that chemical j will actually influence the toxicity of chemical i, and that the influence will be relevant to human health assessment. The classification scheme is given in Table 5.2. The binary weight of evidence classification is then converted into a numerical weight. As with the Mumtaz and Durkin[34] method, positive values indicate synergism, and negative values indicate antagonism. Because the WOE factor is in the exponent of M, positive values increase the apparent magnitude over additivity (e.g., $M^5 = 3.2$) while negative values decrease the apparent magnitude compared to additivity (e.g., $M^{-5} = 1/M^5 = 0.32$).

The factor θ_{ij} reflects the degree to which components i and j are present in equitoxic amounts. The definition of *equitoxic* is based on the ratio of the hazard quotients. The measure of the deviation from equitoxic amounts for the ith and j^{th} components is defined here simply as the ratio (θ_{ij}) of the geometric mean to the arithmetic mean:

$$\theta_{ij} = \frac{\sqrt{HQ_i - HQ_j}}{(HQ_i + HQ_j) \div 2} \quad (12)$$

Thus, the ith and j^{th} components are said to be equitoxic with respect to each other if their HQs are equal. The change in effective magnitude because of this factor is not great. If $HQ_1 = 1$ and $HQ_2 = 10$, then θ is 0.575 and an M of 10 is reduced to 3.76. Note that as HQ_i approaches HQ_j, $\theta_{i,j}$ approaches unity, and as the difference between HQ_i and HQ_j increases, $\theta_{i,j}$ approaches zero. The term $\theta_{i,j}$ is incorporated into the algorithm under the assumption that, for a given total dose of two chemicals, the greatest deviation from additivity will occur when both of the components are present in equitoxic amounts. This assumption is explicit in Finney's model of a deviation from dose additivity.[38]

We attempted, without success, to locate empirical studies relating interaction magnitude to the relative doses of the components in order to check the interactions HI (Equation 10). The difficulty was in locating adequate data. Most binary studies lacked the data to establish the dose-interaction relationship at doses in the low response region, i.e., near the no-observed-adverse-effect level (NOAEL), primarily because of insufficient dose combinations in the low response range, or other inappropriateness of the experimental protocol. Others described multiple end points that were inconsistently monitored over the range of doses used. With mixtures of more than two chemicals, many studies included only a few pairwise interaction results along with a few whole mixture results. We expect that the relatively high cost of interaction studies will inhibit research in this area for some time, so that the support for this approach will be mainly based on the plausibility of the assumptions.

Cautions and uncertainties with component-based assessments

The component-based procedures discussed earlier for dose-response assessment and risk characterization are intended only for simple mixtures of a dozen or so chemicals. The uncertainties and biases for even a small number of chemical components can be substantial. Component-based methods are particularly susceptible to misinterpretation because the listing of chemical components in a mixture is often misconstrued as implying a detailed understanding of the mixture risk. The risk characterization must include a discussion of what is missing or poorly understood, in order to convey a clear sense of the quality and degree of confidence in the risk assessment.

Exposure uncertainties. The general uncertainties in estimating chemical exposure are beyond the scope of this chapter, but need to be considered along with the uncertainties peculiar to mixtures. The risk assessor should discuss these exposure uncertainties in terms of the strength of the evidence used to quantify the exposure. When appropriate, the assessor should also compare monitoring and modeling data and discuss any inconsistencies as a source of uncertainty. For mixtures, these uncertainties may be increased as the number of compounds of concern increases.

If levels of exposure to certain compounds known to be in the mixture are not available, but information on health effects and environmental persistence and transport suggest that these compounds are not likely to be significant in affecting the toxicity of the mixture, then a risk assessment can be conducted based on the remaining compounds in the mixture, with appropriate caveats. If such an argument cannot be supported, no final risk assessment can be performed with high confidence until adequate monitoring data are available. As an interim procedure, a risk assessment may be conducted for those components in the mixture for which adequate exposure and health effects data are available. If the interim risk assessment does not suggest a hazard, there is still concern about the risk from such a mixture because not all components in the mixture have been considered.

In perhaps a worst-case scenario, information may be lacking not only on health effects and levels of exposure, but also on the identity of some components of the mixture. Analogous to the procedure described in the previous paragraph, an interim risk assessment can be conducted on those components of the mixture for which adequate health effects and exposure information are available. If the risk is considered unacceptable, a conservative approach is to present the quantitative estimates of risk, along with appropriate qualifications regarding the incompleteness of the data. If no hazard is indicated by this partial assessment, those partial results should be conveyed to the risk manager, but the risk assessment should not be quantified until better health effects and monitoring data are available to adequately characterize the mixture exposure and potential hazards.

Dose-response uncertainties.   For many simple mixtures for which a component-based approach might be applied, studies on interactions, even pairwise interactions, will be missing. Use of a dose- or response-additive model is easily implemented, but justification for such approaches is largely based on conceptual arguments, not empirical studies. An investigation into available interaction studies[2] found that roughly half did not report any attempt at data analysis or only reported significance levels ("p values") with no indication of the statistics used. As indicated previously, recent studies by Feron et al.[26] show that there are exceptions to most rules regarding interactions, even the common assumption that additivity is acceptable if chemicals target the same organ. Recent studies on dose additivity have focused on very simple mixtures of chemically and metabolically similar chemicals.[6,31] Improvements in experimental design and statistical hypothesis testing for dose additivity, along with better understanding of the chemical characteristics that accompany observed dose additivity, should lead to improved predictive ability and justification for dose addition as a default approach.

Conclusions regarding toxicologic interaction are also weakly supported by empirical studies. Many studies[2] failed to identify what the "no-interaction" hypothesis was, so that any conclusions regarding nonadditive interaction were difficult to interpret. Other studies identified the no-interaction hypothesis but employed incorrect experimental designs, so that the conclusions were not justified. Perhaps the most substantial weakness in the understanding of toxicologic interactions is the lack of studies, models, and concepts for interactions involving more than two chemicals. The key assumption in both of the interaction weight-of-evidence methods described previously[34,35] is that, at least for low doses, the resulting influence of all toxicologic interactions in a mixture is well approximated by the pairwise interactions. No studies have been located to date that investigate that assumption, although two studies are in progress at the U.S. EPA and ATSDR.

Toxicologic understanding of interaction is also limited. Although interaction mechanisms are commonly assumed to involve either pharmacokinetics and metabolism or toxicologic receptors, nearly all studies on mechanisms and modes of interaction focus on pharmacokinetics.[39] Current pharmacokinetic models for interactions usually address two- and three-chemical mixtures. Interest in pharmacodynamic interactions, particularly at the receptor level, is beginning to increase, but slowly. Clearly, more complex interactions research is sorely needed if the use of interactions information in risk assessment is to improve.

Presentation of data and results

The consequence of this early stage of mixtures risk research is that the risk assessor must use considerable judgment along with plausible approaches such as those discussed previously. The results, however, must be presented transparently. Although the procedures described in this section are developed from available concepts and data on simple mixtures, all component-based quantitative mixtures risk assessments should be limited to one significant digit for the risk value, unless substantial justification is given for higher precision.

Conclusions for application to incineration emissions

Component-based methods require identification and quantitative measurement of the main chemicals judged responsible for the health risk to the exposed population. As stated previously, emissions can be highly variable in composition, including total dose as well as relative proportions of the individual chemicals. Further, the number of component chemicals is likely to be large, in excess of 100. Comprehensive measurement of individual chemicals is unlikely to be feasible, and pairwise interaction data are likely to exist on only a small fraction of those components (see Figure 5.4 for the number of combinations as a mixture becomes more complex). Component-based risk methods then are unlikely to prove practical for incineration emissions. They may be considered for validation of other methods, such as those involving whole mixture testing or total emission characterizations grouped by chemical classes.

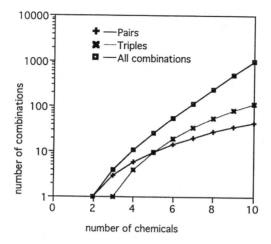

Figure 5.4 Number of combinations versus number of component chemicals in the mixture.

Application to incineration

Introduction

Because of problems associated with land disposal and other waste storage or reduction options, many communities have implemented or considered incineration as a reduction alternative for municipal solid wastes, sludges, and industrial wastes. Medical wastes are also frequently incinerated; combustion of these wastes offers the additional advantage of destroying potentially pathogenic biologic wastes. Hazardous wastes are also combusted at facilities designed specifically to accommodate the special safety precautions required to handle these wastes. While incineration does reduce the volume of waste and may wholly or partially destroy some pollutants through the combustion process or capture some pollutants through the emission control processes, potentially hazardous pollutants may remain in the ash generated during combustion, and potentially hazardous air pollutants may be emitted from smoke stacks of the incinerators. These pollutant emissions to the atmosphere may pose human health risks and are the focus of this section.

In the past, most analyses of health risks associated with atmospheric emissions from combustion sources have focused primarily on single chemical exposure occurring only through the inhalation route. For example, U.S. EPA's human exposure model (HEM) was used to quantitatively predict human inhalation exposures that result from pollutants emitted from utility boilers.[40]

Atmospheric pollutant emissions from combustion facilities have been shown to accumulate in other environmental compartments such as soil[41] and water. It was recognized that accumulation of atmospheric pollutants in environmental compartments could result in exposure through routes other than inhalation;[2,42,43] these exposures were termed "indirect exposure." Indirect exposures occur through contact with, or ingestion of, emitted pollutants that have been transferred by wet and dry deposition to soil, vegetation, and water bodies; these pollutants may also be transferred through the aquatic and terrestrial food chains. Human exposures to mixtures of emitted pollutants may occur through multiple environmental pathways including both direct and indirect exposure routes.

Much of our current knowledge of the environmental fate and exposure concerning combustor emissions has been generated through mathematical models that attempt to predict pathways and exposures.[2,42] Conceptually, the models can be divided into two groups: environmental fate (or sometimes described as pathway models) and exposure models. Environmental fate models contain algorithms that take into account the chemical and physical properties of emitted pollutants as well as the environment that surrounds a facility. These models are used to predict the atmospheric, terrestrial, and aquatic fate of emitted pollutants; they take into account the physical course a chemical or pollutant takes from the source to the organism exposed. Exposure models examine each route through which a chemical or pollutant potentially enters an organism after contact, e.g., by ingestion, inhalation, or dermal absorption. These models then quantify exposure by estimating quantities in the contact medium and contact rates. For example, some models predict human exposure to a pollutant from fish to be the product of the predicted concentration in fish tissue, the quantity of fish consumed over some period of time, and the bioavailability of the pollutant in the human gastrointestinal tract.[44]

Indirect exposure assessments have focused on exposures to individual pollutants or pollutant classes. They have not focused on interactions among pollutants that affect environmental fate, and they have not focused on health effects that result from exposure to mixtures of emitted chemicals. For example, the draft Mercury Study Report to Congress[44] examined the environmental transport and exposure to mercury emitted from a variety of U.S. anthropogenic sources, and the U.S. EPA's draft *Dioxin Assessment*[45] examined the fate and the resulting exposures to emitted dioxin-like congeners. Other studies, for example Cleverly et al.,[46] which assessed exposures to several different pollutants emitted from a single source, did not consider synergistic or antagonistic relationships among the pollutants during the modeling of environmental fate or during the assessment of human exposure.

Emissions of pollutants from smoke stacks are typically assumed to be independent in indirect exposure assessments; that is, they are assumed to exert no influence on the environmental fate of other emitted pollutants. Additionally, they are typically assumed to exert their potential toxicity independently at the point of exposure. As will be described in a later section, the modeling of dioxin-like compounds is an exception to the latter assumption.

Environmental fate of chemical mixtures

Conceptually, the effects of interactions among chemicals can be thought of as affecting the environmental transport of pollutants or the outcome of an exposure event. Environmental transport is a continuous process that may have biotic and abiotic components. It originates at the point of combustion; pollutants are released from combustion materials or formed in the combustion or postcombustion air streams. Those pollutants not captured by pollutant control devices associated with the smoke stacks are released to the atmosphere as vapors, components of particles, or as both vapors and particle components.

In the atmosphere, pollutants are subject to air currents that transport them through the atmosphere. Atmospheric pollutants are also subject to forces that result in their deposition to the earth's surface; for example, gravitational forces result in particulate deposition, and precipitation can scavenge pollutants from the atmosphere. Once deposited, pollutants may adhere to plant leaves or become incorporated into soil or water bodies. Pollutants in soil are subject to runoff and erosion; this results in transfer to a water body. Some pollutants may volatilize to the air from soil or water and in this way cycle through the environment.[44] At any point in the cycle, the presence of one chemical may influence the fate of another chemical. Additionally, high environmental background concentrations can also influence the fate of other chemicals in the environment.

Exposures to emitted pollutants can occur through different media during environmental transport. The exposures may occur at the same time or at different points in time. The results of exposures to two different pollutants may be different than the results of the individual exposures. Additionally, they may occur through several different exposure routes. For example, humans may be exposed to pollutants in soil through both the ingestion and the dermal routes (see Figure 5.5).

Chemical interactions that affect fate and transport

There are chemical interactions that affect the fate of atmospheric pollutants. While these interactions may be important, the data are not comprehensive enough at this time to incorporate them into environmental fate modeling. A better understanding of important physical/chemical abiotic processes and a better understanding of uptake and retention of pollutants by biota will provide a stronger basis for incorporating interactions into assessments. One possible breakthrough is the development of structure-activity relationship models; these models could be employed to increase understanding of some of these important processes and interactions. This increased understanding could in the future result in the incorporation of interactions into the modeling of environmental fate.

An example of such an interaction among pollutants occurs in green plants. Cadmium is a toxic metal[13] (the RfD is 1E-3 mg/kg/day in food) that

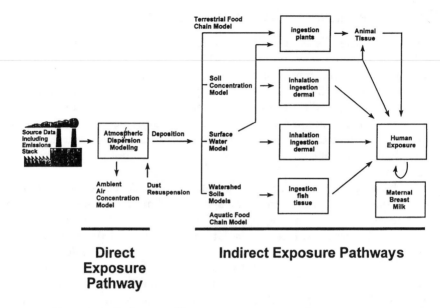

**Direct
Exposure
Pathway**   **Indirect Exposure Pathways**

Figure 5.5 Pathways of human exposure to combustor emissions.

has been measured in the emissions of numerous anthropogenic combustion sources and smelters. A potentially important pathway of cadmium exposure is through the uptake of cadmium by green plants, followed by human consumption of contaminated plants (e.g., garden produce such as lettuce). The human kidney is the target organ for the toxicity.[13]

Studies of municipal solid waste composts that contained high levels of cadmium indicated that high levels of this element could potentially accumulate in garden produce. Further investigation showed that there is a competition between zinc and cadmium for uptake in many plants. When the ratio of cadmium to zinc in the soils was much higher than typical soils (0.005), an accumulation of cadmium in plant tissues could result. The presence of a more normal Cd:Zn ratio reduces cadmium uptake by plants.[47] Elevated zinc concentrations cause phytotoxicity that reduces plant yield, thereby limiting the potential human intake of cadmium from green plants.

Estimating dose for mixtures of similar combustion pollutants

Unlike the interactions data for fate and transport, incorporation of mixtures risk assessment approaches into combustion risk assessment is feasible for some compounds at this time and, as newer methods advance the understanding of risks posed by exposures to combinations of chemicals, will become more and more practical. The characterization of risks posed by mixtures of dioxin-like compounds, which is utilized by U.S. EPA and other governing organizations, attempts to account for the impact of exposure to a mixture of the congeners.[48] A similar approach was recommended for some polychlorinated biphenyl congeners,[49] and a similar yet fundamentally different approach was recommended for some polycyclic aromatic hydrocarbons (PAHs).[50] It must be noted that these procedures are described as interim and that their use confers an additional uncertainty in the risk assessment. These mixtures assessment approaches are useful for classes of compounds with similar structures. The development of assessment approaches for mixtures of dissimilar pollutants is an important research need.

The U.S. EPA prescribes that the environmental transport of emitted dioxin congeners be modeled separately but that at the point of exposure the doses of the congeners are combined through a toxicity equivalence factor (TEF) approach. This is described in the U.S. EPA[48] report, Interim Procedures for Estimating Risks Associated with Exposures to Mixtures of Chlorinated Dibenzo-*p*-dioxins and -dibenzofurans (CDDs and CDFs) and 1989 update. This document cautiously recommended comparing available toxicological data and structure activity relationship information on dioxin class members with those of the best-studied compound, 2,3,7,8-TCDD, to estimate the significance of exposures to the other 209 congeners in this class. The TEF then indicates the toxic potential of the dioxin congener relative to the toxic potential of the index chemical 2,3,7,8-TCDD. The consequence of exposure to each compound is expressed in terms of an equivalent exposure of 2,3,7,8-TCDD by multiplying the concentrations of the individual congeners by their assigned TEF. The resulting 2,3,7,8-TCDD equivalents (TEQ) are then summed to give the equivalent mixture exposure in terms of 2,3,7,8-TCDD. That equivalent exposure is then used to estimate the risk associated with the mixture of these compounds. The TEFs were assigned based on such data as information regarding human carcinogenicity, carcinogenic potency based on animal studies, reproductive effects data, *in vitro* test data, and structure-activity relations. One TEF was assigned to each dioxin congener; these TEFs were assumed to encompass and apply to all health end points and all exposure routes for this class of compounds.

The TEF approach is a form of dose addition, and so relies on the judgment that the congeners have a similar mode of action underlying the toxicity. This application of dose addition to the kinds of data used for the TEFs requires a number of assumptions: the applicability of extrapolation from short-term to long-term health effects, similarities between interspecies metabolism, appropriateness of high-dose to low-dose extrapolations, a common mechanism of action for all members of the class, and the constancy of a TEF across different exposure routes, bioassays (including *in vitro* studies) and health end points.[48] The specific term "TEF" was applied to this class because of the wide acceptance of the approach and the broad applications (i.e., across route and health end points) for which it was designed. To better capture the uncertainty in these assumptions, all TEFs were provided as order of magnitude estimates, and the U.S. EPA regards the results of dioxin TEF application as interim.

After the TEFs were developed for dioxins, seven guiding criteria were developed for the TEF approach.[49,51] It must be noted that a key assumption for the dioxins was that a single TEF could apply to all toxic end points.

This means that, for a given congener, the same TEF would be used to assess cancer risk and to assess potential developmental effects. The criteria are

- Demonstrated need for an interim assessment
- A well-defined group of compounds that occur in environmental samples as mixtures
- TEF based on a broad set of toxicity data covering many end points and many congeners
- Relative congener toxicity generally consistent across many different end points
- Additivity of dose
- A presumed common mechanism for toxic end point of the components
- TEFs are formed through a scientific consensus.

These criteria were developed for specific application to the dioxins and dioxin-like compounds. The criteria listed by Barnes et al.[51] reflect the specific nature of the application to the dioxins and dioxin-like PCBs as discussed below. In particular, the assumption that multiple toxicologic effects are mediated by a common mechanism of action is unusual to the cases listed here.

The Workshop Report on Toxicity Equivalency Factors for Polychlorinated Biphenyl Congeners (EPA/625/3-91/020) reported that certain groups of PCBs appear to act through a common mechanism with 2,3,7,8-TCDD. On this basis TEFs were proposed in this report and others,[52] which related the toxicity of exposure to members of these PCB subclasses to that of 2,3,7,8-TCDD. The same approach to estimating TEQ as was developed for dioxins was advanced for this group.[49] TEFs were proposed only for some members of the class, and the applicability of the proposed TEFs was limited to the health end point. When assessing PCB mixtures, it is important to recognize that both dioxin-like and nondioxin-like modes of action contribute to overall PCB toxicity.[53-55] Because relatively few of the 209 PCB congeners are dioxin-like, dioxin equivalence can explain only part of a PCB mixture's toxicity. TEFs based on action similar to 2,3,7,8-TCDD have been developed for 13 dioxin-like PCB congeners,[52] but no TEFs exist for the nondioxin-like modes of action.[12]

There are over 100 identified PAHs. These are readily formed during combustion, and a wide range of PAHs has been identified in combustor emissions. Very little is known about the toxicity of most members of this class of compounds. Under a weight of evidence scheme, U.S. EPA has judged a small number (i.e., seven) of the class to be "probable human carcinogens."[13] The 1986 Guidelines for Carcinogen Risk Assessment[56] support the calculation of quantitative risk estimates for those compounds that pose a reasonable concern for human health risk.

The Provisional Guidance for Quantitative Risk Assessment of Polycyclic Aromatic Hydrocarbons[50] described an approach for assessing the carcinogenic risks posed by exposures to non-benzo(a)pyrene (B[a]P) PAH that had been judged by the Agency as "B2" substances, i.e., probable human carcinogens.[56] The results of mouse skin carcinogenicity assays for these non-B[a]P B2 PAHs were compared with those of B[a]P to estimate cancer potency. The approach assumed that the B2 PAH had the same cancer slope factor as B[a]P. The ability of these non-B[a]P B2 PAHs to elicit rodent skin tumors was quantitatively compared to that of B[a]P; the results of this quantitative comparison were expressed as an "estimated order of potency." Because this approach was limited to the cancer end point, based on B[a]P exposure from a single (oral) pathway (for the derivation of the slope factor), and considered only a small subset of the PAH, EPA has described it as an estimated order of potency. This naming reflects the uncertainty EPA felt about the application of this type of approach given the current state of science of PAHs. For application, the cancer slope factor for B[a]P was multiplied by the estimated order of potency to estimate cancer risk for that particular PAH. (The U.S. EPA[50] Provisional Guidance Document has been reviewed and cleared but does not represent official Agency policy at the time of this writing.)

Inhalation exposures to chemical mixtures of dissimilar pollutants

One of the more difficult aspects in assessing risks posed by combustor emissions arises from potential temporary increases in emissions that may occur as a result of startup and shutdown in operations, malfunctions, or perturbations in the combustion process or changes in the removal efficiency of the air pollution control equipment. The draft U.S. EPA[42] Addendum document described several procedures for deriving estimates of nonroutine emissions. These include (1) reviewing stack emission testing and engineering reports on the performance and operations of combustion technologies, furnace designs, and air pollution control devices that are most similar in every respect to the facility under review (often the emissions testing is accompanied by records that may give a gross indication of whether the system is functioning according to design specifications; emissions data may exist for some of these periods), and (2) examining tests designed to determine the removal efficiency of the air pollution control devices (APCD) may also provide information regarding temporary increases in the emissions during equipment malfunction. From these procedures and others it is clear that the quantity of pollutants emitted under these upset operating conditions may increase for short periods of time.

Inhalation exposures may increase not only during times of incinerator upset but also under atmospheric conditions that result in decreased atmospheric dispersion of the plume (e.g., inversions, low wind speed). These conditions, though typically short in duration, may result in elevated concentrations in the atmosphere surrounding a combustion facility. While emissions during incinerator perturbations and malfunctions and pollution control device failure are considered in risk assessments, typically, exposures to the emitted pollutants are considered as individual chemical exposures. Additionally, accident scenarios such as on-site fires under conditions of poor dispersion are also considered in combustion facility risk assessments, but the resulting risks are generally predicted for only individual compounds. By not considering the risks from chemical interactions, particularly under some of these high end exposure scenarios, important uncertainties could be introduced into the assessment.

At present, there is not a practical scientific approach that is generally applicable for estimating risks posed by elevated exposures to chemical mixtures under these conditions. The types and quantities of compounds emitted under these conditions may deviate from those normally emitted. The types of compounds emitted under these conditions are thought to be highly variable depending on the conditions of combustion. Additionally, many of the compounds have not been identified and the toxicity of the compounds individually is not well understood at this time. These scenarios are typically described qualitatively as uncertainties within an assessment.

A mixtures approach has been developed in the area of combustion toxicology that may be applicable at some time in the future to risk assessments of combustors, but a great deal of additional data will be needed. It is summarized here because the findings are instructive in that they shed light on the complexities posed by the consideration of toxicity of the mixtures in such an assessment.

Levin[57,58] described a seven-gas model for combustion toxicity. This model predicts toxic potency of combustion products based on the toxicological interactions of a relatively small number (seven) of gases produced during fires. The N-Gas model predicts toxicity for inhalation exposures to carbon monoxide, carbon dioxide, oxygen, hydrogen cyanide, hydrogen chloride, hydrogen bromide, and nitrogen dioxide. The model is based on lethal concentrations tested in animal models. (Animals were exposed in the studies used to develop the model then monitored for 14 days postexposure.) Synergistic and antagonistic effects have been observed among combustion gases and were incorporated into the model. For example, NO_2 is antagonistic with HCN; the protective effect of the combination is thought to be the result of methemoglobin production. Interestingly, the toxicity of NO was increased through concurrent exposure to CO_2; however, when HCN was added to this binary mixture the synergy of $NO + CO_2$ was lost.[57]

For application of this approach to be practical, several modifications will have to occur both in the animal testing system and in the model. Researchers will have to examine other adverse health effects in addition to lethality, which means that the concentrations of the gases used in experimental chambers will have to be decreased. Other compounds that pose risks over short periods of exposure duration through the inhalation route will need to be identified. These experimental findings will then have to be incorporated into the model.

Indirect exposures to chemical mixtures

Pollutants emitted from incinerators are predicted to occur in environmental media and biota around sources. For example, Cleverly et al.[46] predict soil concentrations for all of the pollutants analyzed in their assessment. Emitted pollutants such as mercury and dioxin congeners are predicted to occur in fish and aquatic sediments.[12,59] Simultaneous exposures to mixtures of pollutants through soil ingestion or fish consumption are clearly possible, and interactions among these pollutants are also possible.

Since U.S. combustion practices are generally good, predicted pollutant concentrations are generally low. The highest risks are likely to occur from facilities that emit pollutants that bioaccumulate in the terrestrial or aquatic food webs.

Surf and turf

For example, many facilities emit both mercury and dioxin congeners. Both mercury and dioxin congeners have been shown to bioaccumulate and magnify in the environment. Mercury may be transformed in the environment to methylmercury, and methylmercury bioaccumulates and biomagnifies throughout the aquatic food web. Piscivorous fish typically contain the highest concentrations of methylmercury in the aquatic food webs, and humans typically consume piscivorous fish. Dioxin congeners can also bioaccumulate in fish. Additionally, dioxin congeners may be taken up from the atmosphere by green plants. The green plants may in turn be consumed by foraging livestock and incorporated into their tissues. Humans may be exposed through consumption of beef contaminated through the terrestrial food-chain. It is plausible that people consuming both contaminated fish and contaminated beef from the area around an incinerator may be exposed to levels of both methylmercury and dioxin congeners higher than those experienced by others in the population. The potential health effects from these mixed exposures are unknown at this time. Other exposure scenarios that result in exposure to mixtures of chemicals released from one source could also be constructed. For example, arsenic accumulates in soils.

Other mixtures exposure issues

Areas of the United States may contain high levels of pollutants. These "background" pollutant concentrations may be natural in origin or the result of anthropogenic activities. Most risk assessments currently assess only the incremental exposures that result from the introduction of a new source. Background concentrations may also result in human exposure. In the future some assessments may include the examination of the effects of background concentrations along with the assessment of an individual source. If these types of assessments occur, they should examine the potential effects of exposures to mixtures of pollutants from background and the anthropogenic source of concern.

Some assessments have attempted to examine the cumulative impacts of anthropogenic sources.[44] Models such as RELMAP, which can model the atmospheric transport of pollutants from the multiple sources across a continent, could be utilized to examine anthropogenic source impacts for multiple pollutants. This type of model may be utilized to examine exposure to emitted chemical mixtures.

Conclusions

Ideally, when assessing risks posed by combustor emissions, considerations such as the effects of pollutant interactions in environmental media and impacts of chemicals as mixtures should be included. It is recognized that pollutant interactions may affect pollutant distribution in the environment, bioavailability, and uptake/retention; however, this uncertainty cannot be resolved given the current understanding of environmental fate. It is clearly an uncertainty in assessments that utilize current modeling procedures.

The risks posed by exposure to chemical mixtures can be considered (even if crudely at this time), although this adds an additional element of uncertainty to the assessment. Models exist for examining simple combinations of binary chemical mixtures that could co-occur in environmental media and biota as a result of emissions. One approach would evaluate the individual pollutants to which hypothetical individuals are predicted to have the highest exposure. (The risk assessor could predetermine a regulatory exposure level.) If individuals are predicted to be exposed to more than one pollutant above this level, it would trigger an examination of the toxicology of the chemical pair. A database such as MIXTOX could be utilized to examine such interactions. A second approach for dealing with mixtures issues would begin with an examination of stack test data. The assessor could assess the risks posed by exposure to mixtures of compounds emitted in greatest quantity, then examine the likelihood of exposure to these compounds.

There are many research needs in this field. The most important needs at this time include identification of chemicals in emissions under both routine and nonroutine emission conditions and a better understanding of hazards potentially posed by these compounds both individually and more importantly as chemical mixtures. Better risk assessment methods are needed for estimating the toxicity of mixtures of chemicals emitted from combustion sources. The examination of factors that affect environmental fate of combustion pollutants should be more comprehensive; this should lead to the development of more applicable environmental fate and exposure models.

References

1. NRC (1988). *Complex Mixtures: Methods for In Vivo Toxicity Testing.* National Research Council, National Academy Press, Washington, DC.
2. U.S. EPA. 1990. *Technical Support Document on Health Risk Assessment of Chemical Mixtures.* EPA/600/8-90/064. Office of Research and Development. August.

3. Calabrese, E.J. (ed.). 1991. *Multiple Chemical Interactions*. Lewis Publishers, Chelsea, MI.

4. Yang, R.S.H. (ed.) 1994. *Toxicology of Chemical Mixtures: Case Studies, Mechanisms, and Novel Approaches*. Academic Press, New York.

5. Krishnan, K. and J. Brodeur. 1994. Toxic interactions among solvents and environmental pollutants: corroborating laboratory observations with human experience. *Environ. Health Perspect.*, 102 (Suppl 1): 103–108.

6. Simmons, J.E. 1995. Chemical mixtures: challenge for toxicology and risk assessment. *Toxicology*, 105(2-3): 111–119.

7. Feron, V.J. and H.M. Bolt. (eds.) 1996. *Food Chem. Toxicol.*, 34: 1025–1185.

8. U.S. EPA. 1986. Guidelines for the health risk assessment of chemical mixtures. *Fed. Reg.*, 51(185): 34014–34025.

9. Schlesinger, R.B. 1995. Interaction of gaseous and particulate pollutants in the respiratory tract: mechanisms and modulators. *Toxicology*, 105: 315–325.

10. Gerrity, T.R. 1995. Regional deposition of gases and particles in the lung: implications for mixtures. *Toxicology*, 105: 327–334.

11. U.S. EPA. 1989. *Risk Assessment Guidance for Superfund. Vol. 1. Human Health Evaluation Manual* (Part A). EPA/540/1-89/002.

12. U.S. EPA. 1996. PCBs: *Cancer Dose-Response Assessment and Application to Environmental Mixtures*. National Center for Environmental Assessment. EPA/600/P-96/001F.

13. U.S. EPA. 1997. Integrated Risk Information System (IRIS). Online. National Center for Environmental Assessment, Cincinnati.

14. Albert, R.E., J. Lewtas, S. Nesnow, T.W. Thorslund, and E. Anderson. 1983. Comparative potency method for cancer risk assessment application to diesel particulate emissions. *Risk Anal.*, 3: 101–117.

15. Lewtas, J. 1985. Development of a comparative potency method for cancer risk assessment of complex mixtures using short-term *in vivo* and *in vitro* bioassays. *Toxicol. Ind. Health*, 1: 193–203.

16. Lewtas, J. 1988. Genotoxicity of complex mixtures: strategies for the identification and comparative assessment of airborne mutagens and carcinogens from combustion sources. *Fund. Appl. Toxicol.*, 10: 571–589.

17. Gandolfi, A.J., I.K. Brendel, R.L. Fisher, and J.-P. Michaud. 1995. Use of tissue slices in chemical mixtures toxicology and interspecies investigations. *Toxicology*, 105: 285–290.

18. Nesnow, S. 1990. Mouse skin tumours and human lung cancer: Relationships with complex environmental emissions, in *Complex Mixtures and Cancer Risk*. IARC Scientific Publ. No. 104, p. 44–54.

19. Schoeny, R.S. and E. Margosches. 1989 Evaluating comparative potencies: developing approaches to risk assessment of chemical mixtures. *Toxicol. Ind. Health*, 5: 825–837.

20. Svendsgaard, D.J. and R.C. Hertzberg. 1994. Statistical methods for the toxicological evaluation of the additivity assumption as used in the EPA chemical mixture risk assessment guidelines, in *Toxicology of Chemical Mixtures: Case Studies, Mechanisms, and Novel Approaches*. R. S. H. Yang, ed. Academic Press, New York, 599–642.

21. Pozzani, U.C., C.S. Weil, and C.P. Carpenter. 1959. The toxicological basis of threshold values. 5. The experimental inhalation of vapor mixtures by rats, with notes upon the relationship between single dose inhalation and single dose oral data. *Am. Ind. Hyg. Assoc. J.*, 20: 364–369.

22. Smyth, H.F., C.S. Weil, J.S. West, and C.P. Carpenter. 1969. An exploration of joint toxic action. I. Twenty-seven industrial chemicals intubated in rats in all possible pairs. *Toxicol. Appl. Pharmacol.*, 14: 340–347.

23. Smyth, H.F., C.S. Weil, J.S. West, and C.P. Carpenter. 1970. An exploration of joint toxic action. II. Equitoxic versus equivolume mixtures. *Toxicol. Appl. Pharmacol.*, 17: 498–503.

24. Murphy, S.D. 1980. Assessment of the potential for toxic interactions among environmental pollutants, in *The Principles and Methods in Modern Toxicology*, C.L. Galli, S.D. Murphy, and R. Paoletti, eds., Elsevier/North Holland, Amsterdam.

25. Ikeda, M. 1988. Multiple exposure to chemicals. *Reg. Toxicol. Pharmacol.*, 8: 414–421.

26. Feron, V.J., J.P. Groten, D. Jonker, F.R. Cassess, and P.J. van Bladeren. 1995. Toxicology of chemical mixtures: challenges for today and the future. *Toxicology*, 105:415–427.

27. Plaa, G.L. and M. Vénzina. 1990. Factors to consider in the design and evaluation of chemical interaction studies in laboratory animals, in *Toxic Interactions*, R.S. Goldstein, W.R. Hewitt, and J.B. Hook, eds., Academic Press, New York, 3–30.

28. Kodell, R.L. and J.G. Pounds. 1991. Assessing the toxicity of mixtures of chemicals, in *Statistics in Toxicology*, D. Krewski, and C. Franklin, eds., Gordon and Breach, New York. 559–591.

29. Muska, C.F. and L.J. Weber. 1977. An approach for studying the effects of mixtures of environmental toxicants in whole organism performance, in *Recent Advances in Fish Toxicology*, R.A. Taub, Ed., U.S. EPA, EPA-600/3-77-085. 71–87.

30. Gessner, P.K. 1995. Isobolographic analysis of interactions: an update on applications and utility. *Toxicology*, 105(2–3): 161–179.

31. Gennings C. 1995. An efficient experimental design for detecting departure from additivity in mixtures of many chemicals. *Toxicology*, 105: 189–197.

32. Svendsgaard, D.J. and W.R. Greco. 1995. Session summary: Experimental designs, analyses and quantitative models. *Toxicology*, 105: 156–160.

33. National Research Council. 1988. *Complex Mixtures: Methods for In Vivo Toxicity Testing*, National Academy Press, Washington, DC.

34. Mumtaz, M.M. and P.R. Durkin, (1992). A weight-of-evidence scheme for assessing interactions in chemical mixtures. *Toxicol. Indus. Health*, 8:377–406.

35. Hertzberg RC. 1996. *Risk Assessment of Truly Complex Mixtures Using EPA's Almost Proposed Guidelines*, presented at the Health Effects Institute Annual Conference, Session on Vehicle Emissions as a Complex Mixture, April 28–30, 1996, Asheville, NC.

36. Bliss, C.I. 1939. The toxicity of poisons applied jointly. *Ann. Appl. Biol.*, 26: 585–615.

37. Teuschler, L. and R.C. Hertzberg. 1995. Current and future risk assessment guidelines, policy, and methods development for chemical mixtures. *Toxicology*, 105: 137–144.

38. Finney, D.J. 1971. *Probit Analysis*. 3rd ed. Cambridge University Press, Cambridge, UK.

39. El-Masri H.A., R.S. Thomas, S.A. Benjamin, and R.S. Yang. 1995. Physiologically based pharmacokinetic/pharmacodynamic modeling of chemical mixtures and possible applications in risk assessment. *Toxicology*, 105(2–3): 275–282.

40. U.S. EPA. 1998. *Study of Hazardous Air Pollutant Emissions from Electric Utility Steam Generating Units,* Volume I. Final Report to Congress. EPA/453/R-98-004a. February.

41. Lorber, M., P. Pinsky, P. Gehring, C. Braverman, D. Winters, and W. Sovocool. in press. Relationships between dioxins in soil, air, ash, and emissions from a municipal solid waste incinerator emitting large amounts of dioxins.

42. U.S. EPA. 1993. *Review Draft Addendum to the Methodology for Assessing Health Risks Associated with Indirect Exposure to Combustor Emissions.* Office of Health and Environmental Assessment. Office of Research and Development. EPA-600-AP-93-003. November 10.

43. Rice, G.E., P. McGinnis, and R. Bruins. 1991. *Indirect Exposure: A Comparison of Two Methodologies,* presented at the ASTSWMO 1991 National Solid Waste Forum, July 15, 1991, Las Vegas, Nevada. Session 4A: Waste-to-Energy Facilities and Ash Residue: Problems/Solutions.

44. U.S. EPA, 1997. *Mercury Study Report to Congress Volume IV: An Assessment of Exposure to Mercury Emissions in the United States.* Final Report. Office of Air Quality Planning and Standards and Office of Research and Development. (EPA/452/R-97-006). December.

45. U.S. EPA. 1993. *Estimating Exposure to Dioxin-Like Compounds.* Vol. III: Site-Specific Assessment Procedures. Working Draft.

46. Cleverly, D.H., C.C. Travis, and G.E. Rice. 1992. The analysis of indirect exposure to toxic air pollutants emitted from stationary combustion sources: a case study, presented at the 85th Annual Meeting of the Air and Waste Management Association, June 21–26, 1992, Kansas City.

47. Chaney, R.L. and J.A. Ryan. 1994. *Risk Based Standards for Arsenic, Lead and Cadmium in Urban Soils.* DECHEMA-Fachgesprache Umweltschutz. G. Kreysa and J. Wiesner, eds., Frankfurt.

48. U.S. EPA. 1989. *Interim Procedures for Estimating Risks Associated with Exposures to Mixtures of Chlorinated Dibenzo-p-Dioxins and -Dibenzofurans (CDDs and CDFs) and 1989 Update.* Risk Assessment Forum. EPA/625/3-89/016. March 1989.

49. U.S. EPA 1991. *The Workshop Report on Toxicity Equivalency Factors for Polychlorinated Biphenyl Congeners.* Risk Assessment Forum. EPA/625/3-91/020. June 1991.

50. U.S. EPA. 1993. *Provisional Guidance for Quantitative Risk Assessment of Polycyclic Aromatic Hydrocarbons.* Office of Research and Development, Washington, DC. EPA/600/R-93/089. July 1993.

51. Barnes, D., A. Alford-Stevens, L. Birnbaum, F. Kutz, W. Wood, and D. Patton. 1991. Toxicity equivalency factors for PCBs. *Quality Assurance, Good Practice, and Law.* Vol. 1, No. 1, p. 70–81. October.

52. Ahlborg, U.G., G. Becking, L. Birnbaum, A. Brouwer, H. Derks, M. Feely, G. Golor, A. Hanberg, J. Larsen, D. Liam, S. Safe, C. Schlatter, F. Waern, M. Younes, and E. Yrjanheikki. 1994. *Toxic Equivalency Factors for Dioxin-like PCBs.* Report on a WHO-ECEH and IPCS Consultation, December, 1993. *Chemosphere,* 28(6): 1049–1067.

53. Safe, S. (1994) Polychlorinated biphenyls (PCBs): environmental impact, biochemical and toxic responses, and implications for risk assessment. *Crit. Rev. Toxicol.,* 24(2): 87–149.

54. McFarland, V.A. and J.U. Clarke. (1989) Environmental occurrence, abundance, and potential toxicity of polychlorinated biphenyl congeners: considerations for a congener-specific analysis. *Environ. Health Perspect.,* 81: 225–239.

55. Birnbaum, L.S. and M.J. DeVito. In Press. Use of toxic equivalency factors for risk assessment for dioxins and related compounds. *Toxicology.*

56. U.S. EPA. 1986. Guidelines for carcinogen risk assessment. *Fed. Reg.,* 51(185): 33992–34003.

57. Levin, B. 1996. New research avenues in toxicology: 7-gas N-Gas Model, toxicant suppressants, and genetic toxicology. *Toxicology,* 115: 89–106.

58. Levin, B. 1997. *New Approaches to Toxicology — A Seven-Gas Predictive Model and Toxicant Suppressives.* Abstract presented at the Conference on Issues and Applications in Toxicology and Risk Assessment. Sponsored by Tri-Service Toxicology, Wright-Patterson Air Force Base; U.S. EPA, and ATSDR, April 7–10.

59. U.S. EPA. 1994. *Implementation Guidance for Conducting Indirect Exposure Analysis at RCRA Combustion Units.* Revised draft dated April 22, 1994. Office of Solid Waste, U.S. EPA, Washington, D.C.

60. Albert, R.E. and C. Chen. 1986. U.S. EPA diesel studies on inhalation hazards, in *Carcinogenic and Mutagenic Effects of Diesel Engine Exhaust,* N. Ishinishi, A. Koizumi, R.O. McClellan, and W. Stober, eds., Elsevier, New York, 411–419.

chapter six

Ecological risk assessment issues for hazardous waste incineration

Robert P. DeMott

Contents

Introduction

The aspect of assessing risks from hazardous waste incineration with perhaps the greatest potential for enhancement is the incorporation of specific, quantitative approaches for evaluating risks to the nonhuman components of the environment — ecological risk assessment. In part because of the dominance of concerns about human health, and the frequent assumption that human health risk approaches are conservative enough to provide protection for other components of the environment, evaluations of potential

1-56670-250-X/99/$0.00+$.50
© 1999 by CRC Press LLC

ecological risks from hazardous waste incinerators have been deemphasized. Also, because of the particular challenges of evaluating ecological risks, it is frequently perceived that such analyses produce limited definitive information useful for decision making. However, concern regarding potential ecological risks has increased over time and advancements and refinements have come with use and experimentation. The maturation of ecological risk assessment approaches means that consideration of these important features is more and more feasible.

While growing concern has helped drive the formulation of improved ecological risk assessment methods, compared to human health risk assessment there is much less consensus among scientists about the optimal general scheme, much less the particulars of calculating risk estimates. Consequently, the guidance available from regulatory agencies is much less structured. In terms of hazardous waste incineration, there is limited direction and limited consistency in the consideration that has been given to potential ecological risks. The U.S. Environmental Protection Agency (EPA) is currently in the process of generating an updated general conceptual paradigm for considering ecological risks[1,2] and is expected to prepare a manual incorporating specific protocols for completing ecological risk assessments for Superfund sites. This guidance will likely provide the basis for expanding consideration of ecological risks relating to incineration.

Incineration, with its particular types/forms of environmental releases, creates somewhat specialized ramifications for risk analysis. Incineration is recognized to have the potential for depositing relatively diverse sets of chemicals over relatively large areas. Additionally, there is a growing focus on the potential for chemicals in the environment to affect endocrine processes in wildlife and particular interest in dioxins, also a major concern for incineration. These factors will influence the demand for ecological risk evaluations relating to incinerators. A review of the challenges in evaluating ecological risks associated with incineration will demonstrate how the methods apply to these issues.

Ecological risk assessment and its unique challenges

Development of interest in evaluating ecological risks

On the heels of widespread concern about human health risks, recognition has grown of the potential impacts of chemicals in the environment on nonhuman ecological components. While cancer fears lead to relatively swift advances and formalization of human health risk assessment approaches, ecologists had been observing environmental changes linked to xenobiotics and collecting data on the toxicological responses of various systems, particularly relating to aquatic toxicology, but there was not as much attention focused on formulating methods for providing input to decision making. Clearly, the systems needing consideration were much broader, and the question of what constituted risk was much fuzzier.

At least as a starting point, there was considerable aquatic toxicity data for many key chemicals. To comply with the Federal Insecticide, Fungicide and Rodenticide Act (FIFRA) and Toxic Substances Control Act (TSCA), manufacturers of many chemicals provided toxicity testing data to the EPA relevant to environmental as well as human health.[3,4] Also, researchers studying ecosystem responses to toxic impacts and contaminant transfer in food webs provided laboratory, test pond, and field data. In theory, the available environmental information can serve as the basis for shaping predictive risk analysis methods. In practice, turning retrospective analyses into adequately comprehensive prospective models with acceptable levels of uncertainty has been challenging.

Regulatory background for ecological risk approaches

Under both of the predominant programs for addressing sites affected by hazardous wastes, Superfund (CERCLA/SARA) and RCRA, there have been provisions for including ecological concerns in evaluating potential risks. For Superfund, site-specific information is supposed to determine the need for ecological risk assessment and then the appropriate level of analysis.[5] The primary concern under these programs is evaluating existing contamination, and thus the focus has mainly been on retrospective analyses looking for already realized impacts. A relatively comprehensive sampling and evaluation scheme for retrospective analyses was outlined by the EPA.[6] However, the limited specifics in terms of both approach and default parameters found in guidance documents for prospective ecological risk assessments have led to inconsistencies and provided a convenient rationale for omitting such analyses.

A 1989 review of Superfund or RCRA sites that had included an ecological evaluation found that approximately 75% of those evaluated relied on either qualitative descriptions of impacts or simple comparisons of concentrations in environmental media to EPA-characterized safety levels.[7] At the remaining sites, with one exception, toxicity information obtained from testing various species with contaminated media collected at the site was used to estimate potential effects on the populations at the site.[7] These approaches require existing contamination and samples collected from contaminated media. Only one of the sites reviewed[7] attempted a truly prospective assessment of effects from multipathway exposure via different trophic levels, analogous to a human health risk assessment including indirect exposures through the food chain, and this analysis was limited to one species.

The older approaches are generally strongest as predictors of effects on individual organisms, especially for compounds that are well characterized. The approaches are more difficult to evaluate, and to assign confidence estimates, when higher levels of ecological organization and complex mixtures of compounds are of interest. Concerns about incineration focus on these two topics — the products are complex, diverse, with interactions that are poorly understood, and the potential for generalized ecological health impacts over a wide area is recognized to be an important consideration.

The EPA's updated approach outlined in the *Framework for Ecological Risk Assessment*[1] and developed in the *Proposed Guidelines for Ecological Risk*[2] attempts to make the link stronger and more quantifiable between the overall goals considered to be important (these can be population or ecosystem level concerns) and the specific end points that can be evaluated at a particular site and used to characterize potential risks. The concept is that, by carefully formulating the assessment to consider stressors (chemical or otherwise) and effects that are both significant to larger ecological concerns and can be well characterized, improved inferences about the overall potential risks can be realized. Application of this concept, and perhaps even the specific updated EPA approach, to incinerator-related ecological assessments should allow improved evaluation of diverse receptors and stressors, but not enough assessments are yet available to provide an adequate basis for review.

Specialized issues of ecological risk assessment

The most significant complications of ecological risk assessment compared to human health risk assessment are the need to evaluate potential effects on multiple species with varied responses to particular chemicals and the need to have variable and diverse levels of protection. Where humans are concerned, we need to predict the potential response for one species, and have one level of protection — conservative assurance that individuals are extremely unlikely to realize adverse effects. For various ecological receptors, sensitivities to particular chemicals can be widely disparate. Frequently, it is necessary to consider and extrapolate toxicity information from aquatic and terrestrial vertebrates and invertebrates in characterizing potential risks to the ecological receptors that are of concern in a particular investigation. Also, for some potential receptors, the goal may be to ensure that there are not adverse effects to any individual, but, from an ecological perspective, protection of individuals is often not significant.

Determination of protection goals or assessment end points

While a "healthy" environment seems intuitively obvious as the overall goal for protection, it is much more difficult to determine exactly the factors that are important in a given ecosystem and for particular stressors.[8] Ecological risk assessments must define protection goals based on the particular circumstances being considered. The goals, or units of protection, can be individuals in various species or:

Populations — the interbreeding pool of a certain species in an area
Communities — the populations sharing an area
Ecosystems — the living and nonliving features that characterize an area

Where endangered or threatened species are involved, statutory protection is given to individual organisms. On the other hand, individuals of other species or even the population of particular species may not be of concern for a particular ecosystem if the community balance is not significantly disrupted by localized elimination. Then the pertinent question becomes whether the overall ecosystem is of primary importance or whether all species or all individuals should be protected. While it may seem to be undesirable to acknowledge the potential for local elimination of organisms, this does not necessarily represent an ecological threat, and, especially when the assessment is intended to evaluate various management options, certain tradeoffs may be a part of selecting the best overall approach. For example, where a population of a common sediment-dwelling invertebrate may be at risk because of high sensitivity to a certain sediment contaminant, but the alternative is to dredge the area, likely silting over local fish spawning sites and exposing a wide variety of fry to remobilized chemicals, the preferred alternative from an ecosystem perspective might be to leave the sediments. With multiple scientific answers to the question, "What should be protected in the environment?" the needs and intended uses of the risk assessment guide the appropriate selection for particular evaluations.

The EPA uses the term "assessment end points" to refer to "explicit expressions of the actual environmental value that is to be protected."[2] In selecting assessment end points for an ecological risk assessment, the overall goal is to develop a set of protection goals that can be linked to measurable criteria relating to the anticipated stressors under investigation. Assessment end points should incorporate some ecological entity (e.g., individual largemouth bass or a population of largemouth bass) to be considered and some characteristic related to the entity that is considered significant (e.g., maintenance of breeding.[2] When deciding upon the relevant assessment end points for a particular evaluation, certain criteria should be considered. First, how relevant are the end points to the overall ecology of the area? For example, is the species a critical predator for maintaining community balance or one of a number of species in a particular ecological niche? Second, there should be some aspect related to the assessment end point that is sensitive or susceptible to the stressors anticipated for the evaluation. For example, is the species selected particularly sensitive to dioxin toxicity? Also, while not a theoretical necessity for estimating risks, assessment end points that can be considered in light of the potential alternative strategies are most useful for decision making based on the risk assessment. For example, selecting a fish species that can be evaluated with either a dredging or nondredging plan can be a practical advantage.

Indirect vs. direct toxic effects

Because of the interrelationships within an ecosystem, some species can be faced with indirect effects related to toxic responses elsewhere in the community. For example, there may not be direct toxicity of the stressors of concern on a given species, but its primary food source may be eliminated. Indirect effects can be disruptions of predator–prey balances and also of the competitive balance controlling resources used by many populations.[9] A spider's web analogy is useful, where disturbances at one point are transmitted along various strands of the food web. There is no generalized method for identifying the particular interactions, or web, that applies to a particular stressor. To consider indirect effects in a prospective assessment, information must be developed for each suspected stressor based on the makeup of the potentially impacted environment.

It is possible, using chemical-specific toxicity data and computer simulations of ecosystem interactions, to model the contributions of indirect effects. Such an analysis with chloroparaffin in an aquatic ecosystem suggests that overall effects on zooplankton-consuming fish contain a significant contribution related to toxicity in phytoplankton, several steps down the food chain.[10] This stressor is of interest relating to incineration because it is a complex mixture of chlorinated hydrocarbons that is relatively analogous to some incinerator emissions. The lesson for developing ecological risk assessment strategies is that consideration of just the toxicity of chemicals on certain species of concern may not be sufficient. The critical factor for community health might be an effect on a nonobvious organism several trophic levels lower.

Degree of protection

In human health risk assessments, it is clear that the goal is to completely protect all of the individuals. Where the individual is not the unit of protection in ecological risk assessments, a determination needs to be made of the degree of protection that is required for particular populations. Is it sufficient to protect 50% of the individuals of the most sensitive species from death, or 50% from any effect, or should 95% of all species not show any effect? It is not straightforward from an ecological perspective to determine *a priori* the degree of protection that is important for each receptor and is generally required for ecosystem stability. For species that normally are subject to very high mortality rates, a predicted impact in an ecological risk assessment of 20% mortality might not be distinguishable from the natural occurrence, and would be unlikely to affect the community structure. On the other hand, for long-lived species that invest significantly in each individual offspring (such as bears), the 20% mortality rate could have significant effects on population stability and affect the community dynamics in the long term. Whether a certain mortality rate significantly affects the population, community, and ecosystem is dependent on ecosystem and community structure. Again, situation-specific information is required to determine the best degree of protection for a particular species in a particular evaluation. The key for thorough ecological risk evaluation is identifying the populations that could be subject to particular effects and determining, based on the population's ecological characteristics, how the predicted effect relates to population and community dynamics.

Temporal and spatial scale concerns for ecological communities

Human health risk assessments are based on the one, obvious time scale — length of a human life. When developing ecological risk assessment approaches, what time scales need to be considered? To illustrate the difficulty, contrast a day in the life of an oak tree with the generations of algae produced on the pond nearby during that day. A one-day exposure to a particular stressor could correspond to the entire lifetime of algal cells, including, most importantly, their reproductive stage, and wipe out the algae population of the pond. The same exposure would represent an insignificant fraction of the oak tree's life.

Accounting for the duration of exposure in the analysis of risk for a given population needs to take account of the relative proportion of the life spans represented by the exposure. This issue is complicated by the fact that the proportion of life span relevant to toxicity can vary considerably even within a species depending on life history and the timing of the exposure. Life spans for some organisms vary widely from generation to generation, and many organisms have a complex life cycle including larval or pupal stages that may have different responses to stressors than the adults.[11] Thus, for example, stressor chemicals typically elevated in a river and lake system during spring floods may not be of particular concern for the tadpoles of a frog species that breeds in summer and metamorphoses within a few weeks, but bullfrog tadpoles that overwinter might be significantly impacted because of the timing of their larval life. When examining potential community impacts, the time scale may change again. Many communities exhibit cyclical changes that may be seasonal, annual, or longer[11] that should be considered.

Another issue is the relevance of the time scales used in modeling fate and transport of contaminants through the environment. The commonly used models are most reliable and optimized for time frames relating to human exposure. Biotic transfer of contaminants, in particular, needs to be given close attention since relatively simple food webs are used and models may not be run long enough to detect the eventual appearance of a potentially problematic exposure through bioaccumulation. Slow transfer and bioaccumulation of a persistent compound through long-lived members of a food web, such as some trees, could mean that the potential for toxic exposures could develop many years into the model. Countering such long-term concerns, it is also important to consider how models handle mass balance. Frequently, in the name of conservatism, mass balance is handled loosely, such that potential surficial soil exposures to a volatile contaminant are held steady over time, even though much of the chemical evaporated and more may have been degraded in the soil. Especially for bioaccumulation pathways in long simulations, assuming there is an infinite source for contaminants, leads to artificial overstatements of the potential risks.

Spatial scale is also complicated to account for in ecological risk assessment. Typically for risk assessments involving airborne transport and deposition, such as incinerator emissions, an area of a certain size is modeled as the potentially affected area, and this outlines the ecosystem of interest. The organisms using this area, of course, have no idea of the boundaries, yet the spatial orientation of populations and the communities can affect the potential effects of stressors. Generally, as the level of biological organization increases, the time and space scales expand.[12] Communities can encompass fairly large areas and overlap the impacted area in various ways. Even for individual organisms, the relationship between home range and the impacted area can be complex. The more mobile the organism, the more likely that its exposure will change as it moves in and out of areas with particular levels of contamination, or out of the impacted area entirely. Consideration of spatial issues such as microhabitat preference and mobility is significant when attempting to relate the ecosystems in an area to potential risk factors.[13]

This section has described some of the difficulties in adapting the single-species, human health risk assessment paradigm to considerations of potential ecological risk. Expanding the scope of a risk analysis in this manner expands the complexity of the issues that must be addressed. Some of the questions that must be dealt with are at the basis of the risk analysis. What is it that we wish to protect, and to what degree? Other questions illustrate the interconnected nature of the environment. How do we account for convoluted pathways of indirect exposure? The other topics that have been discussed include the relationship between end points that can be adequately measured and the true protection end points, and differences in scale that can affect multispecies, multilevel analyses. Understanding these issues provides some basis for critically analyzing the risk assessment options and considering optimal strategies for evaluating the potential ecological risks from incinerator emissions.

Incinerator-related concerns in the ecological risk assessment process

Size and diversity of area of concern

The first problematic issue in developing ecological risk assessment plans for incinerators is characterizing the area to be evaluated. At typical hazardous waste sites reasonable boundaries for the area of concern can usually be established based on site history and sampling environmental media.[6,7] Because of the nature of incinerator emissions, the affected area of deposition can potentially be quite large and is not subject to some of the geographic and geological constraints found at typical waste sites. The height at which the emissions are released, air currents, and directions are some of the primary factors determining the potentially contaminated area. In evaluating incineration as an option, the goal is to be able to predict the potential risks that might result, not just estimate risks after the fact; therefore, the method must often be applied in the absence of existing contamination to sample.

This necessitates reliance on air transport models to characterize where deposition is likely to occur, and at what levels, adding to the uncertainty associated with determining both the size of the potentially impacted area and the extent of contamination.

Once the area of deposition has been characterized through air modeling, it can be mapped in terms of ecosystems, habitats, species, and features that are expected to be present. Because of the potential size of the area of concern, incinerator risk assessments may need to consider relatively large ecosystem ranges, and diverse habitats with varied receptor species. This expands the ecological risk assessment task and requires careful consideration of the receptors that should be included. Further, potential effects at the population and community level become a more serious consideration as the size and nature of the area of concern expand. This type of broad characterization of the area of concern is needed to identify assessment end points and goals for protection that are relevant to scale of the potential impacts.

Though the scope of the task is expanded, the factors that need to be considered in characterizing the area of concern are generally the same as those needed in typical site ecological risk assessments, and the factors identified by EPA guidance[5,6] constitute a fairly comprehensive checklist to consider. The type of characteristics that should be described and mapped, where relevant, include

- Distribution of habitat types (grasslands, forest, ponds, etc.)
- Terrestrial features such as soil types, geology
- Aquatic features such as area, depth, flow, clarity, substrate type, water quality parameters
- Topography
- Presence of preexisting sensitive or compromised areas
- Vegetation community description, terrestrial and aquatic
- Faunal presence, including microfauna
- Species profiles (abundance, distribution, description of sensitive stages), especially indicator, endangered, specially valued, or unique species
- Notable species absences or atypical species profiles

Ecological information should be used to map population ranges as they exist in the region, not only within the area identified as potentially impacted by deposition. Accounting for potential population or community level effects requires that the actual extent of the ranges be available.

Identifying chemicals of concern

Having characterized the area for evaluation, the next complication associated with incineration is determining which chemicals should be considered. At typical hazardous waste sites, site history and analytical chemistry are a reasonable basis for selecting chemicals of concern. When preparing an upfront evaluation of the potential risks from incineration, deciding what products are likely to be emitted can be difficult. The set of chemicals released following combustion is heavily dependent on the source materials being incinerated and the characteristics of the facility. Some type of estimates of the chemicals likely to be emitted and the relevant quantities is needed to identify a list of chemicals to be included as potential stressors in the risk assessment.

There are a few classes of compounds that can be considered as at least important candidate stressors since they are often associated with combustion and can pose substantial levels of estimated risk. These include halogenated compounds, such as chlorinated dioxins and dibenzofurans, and volatiles, such as benzene and toluene that are products of incomplete combustion, metals, and acid gases.

The composition of the products of incomplete combustion is highly variable, but dioxins and dibenzofurans in particular are common by-products that are considered important environmental contaminants. These compounds are important because of the relatively high toxicity of some of the particular forms, and their ability to bioaccumulate and move up the food chain. Dioxins have been characterized as potential disrupters of endocrine processes, a growing area of ecological concern, but it remains unclear what type of effects can actually be realized in receptors at environmental levels associated with airborne deposition.[14-16]

The metals, as elements, cannot be broken down during incineration, and thus the entire input of metal is accounted for between the residual solids and emissions from the facility. Metals can be volatilized and released as gases, though this is significant for only a few, but are primarily released into the air as particulates. Mercury and selenium, which are significantly volatile under incineration conditions, arsenic, beryllium, chromium, and lead are candidates to consider. All of these metals have substantial potential for ecotoxicity and are common in combustor emissions. Like the dioxins, mercury has substantial potential to bioaccumulate and should be carefully considered in terms of food chain transfers.

As the list of chemicals of concern is developed for an ecological risk assessment, consideration should be made where possible of background levels and alternative sources. It is reasonable in some cases to exclude chemicals from the evaluation if deposition modeling indicates that additions to background will be insignificant, toxicity is relatively low, and fate and transport models and bioaccumulation potential do not indicate the likelihood of high levels of indirect exposure. This approach was used in the ecological assessment for the Kimball thermal oxidation unit.[17] Excluding chemicals based on background is typically appropriate for naturally occurring inorganic chemicals, though arguments can be made for not considering certain organic chemicals that are ubiquitous, as well. Consideration of alternative sources for particular chemicals within the area of concern is useful for determining whether the incineration facility under evaluation is expected to add substantially to the overall burden of chemicals in the area. In this case, it may be important to consider the potential effects of chemicals that would be considered marginal contributors when just the incinerator-related input was estimated, but where there are substantial contributions from other sources. This type of consideration is most important for chemicals that tend to bioaccumulate since the impacts of relatively low levels can end up magnified in the higher trophic level receptors.

Characterizing exposure

With the area of concern, potential receptors, and stressors identified, the next step in ecological risk assessment is to describe how exposures are anticipated to occur, and what levels of exposure can be expected. Exposures may be through contact or ingestion of media containing contaminants (air, soil, sediment, and/or water depending on habitat), or through ingestion of food sources containing contaminants. In developing a plan for the risk assessment, a matrix can be set up categorizing all of the potential exposure media and the manner in which exposure can occur (exposure pathway) relevant to each identified receptor. Factors to consider include food and water intake rates, seasonal and life-stage changes in feeding habits, life history features relating to contact with particular media (e.g., swimming or burrowing behavior), and the source of prey items. The pathways of exposure that are typically considered include dermal contact and ingestion of soil and water and sediment and biotic transfers of contaminants through prey. For certain classes of chemicals, bioaccumulation will likely drive the exposure levels, especially for high-level predators. Inhalation can be evaluated, but it usually is not a major contributor unless there is a substantial concern about volatile compounds, so these pathways are frequently excluded to simplify the process.

The typical approaches used in ecological risk assessment apply for incinerator evaluations,[5,6] with the twist, however, that concentrations measured in environmental media will not be available when a completely prospective evaluation is the goal. Concentrations in the required environmental media can be estimated from fate and transport models, but there is uncertainty associated with this approach. Concentrations being passed through the food chain can be estimated based on models that account for the bioaccumulative potential of chemicals. Thus, a concentration of the stressor chemicals in prey items for various receptors can be estimated.

For each contaminant, exposure pathways need to be modeled for each population, accounting for their different dietary, habitat, and life history characteristics. To account for exposures along the food web, the analysis needs to begin with the lower trophic levels and progress sequentially so that dietary intake for each trophic level can be considered. The degree of bioaccumulation at each trophic level needs to be factored into the food web as well. Food web analysis is an important focus in ecology, and there are numerous methods available.[18] The choice of a specific approach will depend on the characteristics of the potentially affected area since analysis methods are typically optimized for a particular type of ecosystem.

Attempting to account for all significant exposure pathways for a number of receptors is a cumbersome process. There are data gaps in values needed to estimate contaminant transfer, and the high degree of uncertainty can be introduced through the exposure assumptions. Not surprisingly, an EPA review found that exposure scenarios were frequently oversimplified in ecological risk assessments.[19] While complete characterization of the pathways contributing to exposure for various receptors may not be efficient, or feasible, the usefulness of the evaluation depends on developing scenarios that are reasonable representations of the anticipated exposures. Determining the critical considerations based on receptor and chemical characteristics, and devising approaches to adequately account for these exposures is the key challenge in ecological risk assessment.

Risk characterization

Once the receptors and stressors have been identified and exposure scenarios have been developed, the toxic potential of the stressors is used as the basis for characterizing the potential ecological risks. This results in some form of qualitative or quantitative description of the potential for adverse outcomes to be realized among the receptors. There are no major issues related to risk characterization that are unique to ecological risk assessments evaluating incinerators, but there are some general difficulties deserving mention.

One of the most difficult aspects of environmental risk assessment has been establishing criteria that should be used as the basis for determining whether estimated risks are significant. What effects and what level of estimated impact upon individuals, populations, or communities represent a potential problem? This difficulty is compounded by the challenges of attempting to compare various responses in various receptors. As a consequence, there is considerable uncertainty in establishing priorities in terms of the chemicals and/or effects that are potentially the most important. The difficulty in selecting these revolves around the lack of consensus about what constitutes environmental risk. EPA's 1988 analysis noted that at *none* of the reviewed sites was there a clear statement as to what was considered a "significant" environmental risk.[7] This was identified as a major area for improvement, and the review specifically pointed out the need for methods to evaluate the risk associated with potential emissions from incinerators.[7] In the absence of criteria for determining risk, the typical approach has been to provide some description for each potential hazard identified and to essentially leave open the question of how acceptable are the associated risks. This approach might be more properly termed a hazard presentation than a characterization of risk. Risk managers are essentially left to determine whether these potential hazards are individually important instead of determining whether the likelihood and severity of potential impacts, individually and as a whole, are important. Especially where complex environmental issues are involved, such as with the potential risks relating to incineration, as much critical and synthetic analysis as possible should be provided in the risk characterization to try to provide a tool that is useful to the risk manager attempting to evaluate options. In developing the evaluation, the opportunity resides with the risk analyst to account for relationships operating between different features in the ecology of the area and to consider how stressors can impact directly and indirectly on the environment. Such considerations must influence the risk characterization and be clearly communicated if they are to affect decisions regarding the potential risks.

Summary

Ecological risk assessment procedures are complicated by the need to consider multiple species and their interactions in characterizing potential risks. There are challenging questions regarding what represents potentially significant impacts, what specific features should be protected, and what degree of protection should be provided. The more complicated the ecology being evaluated in a risk assessment, the more difficult it is to extract definitive conclusions from the risk assessment methods. In the past, faced with these challenges, many ecological risk assessments resorted to simplistic approaches which did not produce useful tools for helping support decision making. Now, with increasing emphasis on nonhuman environmental concerns, and increasing attention from the scientific and regulatory communities, the available methods for evaluating ecological risks are becoming more sophisticated.

Evaluating the potential ecological risks associated with hazardous waste incineration can be expected to involve complex environmental factors because of the potential for widespread deposition of emissions and the diversity of the stressors that can be released. Consideration must be given to ecosystems, habitats, communities, populations, and individual organisms that could be potentially impacted. Characterization of the ecological characteristics of a potentially impacted area and evaluation of the particular stressors that are anticipated are the bases for selecting the resources that need to be considered and the assessment end points that can be evaluated. Characterizing the potential impacts of an incineration facility based on relevant ecological features can then provide a tool for decision makers evaluating the risks of incineration or weighing alternative strategies for dealing with hazardous wastes.

This chapter has described some of the complexities relating to ecological risk assessment and identified several unique features relating to incineration that affect the manner in which potential risks need to be examined. The ecological risk assessment process continues to be subject to considerable uncertainty, but a key element in determining the usefulness of such analyses is the manner in which the analysis is shaped and the aspects included for consideration. These tasks generally fall to those developing the risk assessment, and careful selection of the strategies as well as a clear understanding of the limitations and uncertainties can make the difference in providing a reasonable and useful analysis.

References

1. U.S. EPA, *Framework for Ecological Risk Assessment*, U.S. Environmental Protection Agency, Washington, DC, 1992.
2. U.S. EPA, Proposed guidelines for ecological risk assessment, *Fed. Reg.*, 61, 47552, September 9, 1996.
3. Barton, A.L. Ecological risk assessment in the office of pesticide programs. in *Wildlife Toxicology and Population Modeling Integrated Studies of Agroecosystems*. R.J. Kendall and T.E. Lacher, Eds., CRC Press, Boca Raton, FL, 1994, p. 27.
4. Zeeman, M. and Gilford, J. Ecological hazard evaluation and risk assessment under EPA's toxic substances control act (TSCA): an introduction, in *Environmental Toxicology and Risk Assessment*. W.G. Landis, J.S. Hughes, and M.A. Lewis, Eds., ASTM, Philadelphia, 1993, p. 7
5. U.S. EPA, *Supplemental risk assessment guidance for the Superfund program. Part I, guidance for public health risk assessment. Part 2, guidance for ecological risk assessments*, U.S. Environmental Protection Agency, Boston, 1989.
6. U.S. EPA, *Ecological Assessment of Hazardous Waste Sites: A Field and Laboratory Reference*, U.S. Environmental Protection Agency, Corvallis, OR, 1989.
7. U.S. EPA, *Ecological risk assessment methods: a review and evaluation of past practices in the superfund and RCRA programs*, U.S. Environmental Protection Agency, Washington, DC, 1989.
8. Dobson, S. Why different regulatory decisions when the scientific information base is similar?—Environmental risk assessment, *Regul. Toxicol. Pharmacol.*, 17, 333, 1993.
9. Cockerham, L.G. and Shane, B.S. *Basic Environmental Toxicology*, CRC Press, Boca Raton, FL, 1994.
10. Bartell, S.M., Gardner, R.H., and O'Neill, R.V. *Ecological Risk Estimation*, Lewis Publishers, Chelsea, MI, 1992.
11. Burger, J. and Gochfeld, M. Temporal scales in ecological risk assessment, *Arch. Environ. Contam. Toxicol.*, 23, 484, 1992.
12. Gentile, J.H. and Slimak, M.W. Endpoints and indicators in ecological risk assessments, in *Ecological Indicators*, Vol. 2. D.H. McKenzie, D.E. Hyatt, and V.J. McDonald, Eds., Elsevier Science, England, 1992, p. 1385.
13. Pulliam, H.R. Incorporating concepts from population and behavioral ecology into models of exposure to toxins and risk assessment, in *Wildlife Toxicology and Population Modeling: Integrated Studies of Agroecosystems*. R.J. Kendall and T.E. Lacher, Eds., Lewis Publishers, Boca Raton, FL, 1994, p. 13.
14. Peterson, R.E., Theobald, H.M., and Kimmel, G.L. Developmental and reproductive toxicity of dioxins and related compounds: cross-species comparisons, *Crit. Rev. Toxicol.*, 23, 283, 1993.

15. Safe, S., Astroff, B., Harris, M., Zacharewski, T., Dickerson, R., Romkes, M., and Biegel, L. 2,3,7,8-Tetrachlorodibenzo-*p*-dioxin (TCDD) and related compounds as antiestrogens: characterization and mechanism of action, *Pharmacol. Toxicol.*, 69, 400, 1991.
16. Birnbaum, L.S. Endocrine effects of prenatal exposure to PCBs, dioxins, and other xenobiotics: implications for policy and future research, *Environ. Health Perspec.*, 102, 676, 1994.
17. ENSR Consulting and Engineering, Health risk assessment for the Kimball Thermal Oxidation unit, volume I-Technical report, ENSR Consulting and Engineering, Fort Collins, CO, 1992.
18. DeAngelis, D.L. What food web analysis can contribute to wildlife toxicology. in *Wildlife Toxicology and Population Modeling: Integrated Studies of Agroecosystems.* R.J. Kendall and T.E. Lacher, Eds., Lewis Publishers, Boca Raton, FL, 1994, p. 365.
19. U.S. EPA, *A Review of Ecological Assessment Case Studies from a Risk Assessment Perspective*, U.S. Environmental Protection Agency, Washington, DC, 1993.

chapter seven

Evaluation of the literature regarding ecological effects of hazardous waste incinerators and related facilities

Christopher E. Mackay

Contents

Introduction and overview

The concept behind the incineration of hazardous material is to convert compounds that represent a high risk to the environment into safer ones.

1-56670-250-X/99/$0.00+$.50
© 1999 by CRC Press LLC

This is done by combining the hazardous material with atmospheric oxygen at high temperature to break them down to their simplest oxidation products. As with any process, incineration is never completely efficient. The physical and chemical complexity involved in the combustion of large quantities of complex mixtures in various matrices dictates that some of the principal hazardous constituents (PHCs) will not be oxidized. Furthermore, some of the waste material will not achieve complete oxidation and will be manifest as products of incomplete combustion (PICs) that in themselves represent a risk to the environment. These products must either be captured and/or broken down by secondary treatment processes that in themselves contain inefficiencies. Ultimately, some portion of the PHCs and PICs will enter the environment.

Another consideration in the incineration of hazardous wastes is the hazard that the oxidation products may represent. This is of greatest concern when the elemental constituents of the waste stream are themselves hazardous. Examples would be compounds containing lead, mercury, selenium, etc. At the point of combustion, these elements and their oxidation products will enter either the emissions or will remain in the bottom ash. Again, secondary processes are required to contain the majority of this material. However, because of the difficulty of this separation, a small percentage will inevitably be released.

When PHCs, their PICs, or their hazardous elemental constituents are released they have the potential to adversely affect their surroundings. This includes not only humans that may be exposed but also the environment in general. If the release concentrations are high enough or an environmental receptor is sensitive enough, then an adverse ecological effect will incur. An adverse ecological effect could be considered an induced disruption in the energy transport within the system that results in a reduction in the system's mass and/or diversity.

Almost no direct research has been done specifically on the ecological effects of hazardous waste incinerators. The major body of ecological research involves other incineration processes. The concept of the hazardous waste incinerator is relatively new. In the past, hazardous waste was usually combined with other wastes and burned in municipal or industrial facilities. Furthermore, the number of specific hazardous waste incinerators is relatively small, and they tend to be purposely located distant from what would be conventionally considered sensitive ecosystems. The emission products from hazardous waste incinerators tend to differ in quantity but not in kind from other types of incinerators (U.S. EPA 1990). Because of rigorous emission standards and controls, the former is required to be cleaner than the latter (Dempsey 1993). Therefore, it is possible to predict the ecological impact of the incineration of hazardous wastes by examining that of other incinerators and accordingly account for their higher efficiency and thus lower volumes of emissions.

This chapter will review the observed impacts of incinerator products on the surrounding environment with regard to the exposure and potential impact on ecological receptors and systems. Inferences will be made, where possible, to the expected results particular to the quantities and products expected from a hazardous waste facility. This will be summated with a discussion as to the strength of the literature on this issue and what concerns remain to be addressed.

Types and sources of ecological contamination

The hazardous waste incineration process encompasses four potential sources of environmental contamination:

- Contaminants from the initial waste stream
- Contaminants released with gaseous products of the combustion
- Contaminants within the wastewater stream used for cooling and secondary contamination removal processes
- Noncombusted ash residue of the incineration process

The vast majority of the research done on hazardous waste incinerators and their impact on the surrounding environment focuses on the sources and composition of the contaminants. Reviews of typical contamination concentrations from hazardous waste incineration facilities have been reported by Van Buren et al. (1987) for liquid and solid wastes and Oppelt (1986) and Trenholm et al. (1984) for gaseous organic and inorganic emissions, respectively.

Contamination from the waste stream

The first possible source of environmental contamination is the original waste stream. This enters the environment usually by accident or negligence. The only study on the ecological impact of accidental release from hazardous waste incinerators involved incineration at sea. In 1975, the U.S. Environmental Protection Agency (EPA) started issuing test permits to allow the burning of hazardous waste generated in the United States in the Gulf of Mexico and the Pacific (U.S. EPA 1978). At this time, the European Economic Community (EEC) had performed over 350 burns in the North Sea (Sodergren et al. 1990). Although EPA had studied the probability of a "worst case" spill prior to the permitting, Ditz (1988) examined the possibility and ecological impact of noncatastrophic accidental releases. Based on experience with oil tankers, he determined the risk to be 0.6 to 6.1 accidents per 10,000 transits. Furthermore, it was concluded that there is a 5% risk that the spill would be greater than 27,000 kg. Using Mobile Bay, one of the proposed sites ports for these ships, as an example site and PCBs as the product, he demonstrated that a spill of less than 1,000 kg would pose a significant danger to marine organisms based on then-current EPA salt water protection criteria.

Environmental contamination due to exposure to the waste product as the result of accidental release, although catastrophic in nature, is limited in occurrence. The greatest concern for ecological exposure is not from the waste stream but rather through long-term exposure to the by-products of the incineration process. This would include not only the PHCs that pass through into the emissions and effluents but also the PICs and the toxic elemental constituents.

Contamination from flue emissions

Terrestrial deposition

Flue emissions are composed of two separate constituents: gaseous emissions and fly ash. Gaseous emissions are compounds that are in the vapor state at normal ambient temperatures. Fly ash are solids that are carried with the flue gases. They are usually a combination of material dislodged from the combustion bed and materials that condense as the flue gases cool. The impacts of gaseous and fly ash emission from hazardous waste incinerators have been directly examined in a limited number of studies.

Foliar exposure and contamination appear to be of greatest concern with regard to the ecological impact of incinerators. Since this is likely the primary route of introduction, it is important to quantify the contamination and examine the uptake and stability in plant masses. One such study was performed by Bache et al. (1991, 1992) at an unidentified facility. This municipal incinerator burned about 110 T of refuse a day and had no postcombustion emission control systems. Grass samples were taken to a maximum distance of 0.9 km downwind of the facility and analyzed for heavy metal and PCB contamination. Samples taken 100 m downwind had 10-times the lead, mercury, and cadmium concentrations as the corresponding upwind samples. This fell off rapidly except for lead, whose downwind concentration was still twice the upwind levels at 0.9 km. The high local contamination concentrations were likely due to metals carried on fly ash. No significant PCB contamination was found in any of the grass samples taken.

A similar study performed by Carpi et al. (1994) looked at mercury contamination in grass and moss samples associated with a municipal incinerator in New Jersey. This facility burned about 400 tons of municipal waste per day and was equipped with a spray drier and fabric filters. Although stack emission rates were not reported, the investigators found low level contamination on surrounding plant material that was double the remote controls at a distance of 5 km downwind from the facility. The highest mercury concentrations (194 to 231 ppb) were found between 2 and 3 km from the plant. Samples directly adjacent were slightly lower (190 ppb). At 5 km from the plant, the mercury contamination had fallen 63% compared to the highest proximate determinations (129 to 132 ppb). These observations were consistent with local weather patterns, and the authors suggest that this mercury contamination was the result of both wet and dry deposition processes.

A long-term study of the exposure of ecological end points to fall out from municipal incinerators was carried out by Keller et al. (1994). From 1971 to 1990, Cl⁻ and other contaminants were measured in beech tree canopies adjacent to a municipal incinerator. This incinerator, which initially had an electroscrubber, began operation in 1974. In 1987/1988 the facility was retrofitted with a dry scrubber to remove particulate matter. The plots from which the material was taken was on a hillside downwind and directly adjacent to the facility. Most of the plots were at a greater elevation than the top of the stack. Cl⁻ residues increased by a factor of 14 when the plant commenced operations. This was seen to be greatest at the elevation 50 m above the top of the stack (approximately 150 m downwind). The highest levels were measured subsequent to the addition of the dry scrubbers. In the late 1980s, the Cl⁻ concentrations fell to about six times the preoperational levels when this secondary treatment was put in place. Several metals were also analyzed for, including lead and cadmium. Early operations showed a 2.6-fold increase in lead and a 4-fold increase in cadmium in the beech foliage. Postscrubber levels fell to 160% and 233% of the preoperational controls. The authors were also able to correlate rainfall with contaminant content. Using potassium and magnesium as examples of leachable components, and chloride and zinc as nonleachable, they showed a correlation of the latter with precipitation while there was no significant difference in the former. Overall, there was no correlation between metal contamination and precipitation, suggesting that the contaminants were not being washed off the foliage.

A study specific to foliage uptake from fly ash was performed by Gutenmann et al. (1992). Examining a municipal incinerator that possessed both scrubbers and fly ash filtration systems, they found no significant difference in cadmium and lead concentrations between upwind and downwind plant samples within 1,500 m of the facility. Furthermore, these were not significantly different from the background controls.

One of the only examinations of foliar contamination from a hazardous waste incinerator reported was performed by Eduljee et al. (1986) at the Re-Chem International's Bonnybridge facility in Scotland. Soil and plant samples were examined to a range of 3.5 km and analyzed for PCB, polychlorinated dibenzodioxins (PCDD), polychlorinated dibenzofurans (PCDF), and heavy metal contamination. They concluded that there was no significant difference in the organic contamination levels within the test range and control sites. The only noteworthy differences observed were slight season variations in metal contamination within the surrounding grasses. This however could not be linked to the incinerator itself and was concluded to be the result of nonpoint source depositions.

These studies of flue emissions and contamination of plant material seem to indicate that the effect is limited to the vicinity of the facility. Not unexpectedly, it was also observed that, the more stringent the emission controls, the less local contamination was observed. This was particularly true for heavy metal contamination that was drastically curtailed when the fly ash was removed. All the positive contamination results are linked directly to fly ash. Gaseous emissions appear to have little regional effect. This would be expected since the deposition rate is so much greater for the former than the latter. Furthermore, the dilution and transport factors that affect gaseous emissions would make it extremely difficult to link any single point source to a broad exposure situation. Therefore, it appears that the elimination of fly ash through the use of technologies such as electrostatic precipitators and fiber bag filtration eliminates most of the detectable local flueborne contamination produced by incinerators.

Nonterrestrial deposition

No direct field studies have been performed on the effects of hazardous waste incineration on an aquatic system. However, there have been numerous studies examining the dynamics of contamination transfer between gaseous emissions and the aquatic environment. Organic contaminants such as PCBs and PCDD/Fs, appear to have a greater prevalence within aquatic than terrestrial systems (Nassos 1987). This is intuitively contrary based on the hydrophobic nature of these compounds. However, it has been shown that the microlayer that exists at the air/water interface of aquatic systems has the ability to concentrate nonwater soluble compounds (Hardy 1982; Sodergren 1987). This is an important consideration in the exposure of ecological end points to incinerator by-products. Connolly and Thomann (1982) have estimated that 12% of the PCB load in lake trout in the Great Lakes can be attributed to contamination of this microlayer. Sodergren et al. (1990) has examined the interactions between emission products in an aquatic microcosm. They found that hydrophobic deposition tends to accumulate in the microlayer and then diffuse into the subsurface water. From there, it tends to concentrate in the surface organisms rather than being diluted in the bulk water column. This observation is not specific to hazardous waste incinerators. However, when examining the impact of such a facility, this partition characteristic should be taken into account.

The exposure of marine environments to gaseous emissions from ocean incineration was studied by Lohse (1988) by quantifying the variation in the concentration of hexachlorobenzene and octachlorostyrene contamination in the sediments of the North Sea. These PICs are specific to the incineration of hazardous wastes associated with off-shore incineration practices. Lohse found these products in close association with the ocean combustion area. The concentrations of PICs were independent of non-PIC organochlorines such as PCBs, which strongly argues that the source of the PICs was the ocean-going incinerator ships. It also demonstrated that the burning at sea of hazardous waste was not a significant source of PCB contamination. Lohse reported that downwind drift was very limited, suggesting that aerial deposition occurred very close to the proximity of the source and that transit time in the water column is very short. The first observation was likely due to the short ship's stacks. The second mechanism has been examined by Olsen et al. (1982), who demonstrated that compounds with high K_{ow} values are removed rapidly from the surface waters by sinking benthic particles.

Contamination from bottom ash disposal

Hazardous waste incinerator ash must undergo extraction procedure (EP) toxicity testing before it can be deemed nonhazardous waste. This procedure is designed to test the stability of the material against decomposition and leaching expected to occur in the environment. However, consistency in results has been lacking and the procedure has come under considerable controversy (Josephson 1982). Leachability studies performed on municipal bottom ash indicate that the availability of bound metals is pH dependent (Lisk et al. 1988, 1989). Heavy metals such as lead and cadmium have been shown to be reasonably immobile at high pH. Ash products tend to be alkaline in nature due to the high concentrations of alkali earth metals. This may be an important consideration when considering disposal. Leachate from municipal landfills tends to be acidic as the result of microbial decomposition of noncombusted material. Therefore, the environmental bioavailability of heavy metals in bottom ash would be much more bioavailable if disposed of in a typical landfill as opposed to an ash-only dump site (Reimann 1987).

Numerous studies have been published to examine the stability of municipal incinerator bottom and fly ash after disposal in landfills (Feder and Mika 1984; Giordano et al. 1983; Karasek et al. 1987). However, studies on the ecological effects of incinerator products per se have been few. Mika et al. (1985) did examine the effects of municipal incinerator ash that was dumped in a nonlined pit adjacent to a meadow wetland. This emulated a worst-case condition. The surrounding plant community was monitored over a 4-year period. They found no significant adverse effect on either the diversity or productivity of the wetland resulting from the incinerator ash leachate. Contrarily, the ash appeared to have increased both the diversity and abundance in most of the common species within the wetland meadow. Examination of associated soil contamination indicated not only elevated concentrations of heavy metals but also increased concentrations of nutrients that leached from the incinerator ash. This latter consideration may have accounted for the increased productivity found on the test site. No analysis was performed on the metal contamination in the plants so it is not possible to determine the uptake of the more toxic heavy metals in this circumstance.

Exposure and uptake of contaminants by ecological end points

Exposures of sensitive ecological end points to products of hazardous waste incineration will likely occur by two routes. First is by direct exposure to contaminant products. Second, exposure to toxicants through the ingestion of contaminated foodstuffs. Except for accidental release of PHCs from the incineration facility or catastrophic failure in the control of the combustion process, this first route of exposure would represent a risk only to long-lived primary producers such as trees. The most likely route that may represent a significant risk to other ecological end points would be through the food chain. This is because many of the by-products of concern are highly persistent and/or possess high K_{ow} values such that they tend to biomagnify.

The topic of introduction of combustion products into the food chain has sparked some research primarily because of the threat this may represent to humans through the contamination of foodstuffs. One such study by Nicholson et al. (1995) looked at the retention of atmospherically deposited cadmium in grasses and herbage. They found that surface deposition of the metal accounted for 1 to 20% of the total metal load, with the rest incorporated within the mesophyll. The route of uptake appears species and condition specific, with stomatal uptake accounting for 20 to 60% of the total metal uptake and the rest coming from root uptake and translocation (Hovmand 1983). It should be noted that neither of these studies was done in association with an incinerator facility. It could be expected that, in the vicinity of point sources, the surface contamination may represent a larger proportion of the overall metal load in vegetation.

After contaminants have entered the food chain through primary producers, there is a question of bioavailability to higher grazing organisms. Slob et al. (1995) looked at congener-based bioavailability of PCDD/Fs found in grasses associated with a municipal incinerator. It had been shown that cattle grazed near these facilities possess higher PCDD/F concentrations in their milk (Liem et al. 1991). The grass in Slob's study contained 7 to 10 pg I-TEQ per gram dry weight while the resulting milk contained 10 to 11 pg I-TEQ per gram milk fat. Overall, they found a 7.5% of the total PCDD/F load available in the milk. The highest congener bioavailability was found for PCB-126 and PCB-169 (30 to 36%). For dioxin compounds, the highest bioavailability was found with 2,3,7,8-TCDD and 1,2,3,7,8-PCDD (10 to 15%). The bioavailability of the PCDFs was comparatively low, ranging from 0 to 5%. The exception was 2,3,4,7,8-PCDF, which was about 12% bioavailable. Therefore, it was concluded that any biomagnification was countered by the reasonably low bioavailability such that for the milk there was no real bioconcentration.

Ecological impact of incinerator products

Effects on soil microflora

The contamination of soil from incinerator products is usually considered only as a route and not an ecological end point in itself. However, recent research indicates that heavy metals and PCBs can have a detrimental effect on the soil microflora. One study on grassland soils found a negative correlation between PCB contamination rates and soil productivity and biomass (Dusek and Tesarova 1996). Soil adjacent to a municipal incinerator in the Czech Republic was found to contain 14 ng/g soil PCB contamination. This site was examined over a 3-year period and was found to have a significantly lower soil respiration rate and biomass content compared to a control site (4.4 ng/g soil PCB). It appears common that this class of compounds possesses a relatively high toxicity to fast-growing heterotrophs (Blakemore and Carey 1978).

As well as affecting the overall productivity of soil microflora, PCBs have also been shown to selectively disrupt nitrification. Nitrobacters have been shown to be very sensitive to both PCBs as well as heavy metal contamination (Dusek 1995; Baath 1989). Disruption in nitrobacter population results in an increase in soil nitrite concentrations. This has been observed in soils that have been contaminated with high concentrations of heavy metals (Liang and Tabatabai 1978). This is not likely to have a broad effect with regard to nitrite toxicity within the ecosystem. However, it may produce a profound localized effect, reducing the rates of nitrogen mineralization and thus retarding overall nutrient cycling and the carrying capacity of the affected region.

It should be noted that, in both microflora studies described above, the organic and heavy metal concentrations were orders of magnitude higher than that expected from a properly operated hazardous waste incineration facility. However, studies into the effect of low levels of contamination have not been published.

Terrestrial plants

Terrestrial plants have been shown to be the major target for the direct exposure to incinerator emissions both organic and inorganic. However, there is little evidence of adverse effect. Toxicity tests were conducted in which corn plants grown in soil containing 30% refuse bottom ash accumulated cadmium and lead and did show signs of toxicity (Giordano et al. 1983). However, the authors attributed this to the high chloride content of the soil (10 to 12%) and not to the accumulation of heavy metals. Wadge and Hutton (1986) investigated the toxicity of fly ash to barley and cabbage. Although the plants did appear to accumulate lead, cadmium, and selenium, reductions in growth rates were only seen with ash concentrations of 40% (w/w) of the soil medium. Again, the adverse effects appeared to be ionic and not due to any specific heavy metal or organic toxicity.

The only adverse ecological effect of incineration to be reported was the result of acid deposition [see reviews by Nohrstedt (1985), Stumm et al. (1987), and Woodman and Cowling (1987)]. This effect cannot be attributed solely to the incineration of hazardous materials and is more attributable to combustion processes involved in transportation and power generation.

Terrestrial animals

No adverse ecological effect has been demonstrated in higher animals as the result of any waste incineration process. As stated, exposure studies have shown accumulation of heavy metals and PCDD/F within test animals, but no adverse effect on either the individual or community level has been reported.

Toxicity testing performed on products of incineration have concentrated on the effects of fly ash. Direct short-term oral toxicity studies in rats and guinea pigs by Theelen (1992) showed no significant effect on body weight at dosages of 25 g/kg feed. An inhalation study performed by Alarie et al. (1989) demonstrated acute reduction in gas exchange and subchronic manifestations of multifocal pneumoconiosis after a 14-day exposure to 292 mg/m^3 of municipal fly ash. Response appeared to be correlated with heavy metal content in the fly ash, but only two sources were tested. Nevertheless the response demonstrated that municipal fly ash has a relatively low potency with regard to effects of gas exchange. Since fly ash is not emitted from hazardous waste incinerators, this would only be a concern with regard to ash disposal and not directly with the operating facility.

Aquatic organisms and fish

There are numerous studies that have reported toxicity in fish as the result of exposure to either heavy metals or PCDD/Fs (Carpenter 1930; Opperhuizen et al. 1986; Mehrle 1988; Mennear and Lee 1994). However, the only study that examined the toxicity of refuse product (fly ash) directly reported no toxicity in rainbow trout fry (Helder et al. 1982). However, toluene extracts were found to contain high concentrations of PCDD/Fs suggesting that the fly ash's benign nature was the result of the PCDD/Fs being tightly adsorbed to the matrix. A similar study performed by Bache and Lisk (1989) examined the effects of municipal fly ash on goldfish. Although the lead and cadmium concentrations in the ash were measured at 5,334 and 185 ppm, respectively, only 7% of the total salts were dissolved in the water over a 12-day test period. Concentrations in the goldfish were determined to be 3.75 ppm lead and 1.55 ppm cadmium. The fish manifested no signs of toxicity such as abnormal swimming behavior or accumulation of mucus in the gills. The investigators concluded that at mid-pH ranges (7.2 to 7.5), the bulk of heavy metals in fly ash are not available for absorption by teleosts.

Summary

The ecological impact of hazardous waste incinerators has been a topic that has received almost no attention. This is probably because investigators working in the field have found a better environmental model in the municipal waste incinerator. These facilities are more numerous with higher permissible outputs of heavy metals and toxic organic compounds than the rarer and more strictly controlled hazardous waste units. Therefore, in the absence of direct experimental evidence, it is necessary to draw inferences as to the potential risk of hazardous waste incineration from results attained from their municipal counterparts.

Exposure studies appear to indicate that, as point sources, refuse incinerators have a very limited effect. Even in the absence of any secondary emission controls, detectable increases in PCDD/Fs and heavy metals are limited to a few miles downwind from these facilities. The major medium for the transport of this material is via the fly ash. Gaseous emissions appear to dilute rapidly within the environment such that their contribution to environmental deposition cannot be discerned from typical background concentrations. The second major source of contamination results from the disposal of the ash products that are residual to the incineration process. Even in its raw form, this material appears to be relatively environmentally stable. Organic contaminants adsorb to the material tightly and resist leaching. Heavy metals seem only to leach at relatively low pHs.

Except for limited observations on soil microflora, there is no clear example of an adverse ecological effect that can be attributed directly to any incinerator location or processes. Controlled laboratory studies appear to suggest that most ecological end points have a higher acute tolerance to incinerator emission products than would be expected from typical operational exposures. There is still some question as to the potential for chronic, long-term effects. However, these types of adverse responses are very difficult to observe and measure because the dynamics of most ecosystems have a large potential to overcome minor environmental stresses. Furthermore, since the typical toxic products of incineration such as PCDD/Fs and heavy metals are ubiquitous at low concentrations, it is exceedingly difficult to attribute any small change in an observed ecosystem to a single point source or class of producers.

At this time it is not possible to conclude that hazardous waste incinerators have no adverse effect upon surrounding ecosystems. However, the very minor effects noted with municipal incinerators suggest that the impact of a hazardous waste incinerator would be extremely limited and difficult to perceive. The evidence appears to suggest that, given the tight regulatory controls and the required safeguards on such facilities, any impact that they may have will be extremely difficult to detect and relatively negligible in comparison to other types of anthropomorphic practices.

References

Alarie Y, Iwasaki M, Stock MF, Pearson RC, Shane BS, Lisk DJ, Effects of inhaled municipal refuse incinerator fly ash in guinea pig, *J Toxicol Environ Health*, 28, 13–25, 1989.

Baath E, Effects of heavy metals in soil on microbial processes and populations, a review, *Water Air Soil Pollut*, 47, 335–379, 1989.

Bach CA, Lisk DJ, Cadmium and lead accumulation by goldfish exposed to aqueous refuse incinerator fly ash leachate, *Bull Environ Contam Toxicol*, 43, 846–849, 1989.

Bache CA, Elfving DC, Lisk DJ, Cadmium and lead concentration in foliage near a municipal refuse incinerator, *Chemosphere*, 24, 475–481, 1992.

Bache CA, Gutenmann WH, Rutzke M, Chu G, Elfving DC, Lisk DJ, Concentrations of metals in grasses in the vicinity of a municipal refuse incinerator, *Arch Environ Contam Toxicol*, 20, 538–542, 1991.

Blakemore RP, Carey AE, Effects of polychlorinated biphenyls on growth and respiration of heterotrophic marine bacteria, *Appl Environ Microbiol*, 35, 323–328, 1978.

Capri A, Weinstein LH, Ditz DW, Bioaccumulation of mercury by Sphagnum moss near a municipal solid waste incinerator, *Air Waste*, 44, 669–672, 1994.

Carpenter KE, Further researches on the action of metallic salts on fishes, *J Exp Zool*, 56, 407–422, 1930.

Connolly JP, Thomann RV, Calculated contribution of surface monolayer PCB to contamination of Lake Michigan lake trout, *J Great Lake Res*, 8, 367–375, 1982.

Dempsey CR, A comparison of organic emissions from hazardous waste incinerators verses the 1990 toxic release inventory air releases, *J Air Waste Mgt Assoc*, 43, 1374–1379, 1993.

Ditz DW, The risk of hazardous waste spills from incineration at sea, *J Hazard Mater*, 17, 149–168, 1988.

Dusek L, Activity of nitrifying populations in grassland soil polluted by polychlorinated biphenyls (PCBs), *Plant Soil*, 176, 273–282, 1995.

Dusek L, Tesarova M, Influence of polychlorinated biphenyls on microbial biomass and its activity in grassland soil, *Biol Fertil Soils*, 22, 243–247, 1996.

Eduljee G, Badsha K, Scudamore N, Environmental monitoring for PCB and trace metals in the vicinity of a chemical waste disposal facility. II, *Chemosphere*, 15, 81–93, 1986.

Feder WA, Mika JS, Movement of fly ash-generated Pb and Cd through soil leached with acid precipitation, Abstract of Papers of the National Meeting of the American Association for the Advancement of Science, 150, 24–29, 1984.

Giordano PM, Behel AD Jr, Lawerence JE Jr, Solleau JM, Bradford BN, Mobility in soil and plant availability of metals derived from incinerated municipal refuse, *Environ Sci Technol*, 17, 193–198, 1983.

Gutenmann WH, Rutzke M, Elfving DC, Lisk DJ, Analysis of heavy metals in foliage near a modern refuse incinerator, *Chemosphere*, 24, 1905–1910, 1992.

Hagemeyer M, Kahle H, Breckle SW, Waisel Y, *Water Air Soil Pollut*, 29, 347–359, 1986.

Hardy JT, The sea surface film microlayer: Biology, chemistry and anthropogenic enrichment, *Prog Ocanogr*, 11, 307–328, 1982.

Helder T, Stutterheim E, Olie K, The toxicity and toxic potential of fly ash from municipal incinerators assessed by means of a fish early life stage test, *Chemosphere*, 11, 965–972, 1982.

Hovmand MF, Tjell JC, Mosbaek H, *Environ Pollut Ser A*, 30, 27, 1983.

Josephson J, Immobilization and leachability of hazardous wastes, *Environ Sci Technol*, 16, 219A–223A, 1982.

Karasek FW, Charbonneau GM, Reuel GJ, Tong HY, Determination of organic compounds leached from municipal incinerator fly ash by water at different pH levels, *Anal Chem*, 59, 1027–1031, 1987.

Keller TH, Matyssek R, Gunthardt-Goerg MS, Beech foliage as a bioindicator of pollution near a waste incinerator, *Environ Pollut*, 85, 185–189, 1994.

Liang CN, Tabatabai MA, Effects of trace elements of nitrification in soils, *J Environ Qual*, 7, 291–293, 1978.

Liem AKD, Hoogerbrugge R, Kootstra PR, Van der Velde EG, De Jong APJM, *Chemosphere*, 23, 1675–1684, 1991.

Lisk DJ, Environmental implications of incineration of municipal solid waste and ash disposal, *Sci Total Environ*, 74, 39–66, 1988.

Lisk DJ, Secor CL, Rutzke M, Thomas HK, Element composition of municipal refuse ashes and their aqueous extracts from 18 incinerators, *Environ Contam Toxicol*, 42, 534–539, 1989.

Lohse J, Ocean incineration of toxic wastes: A footprint in North Sea sediments, *Mar Pollut Bull*, 19, 366–371, 1988.

Mehrle PM, Buckler DR, Little EE, Smith LM, Petty JD, Toxicity and bioconcentration of 2,3,7,8-tetrachlorodibenzodioxin and 2,3,7,8-tetrachlordibenzofuran in rainbow trout, *Environ Toxicol Chem*, 7, 47–62, 1988.

Mennear JH, Lee CC, Polybrominated dibenzo-p-dioxins and dibenzofurans: literature review and health assessment, *Environ Health Perspect Supl*, 1, 265–274, 1994.

Mika JS, Frost KA, Feder WA, The impact of land-applied incinerator ash residue on a freshwater wetland plant community, *Environ Pollut Ser A*, 38, 339–360, 1985.

Nassos GP, The problem of ocean incineration: a case of modern mythology, *Mar Pollut Bull*, 18, 211–216, 1987.

Nicholson FA, Jones KC, Johnston AE, The significance of the retention of atmospherically deposited cadmium on plant surfaces to the cadmium content of herbage, *Chemosphere*, 31, 3043–3049, 1995.

Nohrstedt HO, Studies of forest floor biological activities in an area previously damaged by sulfur dioxide emissions, *Water Air Soil Pollut*, 25, 301–311, 1985.

Olsen CR, Cutshall NH, Larsen IL, Pollutant-particle associations and dynamics in coastal marine environments; a review, *Mar Chem*, 11, 501–533, 1982.

Oppelt ET, Hazardous waste destruction, *Environ Sci Technol*, 20, 312–318, 1986.

Opperhuizen A, Wagenaar WJ, van-der-Wielen FWM, van-den-Berg M, Olie K, Gabas FAPC, Uptake and elimination of PCDD/PCDF congeners by fish after aqueous exposure to a fly-ash extract from a municipal incinerator, *Chemosphere*, 15, 2049–2053, 1986.

Reimann DO, Treatment of waste water from refuse incineration plants, *Waste Mgt Res*, 27, 147–157, 1987.

Slob W, Olling M, Derks HJGM, de Jong APJM, Congener-specific bioavailability of PCDD/Fs and coplanar PCBs in cows: laboratory and field measurements, *Chemosphere*, 31, 3827–3838, 1995.

Sodergren A, Origin and composition of surface slicks in lakes of differing trophic status, *Limnol Oceanogr*, 32, 1307–1316, 1987.

Sodergren A, Larsson P, Knulst J, Bergqvist C, Transport of incinerated organochlorine compounds to air, water, microlayer and organisms, *Mar Pollut Bull*, 21, 18–24, 1990.

Stumm W, Sigg L, Schnoor JL, Aquatic chemistry of acid deposition, *Environ Sci Technol*, 21, 8–13, 1987.

Theelen RMC, Toxicity of flue ash emissions from waste incinerators, *Chemosphere*, 24, 753–761, 1992.

Trenholm A, Gorman P, Tungclaus G, *Performance evaluation of full-scale hazardous waste incinerator*, Vol. I–IV, EPA/600/2-84/181A–D, 1984.

U.S. EPA, Environmental assessment: at-sea and land-based incineration of orga-
    nochlorine wastes, TRW Inc., EPA-600/2-78- 087, 1978.

U.S. EPA, Cancer risk from outdoor exposure to air toxics, EPA-4501/ 1-90-004(a and
    b), 1990.

Van Buren D, Poe G, Castaldini C, Characterization of hazardous waste incinerator
    residuals, Acurex Corporation, EPA-600/2-87/017, 1987.

Wadge A, Hutton M, The uptake of cadmium, lead and selenium by barley and
    cabbage grown on soils amended with refuse incinerator fly ash, *Plant Soil*,
    407–412, 1986.

Woodman JN, Cowling EB, Airborne chemicals and forest health, *Environ Sci Technol*,
    21, 120–126, 1987.

chapter eight

Ecotoxicological risk assessment for hazardous waste incineration: a case study

Patrick J. Sheehan, Frank J. Dombrowski, Michael J. Ungs, and Charles R. Harman

Contents

Introduction

Most of the recent proposals for hazardous waste incinerators have suggested siting these facilities in rural areas away from major human population centers. The selection of rural locations for these proposed incinerators has shifted some of the emphasis of risk evaluations away from assessing the human health risks of projected emissions to a focus on assessing potential risks to ecological systems. This chapter presents a prospective ecotoxicological risk assessment (ERA) for a proposed hazardous waste incinerator to be located in an area of rich wildlife habitat in rural Mississippi. To enhance the instructive value of the case study, the assessment methods and assumptions are clearly described along with the assessment results.

Background

In 1990, an industrial consortium proposed the construction of a hazardous waste treatment facility including an incinerator in a rural area in Noxubee County, Mississippi. The proposed facility was designed to provide state-of-the-art waste treatment for industrial and commercial generators and to accept waste materials classified as both hazardous and nonhazardous wastes under the Resource Conservation and Recovery Act (RCRA). As part of the permitting process under RCRA, the Mississippi Department of Environmental Quality (MSDEQ) requested an ERA to evaluate the potential for ecological effects which might result from routine operation of the proposed facility.

Risk assessment objectives

The objective of the ERA was to provide estimates of risks to plant, fish, mammal, and bird populations resulting from exposures to projected emissions from the proposed treatment facility during routine operations. Emphasis was placed on evaluating risk to plant species important to local agriculture and forestry, and fish and wildlife species with populations resident in the vicinity of the proposed incinerator.

Scope of the assessment

The conceptual approach for the ERA and specific risk assessment tasks completed are identified in Figure 8.1. The ERA consisted of the 10 tasks indicated. This prospective assessment was conducted using computer models to estimate the concentrations of the chemicals of interest (COIs) in various media (i.e., air, surface soil, and surface water) and to project exposures to indicator species. The models relied on a conservative set of assumptions regarding (1) the projected operating characteristics of the facility

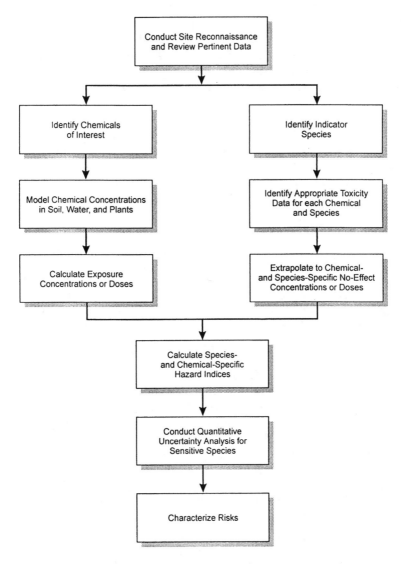

Figure 8.1 Conceptual approach and tasks for the ecotoxicological risk assessment.

(2) the deposition, fate, and transport characteristics of the COIs, and (3) the exposure of the indicators species to the COIs emitted from the facility. The concentrations of COIs estimated in air as a result of proposed facility operations were used to predict the concentration in soil and water that would result from the deposition of projected facility emissions. These predicted concentrations, in turn, were used to model potential exposures to indicator plant and animal populations. Chemical- and species-specific toxicity reference values (TRVs) were developed using established standards and published literature values and appropriate factors, where necessary, to account for possible differences in species sensitivity and laboratory and field conditions. The TRVs then were compared to the estimated media concentrations or doses to calculate chemical and species-specific hazard quotients (HQs). For the most sensitive species identified from this analysis, a quantitative evaluation of uncertainties in exposure was performed using a Monte Carlo technique. Risks were characterized based on a weight-of-evidence analysis.

Problem formulation

In the problem formulation phase of the ERA, the site of the proposed waste treatment facility and the treatment facility were described, chemicals of interest and receptor species of interest were identified, and assessment and measurement end points to be evaluated were selected.

Site characterization

The proposed site for the waste treatment facility is a 500-acre parcel located in a prairie region of east-central Mississippi (Figure 8.2). The area is rural and historically was used as pasture land for cattle. The topography of the site is flat to gently rolling hills. There are two main creeks, Joes Creek and Horse Hunters Creek, located to the west and east, respectively, of the proposed waste treatment facility property boundaries. Intermittent streams originating near the center of the site carry runoff to the nearby creeks during periods of high rainfall. There are also eight man-made ponds created for the purpose of watering livestock located on-site.

The property immediately adjacent to the proposed facility site consists primarily of agricultural and pasture land, with small percentages of land in aquaculture, commercial, and residential uses. Approximately 25% of the land surface in the region is in crop production. Soybean, corn, pasture grasses, and wheat are the primary crops. A 5-acre, commercial catfish pond is located to the south of the site. The pond is fed by groundwater from the owner's private well. About 41% of the land in the region is in managed woodlands. The individual woodlots are classified according to the dominant tree species: loblolly/shortleaf pine forest, commonly found in upland habitat such as slopes and ridges; oak/gum/cypress forest, commonly found in river bottom lands and associated with wetlands; oak/hickory forest,

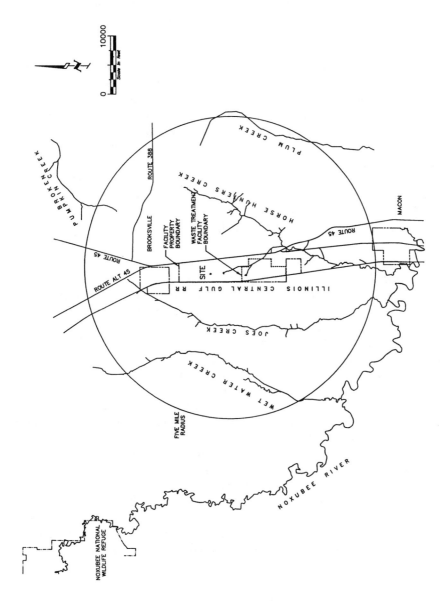

Figure 8.2 Proposed waste treatment facility site and surrounding area.

commonly found in upland habitat; and oak/pine forest, typical of subclimax upland woodland.

Over 90% of the proposed waste treatment facility site is open field dominated by grasses and other herbaceous plant species. Forested areas of the site consist of hedgerows along the borders, scattered isolated woodland pockets, and a pecan grove located in the southeast corner of the property. The site is generally divided into two types of terrestrial plant communities: (1) recovering old field (>90%) and (2) oak/pine forest (5 to 10%). Hedgerows, woodlots, and margins of surface ponds and riparian zones adjacent to stream beds also contain unique vegetation communities, which serve as foraging habitat, cover, and travel corridors for wildlife.

The soils on and immediately surrounding the site are classified by the U.S. Soil Conservation Service as the Vaiden–Okolona–Brooksville complex. A total of eleven soil types were identified on the site, consisting primarily of clays or clay/silt combinations.

The region is characterized by a moderate climate with an average yearly temperature of 63.1°F and an average monthly temperature range from 42.5°F in January to 81.2°F in July. Yearly average precipitation was 56.5 inches, including 1.3 inches of snowfall.

Description of proposed facility

The proposed waste treatment facility is designed for incineration and pozzolanic stabilization. High-temperature rotary kiln incinerator, designed to treat RCRA hazardous wastes, and pozzolanic stabilization, designed to treat the inorganic residues from incineration as well as industrial wastes for which stabilization is specified as the best demonstrated available technology, are the primary technologies to be used.

Other operations proposed at the facility include a landfill for the disposal of stabilized residues, a decontamination unit for trucks, rail cars, large waste containers and equipment, and a landfill leachate treatment unit. The facility is designed to include waste receiving areas, container storage and handling areas, a bulk sludge handling area, tank storage areas, waste staging and feeding areas, and support facilities. The proposed operational lifetime of the facility is projected to be 30 years.

The proposed treatment facility operations are predicted to result in controlled releases of vapor or particulates from 89 discrete sources including the incinerator stack. Several control systems are proposed as part of the facility design to minimize these emissions, such as enclosures and movable hoods in association with exhaust filtration systems, cyclones, fabric filters, and activated carbon adsorption systems.

Species of interest selection

The woodlots and hedgerows are considered important habitat for wildlife on and around the site. The woodlots provide cover and forage habitat for several species of larger mammals. They also provide habitat for a large

number of bird species, such as passerines and owls, and perching and
nesting areas for raptors such as red-tailed hawks. The hedgerows provide
habitat for cover, nesting and foraging and corridors for the migration of
species. The ponds and streams in the area provide habitat for aquatic organisms and piscivorous wildlife such as the black-crowned night heron.

Plant, fish, and wildlife species associated with the proposed facility

As noted, the areas in proximity to the proposed facility include cultivated
agricultural land, open fields, upland woodlots, and riparian zones. This
combination of vegetative characteristics provides excellent habitat for a
variety of game and nongame wildlife species. On-site observational surveys
documented direct and indirect utilization of the area by a number of fish
and wildlife species, including catfish, fathead minnow, and other forage
fish, as well as numerous predatory, gallinaceous, raptorial, and passerine
bird species, and several species of larger mammals.

Three species listed as threatened or endangered (red-cockadid wood-
pecker, bald eagle, and freckbelly madtom) have been identified in the
region. However, observations document no evidence of the occurrence of
these species on-site nor suitable habitat for them in the areas adjacent to
the proposed facility.

Selection of indicator species

Indicator species for the ERA were chosen as representatives of various
functional groups found in the aquatic and terrestrial ecosystems in the
vicinity of the proposed waste treatment facility site. Since the aquatic and
terrestrial habitats on and adjacent to the site support a wide variety of plants
and animals, there was a clear need to define criteria to support the selection
of species of interest for the ERA. In this case, the indicator species were
selected based on (1) role and function in the ecosystem, (2) commercial or
recreational significance, (3) availability of ecotoxicological information, and
(4) MSDEQ permit requirements to address potential effects on agriculture,
aquaculture, forestry, fish, and wildlife. The indicator species selected are
identified in Table 8.1. They include three plant species, two fish species,
three mammals, and two avian species.

The three vegetative plant species selected are the most significant agricultural row crops found in Noxubee County. The selection of these species
addresses the requirements to evaluate the potential for impacts to agriculture.
Further, as there is little in the scientific literature regarding the impacts of
chemicals on native pasture and range grasses, evaluation of the physiologically similar agricultural species (i.e., cultivated grasses) will serve to address
the potential for impacts on native flora. Additionally, the evaluation of wheat
and corn will serve to contrast plants having C3 and C4 photosynthesis.

Loblolly pine was chosen because it is one of the dominant timber species
found in the surrounding area. It is a major forest cover species and, as an
indicator, allows for the evaluation of the potential for economic impacts on

Table 8.1   Justification for Selection of Indicator Species

Indicator species	Justification
Corn (*Zea mays*)	Indicator of agricultural impacts
	Herbaceous plant species
	C4 photosynthesis
Soybean	Indicator of agricultural impacts
(*Glycine max*)	Herbaceous plant species
Wheat	Indicator of agricultural impacts
(*Triticum aestivum*)	Herbaceous plant species
	C3 photosynthesis
Loblolly pine	Indicator of forestry impacts
(*Pinus taeda*)	Woody plant species
Channel catfish	Indicator of aquaculture impacts
(*Ictalurus punctatus*)	Local game fish
	Upper trophic level species (aquatic)
Fathead minnows	Indicator of fisheries impacts
(*Pimephales promelas*)	Mid-lower trophic level species (aquatic)
White-tailed deer	Indicator of wildlife impacts
(*Odocoileus virginianus*)	Major game animal
	Browser
	Inhabits secondary successional habitats
Eastern cottontail	Indicator of wildlife impacts
(*Sylvilagus floridanus*)	Major game animal
	Grazer
	Inhabits grassland habitat
Eastern gray squirrel	Indicator of wildlife impacts
(*Sciurus carolinensis*)	Major game animal
	Inhabits forest habitats
Red-tailed hawk	Raptor
(*Buteo jamaicensis*)	Predator in terrestrial food web
Black-crowned night heron	Piscivore
(*Nycticorax nycticorax*)	Top predator in aquatic food web

forestry resources and ecological impacts on woodland habitats. The choice of this species also allows for the evaluation of the potential for effects on woody species in addition to the herbaceous agricultural species described above.

The fish species selected are considered to be dominant organisms within local aquatic systems. Channel catfish are an upper trophic level fish species, which feed upon a variety of aquatic organism. They serve as prey for a variety of mammalian and avian predators. They are found in streams and ponds throughout the area and are commercially raised at a location immediately south of the proposed waste treatment facility site. The fathead minnow is also a dominant organism in ponds in this area. The fathead minnow is utilized extensively as a forage fish by higher level predators, and it is an important intermediary between the abiotic matrices in the aquatic ecosystem and higher trophic organisms.

The three mammals were chosen as indicator species because they have unique positions within local terrestrial ecosystems. The white-tailed deer, eastern cotton tail, and eastern grey squirrel are grazers or browsers and inhabit different terrestrial habitats. All three species are considered economically and recreationally important as game species within Noxubee County. White-tail deer occupy a niche as a browser in secondary vegetative successional stage communities. The eastern gray squirrel is an arboreal species primarily inhabiting upland woodlot habitats. The eastern cottontail is an important grazing species in grassland and forested plant communities.

Two avian species were selected to represent important predators in local terrestrial and aquatic food webs. The red-tailed hawk feeds primarily on rodents and was observed foraging on the site. The black-crowned night heron feeds primarily on fish and is a top predator in the local aquatic food web.

Chemicals of interest selection

Based on an evaluation of the types of waste materials that the proposed waste treatment facility might receive, an estimated 189 regulated chemical compounds were identified that could be emitted as a result of anticipated operations. Not all of these compounds are highly toxic nor are many expected to be released in large quantities. In addition, it is impractical to evaluate exposures to all 189 chemicals. Therefore, a selection process was developed to identify the subset of chemicals which may be of greatest potential ecological significance.

Typically when selecting chemicals of interest for a baseline risk assessment, measured chemical concentrations in impacted media are compared to established regulatory criteria, toxicity values, or background concentrations.[1] This approach, of course, cannot be applied in a prospective assessment of this type where no media concentration data are available. Therefore, an alternative approach for selecting chemicals of interest was developed. The two-tiered evaluation included (1) an analysis of physiochemical and toxicological characteristics to identify the more persistent, bioaccumulative, and/or toxic chemicals and (2) verification of the availability of chemical-specific data to support fate and transport modeling and risk evaluations.

The parameters used to evaluate a chemical's tendency to persist in the environment or bioconcentrate in the tissues of aquatic or terrestrial organisms included (1) organic carbon partition coefficient (K_{oc}), (2) environmental half-life ($t_{1/2}$), (3) vapor pressure (VP), (4) octanol–water partition coefficient (K_{ow}), and (5) bioconcentration factor (BCF). As a general indicator of a chemical's toxicity to wildlife and fish, two additional parameters also were considered: (1) the ingested dose to rats lethal to 50% of a test population (LD_{50}), and (2) the water concentration lethal to 50% of a fish population (LC_{50}). The potential for ecological impacts was rated as low, medium, or high for each of the above physiochemical and toxicity parameters based on predefined ranges of values as shown in Table 8.2. The chemical-specific values for each of these parameters were then compared to these ranges to

Table 8.2 Physiochemical and Environmental Toxicity Parameters Used
in Selecting Chemicals of Interest for Ecological Risk Assessment

| Parameter | Symbol | Units | Value with respect to potential for ecological effects | | | Reference |
			Low	Medium	High	
Organic carbon partition coefficient	K_{oc}	µg/ml	<1,000	1,000–10,000	>10,000	2
Environ-mental half-life	$t_{1/2}$	days	<30	30–90	>90	2
Vapor pressure	VP	mmHg	>10^{-2}	10^{-2}–10^{-6}	<10^{-6}	2
Octanol water partition coefficient	K_{ow}	unitless	<500	500–1,000	>1,000	2
Bioconcen-tration factor	BCF	unitless	<10	10–10^4	>10^4	3
Rat 50% lethal dose	LD_{50}	mg/kg	>5,000	50–5,000	<50	4
Fish 50% lethal concen-tration	LC_{50}	mg/l	>100	1–100	<1.0	5

rank each chemical with respect to the three general categories of persistence (K_{oc}, $t_{1/2}$, and VP), bioconcentration (K_{ow} and BCF), and environmental toxicity (LD_{50} and LC_{50}). All chemicals for which an overall classification of "medium" or "high" was determined were included as COIs. Based on medium to high ranking in persistence, bioaccumulation potential and toxicity, 29 chemicals were selected as COIs for the ERA (Table 8.3).

A wide variety of data are necessary to support modeling efforts for each COI to derive a quantitative estimate of risk for indicator species. Such data include physiochemical and toxicity parameters, data required for exposure modeling, and toxicity values such as no observed adverse effect levels (NOAELs) for indicator species, or media-specific quality criteria such as ambient water quality criteria (AWQCs). It was therefore necessary when finalizing the list of COIs to consider the availability of such data for each of the chemicals under consideration. It was determined that there was sufficient information available for each of the 29 chemicals initially selected as COIs.

Table 8.3 Chemicals of Interest Selected from the Medium and High Categories Based on Physiochemical and Environmental Toxicity Parameters

Chemical	Symbol	Persistence			Bioconc. potential		Environ. toxicity		Overall classification	
		K_{oc}	$t_{1/2}$	VP	K_{ow}	BCF	LD_{50}	LC_{50}	Persistence/bioconc.	Tox.
Benzo(a)pyrene	BAP	H	H	H	H	M	M	L	H	M
Captan	CAPT	M	L	M	L	L	L	H	M	M
Chlordane	Chlo	H	H	M	H	H	M	M	H	M
2,4-D	2,4-D	L	L	L	M	L	M	M	M	M
DDE/DDT	DDE/DDT	H	H	M	H	H	M	H	H	H
Dibenzofurans	DBF	H	H	NA	H	NA	M	M	H	M
Dichlorobenzene	DiBZ	M	H	L	H	L	M	M	M	M
bis-Ethylhexylphthalate	BIPT	H	L	H	H	M	L	L	H	L
Heptachlor	HeCl	H	H	M	H	H	H	H	H	H
Hexachlorobenzene	HeBz	M	H	M	H	M	L	H	H	M
Hexachlorobutadiene	HeBu	H	H	L	H	L	M	H	H	H
Lindane	LinD	H	H	H	H	M	M	H	H	H
Methoxychlor	MeOX	H	H	M	H	M	M	H	H	H
N-Nitrosodimethylamine	NNDA	L	M	L	L	NA	M	NA	M	M
Parathion	Para	M	L	M	H	L	H	M	M	H
Polychlorinated biphenyls	PCB	H	H	M	H	H	M	M	H	M
Pentachlorophenol	PCPH	H	H	M	H	M	M	L	H	M
Tetrachlorodibenzo-p-dioxin	TCDD	H	H	M	H	M	H	L	H	M
Toxaphene	TOXA	M	H	L	H	H	M	M	H	M
Trifluralin	TRIF	H	M	M	H	H	M	H	H	H
Arsenic	As	NA	H	NA	NA	L	M/H[1]	M	M	H
Cadmium	Cd	NA	H	NA	NA	L	M	H	M	H
Chromium	Cr	NA	H	NA	NA	L	M/H[1]	M	M	H

(continues)

Table 8.3 Chemicals of Interest Selected from the Medium and High Categories
Based on Physiochemical and Environmental Toxicity Parameters (continued)

Chemical	Symbol	Persistence			Bioconc. potential		Environ. toxicity		Overall classification	
		K_{oc}	$t_{1/2}$	VP	K_{ow}	BCF	LD_{50}	LC_{50}	Persistence/ bioconc.	Tox.
Cyanide	hCN	NA	H	L	L	L	M/H[1]	H	M	H
Copper	Cu	NA	H	NA	NA	M	NA	H	H	H
Lead	Pb	NA	H	NA	NA	L	NA	M	M	M
Mercury	Hg	NA	H	M	NA	M	M/H	H	H	H
Nickel	Ni	NA	H	NA	NA	L	M	M	M	H
Selenium	Se	NA	H	NA	NA	L	L	M	M	M

Note: NA: not available, H: high, M: medium, L: low, and 1: dependent upon valence state or chemical form of element.

Assessment and measurement end points

Risk hypotheses proposed for the ERA include assumptions about the behavior of the COIs in environmental media and the potential types of adverse effects to indicator species posed by chronic exposures. The risk hypotheses state the specific question to be addressed in the assessment. These hypotheses are conservatively stated as null hypotheses of the form "projected concentrations of COIs do not exceed levels significantly toxic to biota." The assessment will evaluate whether evidence is sufficient to reject these null hypotheses and to conclude that levels of chemicals sufficient to result in adverse impacts are present. Each null hypothesis is stated generally for the groups of receptors to which it applies.

> *Risk hypothesis #1.* Projected concentrations of chemicals of interest in surface soil in the area do not exceed levels significantly toxic to plants.
> *Risk hypothesis #2.* Projected concentrations of chemicals of interest in surface water in ponds in the area do not exceed levels significantly toxic to fish.
> *Risk hypothesis #3.* Concentrations of chemicals of interest in surface soil, surface water, and/or tissues of forage/prey species in the area are not above levels which pose a significant exposure to wildlife populations.

Assessment end points and associated measurement end points provide information to support or refute the risk hypotheses generated for the ERA. Assessment end points are explicit statements of the characteristics of the ecological system that are to be protected.[6]

Chemicals of interest may be directly toxic to organisms or may be less toxic but have a high potential for bioaccumulation in the food web. Assessment end points, therefore, must address not only the potential for direct adverse effects to lower trophic levels, but also the potential for indirect effects that might occur as a result of bioaccumulation through the food web.

A conceptual model of relevant contaminant migration and exposure pathways that includes potential direct and indirect food web exposures is presented in Figure 8.3. Measurement end points represent quantifiable ecological characteristics that can be measured, interpreted, and related to the valued ecological component(s) chosen as the assessment end point.[6] The following assessment measurement end points were adopted to interpret the questions raised by the stated risk hypotheses.

> *Assessment end point #1.* Adverse effects on the plant community resulting from exposure to COIs in soil.
> - Comparison of the projected concentrations of COIs in soil with toxicity data for plants available from the published literature.
> *Assessment end point #2.* Adverse effects on the fish community resulting from exposure to COIs in surface water.
> - Comparison of the projected concentrations of COIs in surface water with the most recent EPA national ambient water quality criteria (NAWQC).

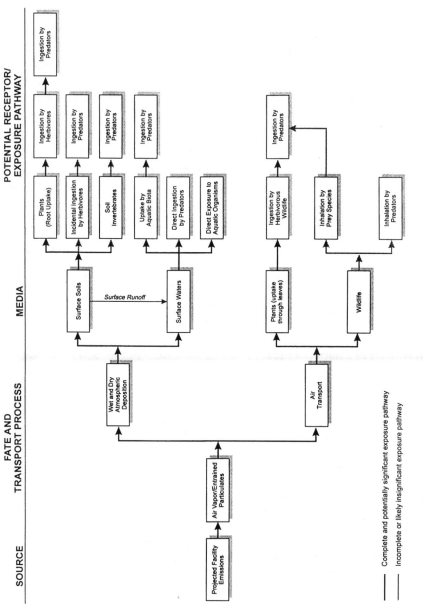

Figure 8.3   A conceptual model for the ecotoxicological risk assessment.

- Comparison of the projected concentration of COIs in surface water with concentration-response data for fish available from the published literature.

Assessment end point #3. Adverse effects on bird and mammal populations resulting from exposures to COIs in surface soil, surface water, and/or forage/prey.

- Comparison of predicted average daily doses of COIs for avian and mammalian indicator species to toxicity reference values for these species.

Exposure assessment

The potential exposures of ecological receptors to COIs was characterized through the application of a three-step modeling process, consisting of air dispersion and deposition modeling, chemical fate and transport modeling, and exposure point concentration/dose modeling, for the indicator species.

Air dispersion and deposition modeling

The industrial source complex-short term (ISC-ST) air dispersion model[7] was used to characterize the movement and behavior of potential atmospheric releases from the proposed facility and to generate input data necessary for subsequent modeling activities. The air dispersion modeling procedure considered projected release rates of chemical emissions during routine operations over flat terrain, which is consistent with the topographic conditions that exist at the proposed facility site and surrounding area.

The modeling inputs were based on engineering and design specifications for the proposed waste treatment facility and local meteorological data. Specific input parameters included emission rates; location, height, exit temperature, and velocity and diameter of stack releases; source area parameters; average annual wind speed and predominant wind direction; ambient air temperature; flagpole receptor height; and building downwash parameters. Meteorological data used for the modeling were from historical measurements for the Meridian, Mississippi/Centerville, Alabama, area.

One model output was created for each year of meteorological data from 1985 through 1989, and the results from the five runs were averaged prior to subsequent calculations. This approach is less conservative than using the "worst" years data to represent wind speed and direction. However, this approach provided a more accurate depiction of weather conditions over the projected operational life of the facility and is therefore more consistent with the objective of evaluating the potential for long-term ecological impacts. The theoretical exposure point concentrations in air for COIs were read from the ISC-ST model output at 500-meter grid intervals at discrete locations within a five-mile radius of the proposed waste treatment facility.

Figure 8.4 illustrates the receptor grid locations, zones of interest, and representative surface water receptor points. The locations of interest that

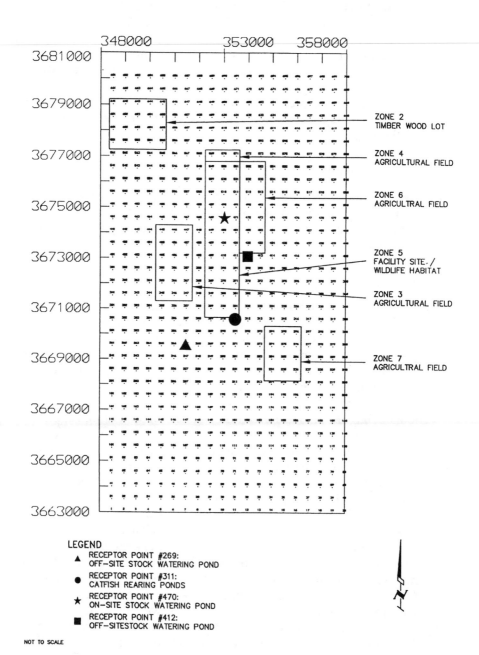

Figure 8.4 Areas of interest for the ecotoxicological risk assessment.

were selected for analysis encompass representative habitat types for all indicator species and are representative of zones of elevated deposition rates. There locations were confirmed by aerial photography and on-site reconnaissance. Zone 1 includes a timber woodlot in an area utilized for forestry operations. Zones 2, 3, 4, and 6 include areas identified as being active agricultural fields. Zone 5 includes all receptor points found within the proposed facility boundaries. Receptor points #269, #311, #470, and #412 are nearest to surface water bodies located close to the proposed facility. Receptor point #311 includes a catfish rearing pond located immediately south of the facility boundary. Receptor point #269 is an off-site stock watering pond; receptor point #470 is a stock watering pond on the proposed facility property, and #412 is a surface water pond located immediately east of the proposed facility.

Estimation of media concentrations

The second step in the modeling process involved the use of the assessment of chemical exposure (ACE) model originally developed by the State of California for use in compliance with the Air Toxics Hot Spots Information and Assessment Act of 1987[8] to estimate chemical concentrations in surface soils and water. The model is an integrated, multistage, Fortran-based program that provides estimates of exposure point concentrations, multimedia doses, and probability of adverse human health effects at specified receptor points. The procedures and algorithms included in the guidance and associated modeling programs are consistent with those recommended in EPA guidelines and other standard risk assessment technical specifications and requirements.[8]

The output from the ISC-ST model provides air concentrations attributable to each source for each COI, assuming a generic or unit emission rate of 1.0 g/sec. ACE utilizes the ISC-ST output to predict the total air concentration of each chemical at each receptor point by multiplying the source-specific emission rates for the facility and summing the results. Chemical concentrations predicted by ACE for other media of interest are generally directly proportional to ambient air concentrations at each receptor point.

The intended goal of ACE in its original form is to assess potential human health risks associated with multipathway exposures to facility emissions. Because the focus of this assessment is to evaluate risks to ecological receptors, some slight modifications to the ACE model were necessary. In the process of calculating the multipathway transport of chemicals, ACE inherently calculates concentrations in other media (e.g., soil and vegetation). Therefore, no changes were necessary to the algorithms to utilize these intermediate calculations.

To reduce the possibility of interfering with the underlying ACE code logic, an additional subroutine was created and attached to the code that performs all of the media concentration and dose calculations. This subroutine reads in all necessary site- and chemical-specific parameters and then simply accesses unaltered ACE algorithms to predict media concentrations and create a file of all results of predicted media concentrations.

An additional modification to the ACE model involved changing the algorithm used for estimating chemical concentrations in surface water. The original algorithm based the water calculation on the relative sum of two concentrations, one from runoff and the other from direct dry deposition onto the water surfaces. This approach is incorrect for two reasons: (1) chemical concentrations cannot be added arithmetically when the water flow rates from different sources (i.e., direct rainfall and runoff) are different, and (2) it assumed that the total mass of estimated facility emissions are deposited on the watershed. It is more accurate to add the mass rate of chemical contribution from direct deposition and runoff and divide this by the total volumetric water flow rate. In addition, only that fraction of the source which falls directly onto the watershed via dry deposition should contribute to runoff transport. The following algorithm was used to modify ACE to compute chemical concentrations in surface water. This equation takes the total mass rate from atmospheric dry deposition over the watershed and the water body and divides by the volumetric rate of water flow from the runoff and direct rainfall sources.

$$C_W = \frac{\left[\left(D_{cp} * W_{sia} * W_c * R_{of} * CF_1 \right) + \left(D_{cp} * S_a * CF_1 \right) \right]}{\left[\left(R_{of} * R_f * W_{sia} * R_{oc} * CF_2 \right) + \left(W_v * V_c \right) + \left(R_f * S_a * CF_2 \right) - \left(E_{vap} * S_a * CF_2 \right) \right]} \tag{1}$$

where C_w = concentration in surface water (µg/kg); D_{cp} = mass of chemical deposited onto ground or water surface (µg/m^2-day); W_{sia} = area of watershed contributing runoff to the surface water body (m^2); W_c = wash coefficient, the fraction of deposited chemical that is incorporated into the runoff water (unitless); CF_1 = conversion factor for days/year (365); S_a = surface area of the water body (m^2); R_{of} = fraction of runoff water assumed to enter surface water body (unitless); R_f = annual precipitation rate (m/year); R_{oc} = fraction of precipitation assumed available for surface water runoff (unitless); CF_2 = conversion factor for water volume (1000 kg/m^3), assuming rain is essentially water with an approximate density = 1 kg/L; W_v = water volume in surface water body (kg); V_c = annual water turnover rate in surface water body (volumes/year) from non-impacted sources (e.g., groundwater); and E_{vap} = annual evaporation rate (m/year).

ACE model input parameters

The mass of chemical deposited onto ground (or water) surface is a function of the ambient air concentration estimated by ACE and the deposition velocity assumed for each chemical or source. Deposition velocities were developed separately for the incinerator stack and for all other sources. Based on engineering design parameters, it was assumed that the most likely particulate size for incinerator stack emissions will be 0.3 µm. For all other sources a particular size of 1.0 µm was assumed. Based on particle and dry deposition research conducted by Sehmel,[9] an estimate for deposition velocity (V_d) was developed using highly conservative assumptions regarding roughness height and friction velocity. By this method, a V_d of 0.003 m/sec from the incinerator stack and a V_d of 0.01 m/sec from all other sources were estimated for chemicals emitted. Several chemical-specific parameters influence the fate and transport of deposited chemicals. The following chemical-specific parameters were used as input for the ACE model: (1) K_{ow}, (2) K_{oc}, (3) half-life in soil $t_{1/2}$ (days), (4) plant root uptake factors (mg/kg per g/m^2), and (5) plant leaf deposition factor (mg/kg per g/m^2). Site-specific parameters also were used in the ACE model to account for unique features of the study area and their influence on the fate and transport of deposited chemicals. For the important local surface water features, these included the watershed area (m^2) and pond surface area (m^2), pond volume (kg), and pond turnover rate site (volumes/year). Other site-specific parameters were identified in Equation 1. These include the wash coefficient, watershed run-off coefficient, and annual evaporation rate.

Post-processing of ACE model output

Because the ACE model will not accommodate source-specific inputs regarding particulate size, it was necessary to use two ACE runs to develop estimates of media concentrations, one based on projected emissions from the incinerator stack source, and one based on emissions from all other sources. The results of these two runs were then summed to yield estimates of total media concentrations associated with potential chemical emissions from the proposed facility.

The media concentrations from ACE were performed for each of the 29 COIs at each of the 700 individual receptor points on the grid (Figure 8.4) and saved in a single large data file. A post-processing program was written specifically to manage the large amount of information contained in this file. Through the use of this program, it is possible to generate basic summary statistics (e.g., minimum, maximum, mean, median, and standard deviation) for chemical concentrations for receptor points located within a specified zone of interest (e.g., Zone #1, Loblolly Pine Timber Woodlot).

The final aspect of post-processing of ACE input involved the consideration of chemical partitioning in aqueous systems. It is known that many metals and organic chemicals having high K_{oc} and K_{ow} values tend to sorb tightly to soil particulates, sediments, and other colloidal matter under normal field conditions. These properties are common to most chemicals included as COIs for this analysis. This partitioning phenomenon makes such chemicals in the particulate phase generally unavailable to biota in the water column. Indeed, aquatic bioconcentration factors reported in the literature are based on the dissolved concentrations, since that is the portion of chemical that is bioavailable via exchange surfaces in aquatic biota. As such, the implicit assumption in the ACE model that the entire mass of chemical deposited or transported to the surface water body can contribute to the water concentration is unrealistically conservative.

To obtain a realistic yet still conservative estimate of chemical concentrations in water, partitioning factors were developed through review of the literature for all indicator chemicals, per the methods described below. These factors are then applied to the ACE-generated water concentrations to yield final estimates of water concentrations for the assessment of potential exposures and effects to indicator fish, mammals, and birds.

The partitioning of organic chemicals from soil or sediment to water is related to chemical-specific properties (e.g. solubility), as well as soil/sediment-specific properties (e.g., particle size, pH, and organic carbon content).[10] It has been shown that a reasonable estimate of chemical sorption behavior to aquatic sediments and/or suspended particulates can be made if the organic carbon content of the materials is known.[11] The use of organic carbon to predict soil adsorption has become widely accepted.[12] Thus, the percent of chemical which would be present as dissolved compound in the water column can be estimated using the following equation:

$$K_p = K_{oc} * f_{oc}$$

where K_p = sediment/water partition coefficient;
$\quad\quad K_{oc}$ = organic carbon partition coefficient in l/kg organic carbon; and
$\quad\quad f_{oc}$ = fraction of organic carbon in sediments (dimensionless).

The soil/water partition coefficient can further be defined as

$$K_p = \frac{C_s}{C_w}$$

where C_s = chemical concentration in sediment and
$\quad\quad C_w$ = dissolved chemical concentration in water.

Given the relationship of K_{oc} to partitioning in water, the percent of an organic compound that would partition to the water was determined for ponds on and near the proposed waste treatment facility. The f_{oc} for site sediments was conservatively assumed to be 1%. This value represents the lower range of values reported for soils in the Brooksville series.[13] Thus, using the above equation for a compound with a K_{oc} of 10,000, a K_p of 100 is calculated indicating that the C_s would be 100 times that of C_w. Stated another way, for a chemical compound with a K_{oc} of 10,000 or greater, 99% of the mass of a compound deposited into a surface water body at the proposed waste treatment facility would be expected to partition into sediments, and 1% would be expected to remain dissolved in the water column. On this basis, the following criteria were developed for the organic compounds to approximate the percentage of the compound which would be expected to partition to sediments:

- $K_{oc} > 10,000 = 99\%$ partitioning to sediments, 1% remaining in water
- $K_{oc} > 1,000$ and $< 10,000 = 90\%$ partitioning to sediments, 10% remaining inwater
- $K_{oc} < 1,000 = 0\%$ partitioning to sediments, 100% remaining in water

For most of the inorganics, several literature references were located which documented the percent partitioning into the water column or percent partitioning into the sediment.[14-18] The majority of the inorganics that were evaluated were found to be tightly bound to the sediment with less than 0.1% partitioning into the water column.[14-19] For the purposes of this assessment, a partitioning factor of 1% was conservatively assumed for these chemicals. Hydrogen cyanide, arsenic, and selenium do not follow the same partitioning trends as the other inorganic chemicals. For example, an EPA[20] technical guidance document cited a partitioning factor of 99.9% in water for hydrogen cyanide. For arsenic and selenium, partitioning properties can vary greatly according to basic water chemistry parameters (e.g., pH, Eh, dissolved oxygen content, f_{oc} of sediments, hardness, etc.).[21,22] To be conservative, an assumed partitioning factor of 100% was adopted. However, it is considered unlikely that 100% of either arsenic or selenium would partition into the water column under field conditions. It should be noted that an additional factor of 0.3 was applied to the water concentration predicted by ACE for mercury. This is based on research conducted by Schintu et al.[23] and the National Research Council of Canada,[24] which suggests that approximately 30% of total mercury measured in water in eutrophic lakes and rivers is in the organic, methylated form. This value represents the highest reported percent methylmercury measured in aquatic systems and is therefore, for this evaluation, considered a conservative estimate of the methylated fraction of mercury expected to be formed in aquatic systems.

Results of ACE modeling

As noted above, the ACE model generates results in the form of predicted media concentrations at discrete receptor locations throughout a 9×16 km grid pattern surrounding the proposed waste treatment facility (Figure 8.4). These data are then used to provide estimates of results for individual receptor locations and aggregate average results for the various zones of interest. In addition, these data can be used to generate a graphic representation of likely chemical dispersion in the form of chemical isopleths in various media. Figure 8.5 presents isopleths of mercury concentrations in soil. This figure clearly shows the areas where the highest chemical concentrations are likely to occur as well as the obvious decrease in chemical concentrations in soil that occur as distance from the projected source areas increases.

Table 8.4 presents the estimated annual average concentrations of COIs in ambient air within each specified zone of interest and at the four receptor point locations representative of surface water habitats. Table 8.5 presents estimated annual average concentrations of these chemicals in soil or surface water at each representative on-site and off-site receptor point location. As noted above, the ACE output presents water concentrations in a mass/mass basis (i.e., $\mu g/kg$). Under the assumption the density of water is 1 kg/l, these values have been converted directly to mass/volume units (i.e., $\mu g/l$). Tables 8.6A and B present estimated chemical concentrations in plant tissues within the woodlot, agricultural field, and wildlife habitat zones of interest.

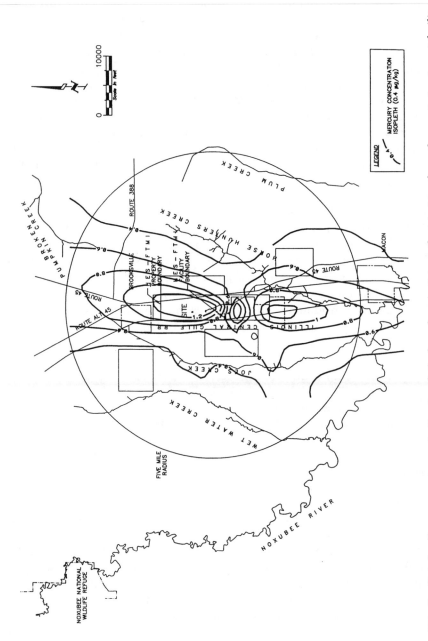

Figure 8.5 Predicted soil mercury concentration isopleths based on anticipated emissions from the waste treatment facility.

Estimated chemical doses to wildlife species

For the three mammalian and two avian indicator species, it was necessary to derive location-specific estimates of average daily doses of chemicals to compare to estimated acceptable doses or "no-effect" criteria in evaluating the potential for adverse effects. The media concentrations of COIs estimated by the ACE model served as the basis for development of estimated average daily doses. The modeled concentrations in surface water or plant tissues were used as the basis for estimation of chemical intakes via direct ingestion of impacted plant tissue and/or surface water, or of secondary ingestion of prey residing in impacted surface water or utilizing impacted plant tissue as a food source. The basis algorithm used in the calculation of chemical doses for indicator species can be expressed as follows:

$$ADD = \frac{\left[\left(C_f * IR_f\right) + \left(C_w * IR_w\right)\right] * CF}{BW}$$

where ADD = average daily dose of chemical (mg/kg-day); C_f = chemical concentration in food source (µg/kg); IR_f = food ingestion rate (kg/d); C_w = chemical concentration in water (µg/kg); IR_w = water ingestion rate (kg/d), assuming water density of 1 kg/l; CF = conversion factor (0.001 mg/µg); and BW = animal body weight (kg).

Chemical concentration in food source (C_f)

For herbivorous wildlife indicator species (i.e., white-tail deer, cottontail rabbit, and grey squirrel) this term refers to the average plant tissue concentration predicted by the ACE model for zone 5 (proposed waste treatment facility site). For each chemical, the highest average concentration predicted for each category of plant (i.e., leafy, vine, and root) was conservatively selected as the value for the analysis (Table 8.6B).

To evaluate exposures to the red-tailed hawk, the COI concentrations in prey (rodents) were estimated by multiplying the projected plant tissue concentrations (Table 8.6B) by a chemical-specific bioaccumulation factor. Based on reported literature values, a bioaccumulation factor of 1 was assumed for inorganic COIs, with the exception of mercury for which a factor of 50 was assumed based on uptake of methylmercury. For all persistent organic compounds, a bioaccumulation factor of 25 was assumed.

To evaluated exposures to the black-crowned night heron, the COI concentrations in prey (fish) were estimated by multiplying the estimated water concentration in the identified ponds (Table 8.5) by a chemical-specific bioconcentration factor. The bioconcentration factors for lead, mercury (methylmercury), and selenium were assumed to be 49, 36,000, and 16, respectively. The bioconcentration factors for DDE, PCBs, and TCDD were assumed to be 51,000, 100,000, and 5,000, respectively.

Table 8.4 Estimated Annual Average Chemical Concentrations in Ambient Air in each Zone or Receptor Location of Interest

	Average chemical concentrations at receptor locations ($\mu g/m^3$)									
Chemical	Zone 1 timber woodlot	Zone 2 agri. field	Zone 3 agri. field	Zone 4 agri. field	Zone 5 proposed facility site	Zone 6 agri.field	Rec. loc. #269 off-site stock pond	Rec. loc. #311 catfish pond	Rec. loc. #412 off-site pond	Rec. loc. #470 on-site stock pond
2,3,7,8-TCDD	8.8E-12	3.3E-11	4.0E-11	5.6E-11	2.4E-10	1.1E-10	6.0E-11	1.5E-10	1.5E-10	1.7E-10
2,4-D	1.1E-05	2.2E-05	2.0E-05	4.3E-05	3.5E-05	2.8E-05	4.7E-05	3.0E-05	2.0E-05	5.7E-05
Arsenic	9.9E-07	3.9E-06	3.2E-06	6.7E-06	2.1E-05	7.7E-06	6.6E-06	9.4E-06	6.1E-06	2.4E-05
Benzo(a)pyrene	1.1E-07	2.3E-07	2.3E-07	4.6E-07	5.0E-07	3.5E-07	5.0E-07	4.0E-07	3.0E-07	6.6E-07
BEHP	9.5E-07	2.2E-06	2.1E-06	4.2E-06	6.3E-06	3.6E-06	4.5E-06	4.2E-06	3.1E-06	7.4E-06
Cadmium	1.1E-06	3.7E-06	3.1E-06	6.5E-06	1.7E-05	6.8E-06	6.5E-06	8.3E-06	5.3E-06	2.0E-05
Captan	1.1E-07	2.3E-07	2.3E-07	4.6E-07	5.0E-07	3.5E-07	5.0E-07	4.0E-07	3.0E-07	6.6E-07
Chlordane	1.1E-07	2.3E-07	2.3E-07	4.6E-07	5.0E-07	3.5E-07	5.0E-07	4.0E-07	3.0E-07	6.6E-07
Chromium	5.1E-05	1.2E-04	1.1E-04	2.2E-04	2.8E-04	1.7E-04	2.4E-04	1.9E-04	1.2E-04	3.9E-04
Copper	2.8E-08	1.6E-07	1.1E-07	2.4E-07	9.1E-07	3.0E-07	2.3E-07	3.8E-07	2.4E-07	1.0E-06
Cyanide	2.9E-06	1.6E-05	1.2E-05	2.6E-05	1.1E-04	3.5E-05	2.3E-05	4.2E-05	2.8E-05	1.1E-04
DDE	1.1E-07	2.5E-07	2.3E-07	4.8E-07	5.7E-07	3.7E-07	5.1E-07	4.3E-07	3.2E-07	7.3E-07
Dibenzofuran	6.9E-12	2.9E-11	3.7E-11	4.8E-11	2.4E-10	1.1E-10	5.2E-11	1.4E-10	1.4E-10	1.6E-10
Dichlorobenzene	7.8E-06	1.7E-05	1.6E-05	3.2E-05	3.6E-05	2.5E-05	3.5E-05	2.7E-05	2.1E-05	4.7E-05
Heptachlor	1.1E-07	2.3E-07	2.3E-07	4.6E-07	5.0E-07	3.5E-07	5.0E-07	4.0E-07	3.0E-07	6.6E-07
Hexachlorobenzene	3.9E-06	8.0E-06	7.4E-06	1.6E-05	1.3E-05	1.0E-05	1.7E-05	1.1E-05	7.4E-06	2.1E-05
Hexachlorobutadiene	2.9E-06	5.9E-06	5.5E-06	1.2E-05	9.7E-06	7.6E-06	1.3E-05	8.3E-06	5.5E-06	1.5E-05
Lead	1.2E-04	2.9E-04	2.6E-04	5.5E-04	8.8E-04	4.5E-04	5.8E-04	5.2E-04	3.4E-04	1.1E-03
Lindane	1.1E-07	2.4E-07	2.3E-07	4.6E-07	5.0E-07	3.5E-07	5.0E-07	4.0E-07	3.0E-07	6.7E-07

Mercury	4.7E-05	9.6E-05	8.9E-05	1.9E-04	1.6E-04	1.2E-04	2.1E-04	1.3E-04	8.8E-05	2.5E-04
Methoxychlor	1.1E-07	2.3E-07	2.3E-07	4.6E-07	5.0E-07	3.5E-07	5.0E-07	4.0E-07	3.0E-07	6.6E-07
N–Nitrosodimethylamine	2.5E-07	1.0E-06	8.2E-07	1.7E-06	5.3E-06	2.0E-06	1.7E-06	2.4E-06	1.6E-06	6.1E-06
Nickel	9.0E-06	2.5E-05	2.2E-05	4.6E-05	9.2E-05	4.1E-05	4.8E-05	4.9E-05	3.2E-05	1.1E-04
Parathion	1.1E-07	2.3E-07	2.3E-07	4.6E-07	5.0E-07	3.5E-07	5.0E-07	4.0E-07	3.0E-07	6.6E-07
Pentachlorophenol	6.0E-08	2.7E-07	3.0E-07	4.4E-07	2.1E-06	8.7E-07	4.6E-07	1.1E-06	1.0E-06	1.6E-06
PCBs	8.4E-10	3.5E-09	4.5E-09	5.8E-09	2.9E-08	1.3E-08	6.3E-09	1.7E-08	1.7E-08	1.9E-08
Selenium	5.8E-08	1.4E-07	1.2E-07	2.7E-07	3.9E-07	2.1E-07	2.8E-07	2.4E-07	1.5E-07	5.2E-07
Toxaphene	1.1E-07	2.3E-07	2.3E-07	4.6E-07	5.0E-07	3.5E-07	5.0E-07	4.0E-07	3.0E-07	6.6E-07
Trifluralin	1.1E-07	2.3E-07	2.3E-07	4.6E-07	5.0E-07	3.5E-07	5.0E-07	4.0E-07	3.0E-07	6.6E-07

Table 8.5 Estimated Annual Average Chemical Concentrations in Soils in Each Zone and Surface Water at Pond Locations of Interest

							Average chemical concentrations at receptor locations (µg/kg)			
Chemical	Zone 1 timber woodlot	Zone 2 agri. field	Zone 3 agri. field	Zone 4 agri. field	Zone 5 proposed facility site	Zone 6 agri. field	Rec.loc. #269 off-site stock pond water	Rec.loc. #311 catfish pond water	Rec.loc. #412 off-site pond water	Rec.loc. #470 on-site stock pond water
2,3,7,8-TCDD	7.3E-08	3.4E-07	4.4E-07	5.5E-07	3.0E-06	1.3E-06	1.9E-10	4.2E-10	2.6E-10	3.0E-10
2,4-D	3.7E-05	7.8E-05	7.3E-05	1.5E-04	1.4E-04	1.1E-04	5.2E-03	2.7E-03	1.1E-03	3.4E-03
Arsenic	1.1E-02	6.3E-02	4.9E-02	1.0E-01	4.2E-01	1.4E-01	1.9E-03	2.5E-03	1.0E-03	4.4E-03
Benzo(a)pyrene	1.6E-05	5.4E-05	7.0E-05	9.4E-05	3.8E-04	1.9E-04	6.2E-07	5.5E-07	2.9E-07	4.9E-07
BEHP	1.7E-05	1.7E-05	7.6E-05	1.3E-04	5.5E-04	2.2E-04	6.8E-06	7.4E-06	3.7E-06	8.2E-06
Cadmium	8.9E-03	4.9E-02	3.8E-02	8.0E-02	3.2E-01	1.1E-01	1.6E-05	2.1E-05	8.0E-06	3.4E-05
Captan	4.5E-07	1.5E-06	1.9E-06	2.5E-06	1.0E-05	5.2E-06	6.2E-06	5.5E-06	2.9E-06	4.9E-06
Chlordane	3.7E-05	1.2E-04	1.6E-04	2.1E-04	8.6E-04	4.3E-04	6.2E-07	5.5E-07	2.9E-07	4.9E-07
Chromium	9.7E-02	4.4E-01	3.6E-01	7.4E-01	2.6E+00	9.0E-01	3.4E-04	1.0E-02	1.1E-04	4.0E-04
Copper	5.8E-04	3.3E-03	2.3E-03	5.1E-03	1.9E-02	6.3E-03	8.5E-07	1.2E-06	4.4E-07	2.0E-06
Cyanide	2.7E-03	1.5E-02	1.2E-02	2.5E-02	1.1E-01	3.5E-02	8.6E-03	1.3E-02	5.2E-03	2.2E-02
DDE	1.2E-04	4.5E-04	5.1E-04	7.7E-04	3.2E-03	1.4E-03	6.8E-07	6.4E-07	3.2E-07	6.4E-07
Dibenzofuran	7.2E-08	3.4E-07	4.4E-07	5.5E-07	3.0E-06	1.3E-06	1.8E-10	4.1E-10	2.6E-10	2.9E-10
Dichlorobenzene	4.9E-04	1.7E-03	2.0E-03	2.9E-03	1.1E-02	5.5E-03	4.4E-03	3.8E-03	2.0E-03	3.7E-03
Heptachlor	1.0E-05	3.4E-05	4.4E-05	5.9E-05	2.4E-04	1.2E-04	6.2E-07	5.5E-07	2.9E-07	4.9E-07
Hexachlorobenzene	1.1E-03	2.2E-03	2.1E-03	4.4E-03	4.6E-03	3.3E-03	1.9E-04	1.0E-04	4.2E-05	1.2E-04
Hexachlorobutadiene	9.1E-05	2.0E-04	1.9E-04	3.9E-04	5.0E-04	3.2E-04	1.4E-05	7.8E-06	3.2E-06	9.4E-06

Lead	3.6E–01	1.8E+00	1.4E+00	3.0E+00	1.2E+01	3.8E+00	9.6E–04	9.2E–04	3.7E–04	1.4E–03
Lindane	8.7E–06	2.9E–05	3.7E–05	5.1E–05	2.0E–04	1.0E–04	6.3E–05	5.5E–05	2.9E–05	5.1E–05
Mercury	3.0E–02	6.1E–02	5.6E–02	1.2E–01	1.0E–01	7.9E–02	6.9E–06	3.6E–05	1.5E–05	4.5E–05
Methoxychlor	1.3E–05	4.2E–05	5.4E–05	7.3E–05	2.9E–04	1.5E–04	6.2E–07	5.5E–07	2.9E–07	4.9E–07
N-Nitrosodimethylamine	4.3E–05	2.4E–05	1.9E–04	3.9E–04	1.5E–03	5.2E–04	4.9E–04	6.7E–04	2.6E–04	1.1E–03
Nickel	4.2E–02	2.2E–01	1.8E–01	3.6E–01	1.4E+00	4.8E–01	9.2E–05	1.0E–04	4.0E–05	1.7E–04
Parathion	3.7E–07	1.2E–06	1.6E–06	2.1E–06	8.4E–06	4.3E–06	6.2E–06	5.5E–06	2.9E–06	4.9E–06
Pentachlorophenol	4.8E–05	2.4E–04	2.7E–04	3.9E–04	1.9E–03	8.1E–04	1.6E–04	3.2E–04	1.9E–04	2.9E–04
PCBs	8.0E–06	3.7E–05	5.0E–05	6.1E–05	3.3E–04	1.5E–04	2.2E–08	5.1E–08	3.2E–08	3.5E–08
Selenium	1.6E–04	8.0E–04	6.1E–04	1.3E–03	4.7E–03	1.6E–03	4.4E–05	4.0E–05	1.6E–05	6.3E–05
Toxaphene	9.0E–05	2.9E–04	3.8E–04	5.1E–04	2.1E–03	1.0E–03	6.2E–06	5.5E–06	2.9E–06	4.9E–06
Trifluralin	1.8E–06	6.0E–06	7.7E–06	1.0E–05	4.2E–05	2.1E–05	6.2E–07	5.5E–07	2.9E–07	4.9E–07

Table 8.6A   Estimated Annual Average Chemical Concentrations in Plant Tissue within each Zone of Interest

| | Average chemical concentration at receptor locations (μg/kg) | | | | | | | | |
| | Zone 1 (timber wood lot) | | | Zone 2 (agricultural fields) | | | Zone 3 (agricultural fields) | | |
Chemical	Root	Leafy	Vine	Root	Leafy	Vine	Root	Leafy	Vine
2,3,7,8-TCDD	9.9E-09	1.7E-06	1.3E-08	4.6E-08	7.7E-08	6.2E-08	6.0E-08	1.0E-07	8.0E-08
2,4-D	1.9E-04	5.7E-04	3.8E-04	4.0E-04	1.2E-03	7.9E-04	3.8E-04	1.1E-03	7.4E-04
Arsenic	1.0E-05	1.0E-05	1.0E-05	5.7E-05	5.7E-05	5.7E-05	4.5E-05	4.5E-05	4.4E-05
Benzo(a)pyrene	1.3E-05	2.1E-05	1.7E-05	4.3E-05	6.8E-05	5.6E-05	5.6E-05	8.8E-05	7.2E-05
BEHP	4.5E-06	1.3E-04	6.9E-05	2.0E-05	5.9E-04	3.1E-04	2.0E-05	6.1E-04	3.1E-04
Cadmium	1.8E-04	1.9E-04	1.8E-04	9.8E-04	1.0E-03	1.0E-03	7.6E-04	8.1E-04	7.9E-04
Captan	1.9E-08	7.7E-06	3.9E-05	6.3E-08	2.5E-05	1.3E-05	8.2E-058	3.3E-05	1.6E-05
Chlordane	3.1E-07	8.0E-06	4.2E-06	1.0E-06	2.6E-05	1.4E-05	1.3E-096	3.4E-05	1.8E-05
Chromium	5.8E-05	6.0E-05	5.9E-05	2.78E-04	2.7E-04	2.7E-04	2.2E-04	2.2E-04	2.2E-04
Copper	4.6E-06	7.5E-06	6.1E-05	2.7E-05	4.3E-05	3.5E-05	1.9E-05	3.0E-05	2.4E-05
Cyanide	1.6E-06	2.3E-06	2.0E-05	9.1E-06	1.3E-05	1.1E-05	7.3E-06	1.0E-05	8.7E-06
DDE	2.0E-05	3.0E-05	2.5E-05	7.6E-05	1.1E-04	9.5E-05	8.6E-05	1.3E-04	1.1E-04
Dibenzofuran	3.2E-06	3.2E-06	3.2E-06	1.5E-05	1.5E-05	1.5E-05	1.9E-05	2.0E-05	2.0E-05
Dichlorobenzene	5.3E-04	1.1E-03	8.2E-04	1.9E-03	3.9E-03	2.9E-03	2.2E-03	4.6E-03	3.4E-03
Heptachlor	6.4E-06	1.4E-05	1.0E-05	2.1E-05	4.6E-05	3.4E-05	2.7E-05	6.0E-05	4.3E-05
Hexachlorobenzene	8.7E-03	6.8E-03	8.7E-03	1.8E-02	1.9E-02	1.8E-02	1.8E-02	1.8E-02	1.8E-02
Hexachlorobutadiene	4.5E-05	1.5E-04	9.9E-05	1.0E-04	3.4E-04	2.2E-04	9.6E-05	3.2E-04	2.1E-04
Lead	3.6E-04	3.8E-04	3.7E-04	1.8E-03	1.9E-03	1.9E-03	1.4E-03	1.5E-03	1.5E-03
Lindane	2.1E-06	1.0E-05	6.1E-06	7.2E-06	3.4E-05	2.0E-05	9.2E-06	4.3E-05	2.6E-05

Mercury	8.9E-04	9.0E-04	8.9E-04	1.8E-03	1.8E-03	1.8E-03	1.7E-03	1.7E-03	1.7E-03
Methoxychlor	1.4E-06	9.1E-06	5.3E-06	4.7E-06	3.0E-05	1.7E-05	6.1E-06	3.9E-05	2.3E-05
N-Nitrosodimethylamine	3.1E-04	4.7E-04	3.9E-04	1.7E-03	2.6E-03	2.1E-03	1.3E-03	2.1E-03	1.7E-03
Nickel	3.3E-04	5.4E-04	4.4E-04	1.8E-03	2.9E-03	2.3E-03	1.4E-03	2.2E-03	1.8E-03
Parathion	2.0E-07	7.9E-06	4.0E-06	6.7E-07	2.6E-05	1.3E-05	6.6E-07	3.4E-05	1.7E-05
Pentachlorophenol	1.9E-05	7.7E-05	4.8E-05	9.6E-05	3.8E-04	2.4E-04	1.1E-04	4.3E-04	2.7E-04
PCBs	2.0E-06	2.8E-06	2.4E-06	9.5E-06	1.3E-05	1.1E-05	1.3E-05	1.8E-05	1.5E-05
Selenium	1.5E-07	1.5E-07	1.5E-07	7.2E-07	7.3E-07	7.3E-07	5.5E-07	5.6E-07	5.6E-07
Toxaphene	1.9E-04	2.0E-04	2.0E-04	6.4E-04	6.6E-04	6.5E-04	8.2E-04	8.6E-04	8.4E-04
Trifluralin	5.1E-06	1.3E-05	8.9E-06	1.7E-05	4.2E-05	2.9E-05	2.2E-05	5.4E-05	3.8E-05

Table 8.6B Estimated Annual Average Chemical Concentrations in Plant Tissue within each Zone of Interest

| | Average chemical concentrations at receptor locations (μg/kg) | | | | | | | | |
| | Zone 4 (agricultural fields) | | | Zone 5 (facility site/wildlife habitat) | | | Zone 6 (agricultural fields) | | |
Chemical	Root	Leafy	Vine	Root	Leafy	Vine	Root	Leafy	Vine
2,3,7,8-TCDD	7.5E-08	1.3E-07	1.0E-07	4.0E-07	6.7E-07	5.4E-07	1.8E-07	3.1E-07	2.4E-07
2,4-D	8.0E-04	2.3E-03	1.6E-03	7.2E-04	2.1E-03	1.4E-03	5.5E-04	1.6E-03	1.1E-03
Arsenic	9.2E-05	9.4E-05	9.3E-05	3.8E-04	3.8E-04	3.8E-04	1.2E-04	1.2E-04	1.2E-04
Benzo(a)pyrene	7.5E-05	1.2E-04	9.7E-05	3.0E-04	4.8E-04	3.9E-04	1.5E-04	2.4E-04	2.0E-04
BEHP	3.3E-05	9.9E-04	5.1E-04	1.4E-04	4.3E-03	2.2E-03	5.7E-05	1.7E-03	8.9E-04
Cadmium	1.6E-03	1.7E-03	1.6E-03	6.4E-03	6.8E-03	6.6E-03	2.1E-03	2.2E-03	2.2E-03
Captan	1.1E-07	4.4E-05	2.2E-05	4.4E-07	1.8E-04	8.9E-05	2.2E-07	9.0E-05	4.5E-05
Chlordane	1.8E-06	4.6E-05	2.4E-05	7.1E-06	1.8E-04	9.6E-05	3.6E-06	9.3E-05	4.8E-05
Chromium	4.4E-04	4.5E-05	4.5E-04	1.6E-03	1.6E-03	1.6E-03	5.4E-04	5.5E-04	5.5E-04
Copper	4.1E-05	6.6E-05	5.4E-05	1.5E-04	2.5E-04	2.0E-04	5.0E-05	8.2E-05	6.6E-05
Cyanide	1.5E-05	2.1E-05	1.8E-05	6.3E-05	8.8E-05	7.6E-05	2.1E-05	2.9E-05	2.5E-05
DDE	1.3E-04	1.9E-04	1.6E-04	5.3E-04	8.0E-04	6.6E-04	2.4E-04	3.5E-04	2.9E-04
Dibenzofuran	2.5E-05	2.5E-05	2.5E-05	1.3E-04	1.3E-04	1.3E-04	6.0E-05	6.0E-05	6.0E-05
Dichlorobenzene	3.1E-03	6.5E-03	4.8E-03	1.2E-02	2.6E-02	1.9E-02	6.0E-03	1.2E-02	9.2E-03
Heptachlor	3.6E-05	8.0E-05	5.8E-05	1.5E-04	3.2E-04	2.4E-04	7.4E-05	1.6E-04	1.2E-04
Hexachlorobenzene	3.6E-02	3.7E-02	3.6E-02	3.8E-02	3.9E-02	3.8E-02	2.7E-02	2.7E-02	2.7E-02
Hexachlorobutadiene	1.9E-04	6.5E-04	4.2E-04	2.5E-04	8.3E-04	5.4E-04	1.6E-04	5.3E-04	3.4E-04
Lead	3.0E-03	3.1E-03	3.1E-03	1.2E-02	1.2E-02	1.2E-02	3.8E-03	4.0E-03	3.9E-03
Lindane	1.2E-05	5.8E-05	3.5E-05	5.0E-05	2.4E-04	1.4E-04	2.5E-05	1.2E-04	7.1E-05

Mercury	3.6E-03	3.7E-03	3.6E-03	3.0E-03	3.0E-03	3.0E-03	2.4E-03	2.4E-03	2.4E-03
Methoxychlor	8.2E-06	5.2E-05	3.0E-05	3.3E-05	2.1E-04	1.2E-04	1.7E-05	1.1E-04	6.1E-05
N-Nitrosodimethylamine	2.8E-03	4.3E-03	3.5E-03	1.1E-02	1.6E-02	1.4E-02	3.7E-03	5.7E-03	4.7E-03
Nickel	2.9E-03	4.6E-03	3.8E-03	1.1E-02	1.9E-03	1.6E-02	3.8E-03	6.2E-03	5.0E-03
Parathion	1.2E-06	4.5E-05	2.3E-05	4.7E-06	1.8E-04	9.3E-05	2.4E-06	9.2E-05	4.7E-05
Pentachlorophenol	1.6E-04	6.2E-04	3.9E-04	7.8E-04	3.1E-03	2.0E-03	3.3E-04	1.3E-03	8.1E-04
PCBs	1.5E-05	2.2E-05	1.9E-05	8.4E-05	1.2E-04	1.0E-04	3.9E-05	5.4E-05	4.6E-05
Selenium	1.2E-06	1.2E-06	1.2E-06	4.3E-06	4.3E-06	4.3E-06	1.5E-06	1.5E-06	1.5E-06
Toxaphene	1.1E-03	1.1E-03	1.1E-03	4.4E-03	4.6E-03	4.5E-03	2.2E-03	2.3E-03	2.3E-03
Trifluralin	2.9E-05	7.3E-05	5.1E-05	1.2E-04	2.9E-04	2.1E-04	5.9E-05	1.5E-04	1.0E-04

Daily food ingestion rate (IR_f)

This parameter pertains to the average mass of food material consumed per day by a species. For white-tail deer the daily food ingestion rate was assumed to be 1.81 kg/day.[25] For rabbits the value was assumed to be 0.3 kg/day.[26] For squirrels, this food ingestion rate was assumed to be 0.064 kg/day.[27]

The food web indicator species average mass of prey consumed per day was assumed to be 0.3 kg/day for red-tailed hawks[28] and 0.21 kg/day[29] for black-crowned night herons. It should be noted that in deriving dose estimates for wildlife, it was conservatively assumed that all forage, water, or prey consumed was obtained from a very restricted area (i.e., either at the receptor location or within the zone of interest). As such, no consideration was given to variations in concentrations of chemicals in food items that would occur over the area of an animal's home range. For small mammals, the effects that this may have on the results is probably not significant since the average plant tissue concentration over the entire area of the proposed facility (i.e., an area probably larger than their normal home ranges) was used. For the deer and avian indicator species, however, a significant portion of their diet and/or water intake could occur in areas that would not be subject to hypothetical impact from the proposed facility.

Chemical concentration in water (C_w)

For wildlife indicator species (i.e., white-tail deer, rabbits, and squirrels), water concentrations (including sediment:water partitioning factors) predicted for the on-site stock pond (receptor location #470) were used for this evaluation. As noted above, only chemicals with water:sediment partitioning factors assumed to be 100% were included in the analysis. Receptor location #470 represents an area of maximal potential for impacts and where habitat utilization by wildlife would be closest to the proposed facility. In addition, assuming that water is obtained solely from that pond throughout the animal's life span adds an additional measure of conservatism to this portion of the analysis.

Average daily ingestion rate for water (IR_w)

It was assumed that white-tail deer ingest approximately 1.4 kg water per day.[25] Daily water ingestion rates for rabbits and squirrels were estimated based on assumed average body weights (BW) using a relationship presented by Suter.[30] This relationship is expressed as

$$IR_w = 0.90 \, BW^{1.2044}$$

Using this method, average daily water ingestion rates for rabbits and squirrels were estimated at 0.18 and 0.056 kg/day, respectively.

Body weight (BW)

Based on life history information compiled for each species, average body weights for each indicator species were estimated. These are as follows: white-tail deer 68.9 kg;[31] cottontail rabbit 1.8 kg;[32] grey squirrel 0.68 kg;[32] red-tailed hawk 1.14 kg;[33] and black-crowned night heron 0.85 kg.[34]

Assumptions regarding pathways of exposure

It should be noted that the portion of daily intake assumed to be attributable to inhalation is not considered in the analysis. This is considered appropriate for two reasons. First, the toxicity criteria to which calculated doses are to be compared are derived solely from ingestion studies. Second, the percentage of total intake attributable to inhalation is typically less than 5% for the wildlife indicator species in zone 5, the proposed waste treatment facility site. Water ingestion was considered significant for the wildlife indicator species for those chemicals having water sediment partitioning factors of 100%, indicating that the entire mass of chemical deposited in the water column was assumed to partition into the dissolved phase. The doses calculated for those chemicals include the contribution associated with water ingestion. For all other chemicals, the portion of the intake attributable to pathways other than food ingestion is considered negligible.

For example, the total dose of arsenic via food ingestion calculated for deer, rabbits, and squirrels in zone 5 were 1.0^{-8}, 6.3×10^{-8}, and 3.6×10^{-8} mg/kg-day, respectively. The total arsenic doses for these species via water ingestion, assuming all water is obtained from the on-site stock pond (receptor location #4) were estimated at 8.9×10^{-8}, 4.4×10^{-7}, 3.6×10^{-7} mg/kg-day, respectively. Using estimated inhalation rates for wildlife[30] and an average ambient air concentration for arsenic in zone 5 of 2.1×10^{-5} µg/m^3, the arsenic dose via inhalation to deer, rabbits, and squirrels were estimated at 5.5×10^{-9}, 1.5×10^{-8}, and 1.8×10^{-8} mg/kg-day, respectively. The inhalation dose thus corresponds to approximately 5.2%, 3.1%, and 4.2% of the total dose via all pathways for these three species. Because air concentration for all other evaluated chemicals are significantly lower than concentration in other media, it is assumed that a similar relationship would exist. As such, the contribution to total intake from inhalation exposure was assumed to be negligible and was not included in the analysis.

Similar assumptions were made for the food web indicator species. Although the potential routes of exposure for the black-crowned night heron and the red-tailed hawk are inhalation of airborne chemicals, ingestion of contaminated water, and ingestion of contaminated prey, toxicity data for inhalation route of exposure are lacking. It is considered highly unlikely that inhalation of chemicals would contribute significantly to total intake relative to food ingestion. In addition, compared to ingestion of prey, ingestion of chemicals in water is expected to contribute a negligible amount to total chemical intake and was therefore not included in the quantitative analysis.

Incidental ingestion of soil for predatory species is also considered insignificant relative to ingestion of prey.

To confirm the assumptions, a sensitivity analysis was performed for water ingestion for selenium for the red-tailed hawk at receptor point #470 (on-site stock pond). For selenium, the maximum estimated concentration in plant tissues of approximately 4×10^{-4} μg/kg yielded an estimated tissue concentration of 4×10^{-7} mg/kg in rodents. Assuming a daily ingestion rate of 0.3 kg rodent tissue/day for hawks, the total selenium intake via food ingestion would be approximately 1.2×10^{-4} mg/day. In contrast, the comparatively daily intake potentially associated with water ingestion is negligible. Given a selenium concentration in water of 6.3×10^{-8} mg/kg (see Table 8.5) assuming that a water ingestion rate for the hawk would be comparable to that of an osprey of 0.77 kg/day,[28] the chemical intake via water ingestion is 2.8×10^{-7} or 4.8×10^{-9} mg/day, approximately 0.004% of the total daily chemical intake of selenium via ingestion. Thus, impacts from ingestion of water were not included in the quantitative evaluation.

Effects assessment

To estimate risks in the ERA, site-specific estimates of doses or media concentrations for COIs are compared to chemical and species-specific criteria that are assumed to represent COI "no-effect" exposure levels. For fish and plant indicator species, threshold criteria are expressed in the form of chemical concentrations in environmental media (i.e., water and soil, respectively) that are considered to have no adverse effects. For mammalian and avian indicator species, threshold criteria are expressed in terms of a chronic average daily dose below which no adverse effects are expected.

The process of developing chemical- and species-specific toxicity reference values (TRVs) parallels the development of reference dose (RfD) values for non-carcinogenic health effects in humans. This is essentially a process of characterizing the relationship between the concentration/dose of an agent and the anticipated incidence of an adverse health effect in an exposed population.[35] However, since no such RfDs have been established for use in performing an ERA, TRVs are developed independently for each chemical and each indicator species (for which adequate toxicological data are available in the literature) according to the following equation:

$$TRV = \frac{Literature\ Toxicity\ Value}{Uncertainty\ Factor(s)}$$

Toxicity values obtained from the literature may be in several different forms. For example, criteria and standards that have been established by regulation specifically for protection of a resource were available for some chemicals and indicator species (e.g., ambient water quality criteria for fish). Alternatively, concentrations or doses shown to be lethal to 50% of test populations ($LC_{50}s/LD_{50}s$), no observed adverse effect levels (NOAELs), or lowest observed adverse effect levels (LOAELs), or other values developed through empirical research may be available. The applicability of literature-based toxicity values depends upon how closely the test conditions under which the values were developed correspond to the intended use of the data. Factors that may influence the applicability of a data set include similarity of test species to indicator species, appropriateness of toxicological end points of the study, route of administration of chemical, and duration of exposure. In extrapolating toxicity test data to indicator species, one or more uncertainty factors are applied to the results to ensure conservatism and protectiveness of the final value identified as a threshold criteria. Using this general approach, threshold criteria for each of the indicator chemicals and indicator species were developed.

The specific approach to developing TRVs varied between species. Therefore, the methods used to develop threshold criteria for each of the chemicals of concern are discussed separately for the fish, plants, mammals, and birds.

Chemical-specific TRVs criteria for fish indicator species

Wherever available, chronic freshwater AWQCs[36] were used as the threshold criteria for the fathead minnow and channel catfish. AWQCs represent freshwater concentrations that, except in cases where a locally important species may be unusually sensitive, freshwater aquatic organisms and their uses should not be affected unacceptably if the four-day average concentration of the chemical does not exceed the criteria value more than every three years on average.[36] For chemicals for which a LOAEL was used as the basis for the AWQC, an uncertainty factor of 10 was applied to develop the threshold criteria. For chemicals without an established AWQC, toxicity data from the literature were used in descending order of precedence:

- An acute or chronic study with a sublethal end point using the indicator species as the test species
- LC_{50} value for the indicator species
- LC_{50} value for a species similar to the indicator species

For the purpose of developing threshold criteria, guppies and bluegill sunfish were considered to be comparable to the fathead minnow, and the carp were considered comparable to the channel catfish. For benzo(a)pyrene, (B(a)P), the only aquatic toxicity value which was located was an LC_{50} value for a sandworm. Therefore, an additional uncertainty factor of 10 (for a total uncertainty factor of 100) was applied to account for potential differences in sensitivity to B(a)P in fish.

All acute LC_{50} data were divided by an uncertainty factor of 100 to account for nonlethal effects, and an additional factor of 100 to account for acute versus chronic exposure duration.

Chemical-specific TRVs for plant indicator species

For all inorganic COIs except cyanide, maximum acceptable concentrations (MACs) in soil for agricultural crops were used as the TRVs for agricultural crops (i.e., soybean, corn, and wheat). To derive a TRV for the loblolly pine, the soil MACs for agricultural crops were divided by an uncertainty factor of 10. For organic chemicals, the PHYTOTOX database[37] was searched for data on each COI and indicator species. Adequate data were available to develop TRVs for only a subset of the COIs for plant species. The rationale for selection of data to develop the TRVs was as follows, in descending order of precedence:

- A soil concentration was reported for the COI and indicator plant species that represented either a LOAEL or NOAEL (e.g., the chemical was applied to the growth medium or was reported as an application rate for the specific plant of concern).
- A soil concentration was reported for the chemical in an appropriate surrogate plant species that represented either a LOAEL or NOAEL.
- A water concentration was reported for plant or seed exposures that represented a LOAEL or NOAEL. A soil concentration was estimated from the water concentration using the partitioning relationship described previously.

Toxicity data for chemicals which were applied as chemical spray were not considered in this assessment, since projected site soil concentrations were used as the basis for comparison.

All LOAEL values were divided by an uncertainty factor of 10 to adjust the toxicity criteria to a NOAEL. Due to potential differences in species sensitivity, the reported toxicity value was divided by an uncertainty factor of 10 when the test species differed markedly from the indicator species. Because the PHYTOTOX database does not report the number of chemical applications, an uncertainty factor of 10 was applied to account for the fact that modeled concentrations represent average values over an extended time period.

Chemical-specific TRVs for mammalian indicator species

Toxicity data reported for domestic livestock were divided by an uncertainty factor of 10 when developing threshold criteria for the deer to account for potential differences in species sensitivity. Likewise, for some COIs, toxicological data were lacking for ruminants or other large mammals. In such cases, rabbit and rodent data were divided by an uncertainty factor of 100 to account for potential interspecies difference. For the cottontail rabbit, the laboratory rabbit was assumed to be comparable. Rodent data were divided by an uncertainty factor of 10 to account for potential differences in species sensitivity with the rabbit. The gray squirrel was assumed to be comparable to rats and guinea pigs. Rabbit and mouse toxicity data were divided by an uncertainty factor of 10 to account for potential differences in species sensitivity with the squirrel.

The rationale for the selection of a representative study or reported dietary concentration to derive the threshold criteria for mammals was as follows, in descending order of precedence:

- NOAEL or LOAEL was reported in a chronic or subchronic study for a sublethal end point using the indicator species as the test species.
- A recommended dietary concentration criterion was reported by the U.S. Fish and Wildlife Service for the species of concern or a similar species.
- An LD_{50} value was available using the indicator species as the test species.
- A NOAEL or LOAEL was reported in a chronic or subchronic study for a sublethal end point using a species similar to the indicator species.
- An LD_{50} value was available using a species similar to the indicator species as the test species.

LOAELs were divided by an uncertainty factor of 10 to adjust the threshold criteria to a NOAEL. Data reported from subchronic studies were divided by an uncertainty factor of 10 to convert to a chronic exposure duration. All LD_{50} data were divided by an uncertainty factor of 100 to account for nonlethal effects and an additional factor of 100 to account for acute versus chronic exposure.

Chemical-specific TRVs for avian indicator species

Adequate data were available to develop TRVs for a subset of the COIs having the highest potential for bioaccumulation in the aquatic and terrestrial food webs. For the purposes of the food web assessment, the rationale for the selection of a representative study or reported dietary concentration to derive the TRVs was as follows, in descending order of precedence:

- A chronic toxicity study for the chemical of concern was available for the indicator species using a sensitive and relevant end point (e.g., impaired reproduction).
- A chronic toxicity study for the chemical of concern was available using a closely related species with similar dietary preferences using a sensitive end point.
- A recommended criterion for dietary intake in birds has been developed by the U.S. Fish and Wildlife Service using toxicity data for the indicator species or similar species.
- Toxicity data for the indicator chemical were available for a surrogate species using a nonlethal end point.

Like the mammalian indicator species, LOAELs were divided by an uncertainty factor of 10 to adjust the threshold criteria to a NOAEL. Data reported from subchronic studies were divided by an uncertainty factor of 10 to convert to a chronic exposure duration. All LD_{50} data were divided by an uncertainty factor of 100 to account for nonlethal effects and an additional factor of 100 to account for acute vs. chronic exposure duration. An example of the TRV derivation process for the black-crowned night heron is provided in Table 8.7.

Conservation in derived TRVs

It is important to note that these TRVs are screening-level values developed specifically for use in this ERA and should not necessarily be interpreted as having wider significance or application in any other context. For example, in some cases the TRV developed for certain chemicals and species may be within or below naturally occurring background levels or practical analytical quantitation limits due to the application of uncertainty factors according to the rules outlined above. The TRV derived for mercury for loblolly pine, for example (i.e., 0.05 mg/kg in soil), is in the low end of the range of naturally occurring levels in soils in Mississippi of 0.03 mg/kg to 0.72 mg/kg[38] and below the mean level of 0.104.[38] This is due to the highly conservative approach used in selecting the uncertainty factors used in developing the TRVs.

Thus, while it can be stated with a high level of confidence that estimated exposure at levels below the TRV will not result in any adverse effects to receptors, any calculated doses or media concentrations at levels that may exceed criteria developed through this methodology should not be interpreted as an absolute indication that adverse effects to the specified species would occur.

Risk characterization

The potential of each of the indicator chemicals to produce adverse effects was evaluated for each of the species of concern through the calculation of a hazard quotient. Because the TRVs are based on different end points, the terms applied in the equation for the hazard quotient (HQ) vary to some extent for different species. However, the general approach is the same and can be expressed as follows:

$$HQ = \frac{Average\ Daily\ Dose/Media\ Concentration}{TRV}$$

In cases where individual chemicals potentially act on the same organs or result in the same health end point, additive effects may occur. The following equation was used to calculate a hazard index (HI) for potential effects in an indicator species as a result of concurrent exposure to multiple chemicals:

$$HI = \sum_{i=1}^{n} HQ_i$$

For the purposes of this ERA, HQs for all the chemicals of concern were added regardless of the target organ or effect. Although this screening level approach would likely overestimate the potential for adverse effects, an HI less than or equal to 1.0 would indicate that adverse effects are not anticipated, even when considering exposure to multiple chemicals.

Plant species

The chemical-specific HQ for each plant indicator species was calculated using the following equation:

$$HQ_{plant} = \frac{Calculated\ Soil\ Concentration}{TRV_{Plant}}$$

The calculated soil concentrations for corn, soybeans, and wheat were the highest average modeled concentrations for either zone 3 or zone 4 agricultural fields (Table 8.8). The maximum average chemical concentration predicted in soil was located in zone 3 for all chemicals, except mercury and 2,4-D, which had higher concentrations in soil in zone 4. Potential impacts to agricultural crops were assessed based upon predicted soil concentrations in these areas. The calculated soil concentrations for loblolly pine were taken from zone 1 (timber woodlot), the closest location in which commercial forestry may occur. Estimated chemical concentrations in plant tissues from zones 1 to 6 are summarized in Tables 8.6 A and B.

As indicated in Table 8.8, the HI calculated for corn is 0.12. Similarly, the HI's calculated for soybeans and wheat are 0.10 and 0.051, respectively. The HI calculated for loblolly pine is 0.01 (Table 8.9). These results indicate that between a 10- and 100-fold margin of protection would exist for plant species exposed to projected chemical emissions from the proposed waste treatment facility. Therefore, no adverse impacts for any of the agricultural plant species or the loblolly pine are expected due to chemical emissions from the routine operation of the proposed waste treatment facility.

Table 8.7 Toxicity Reference Value Derivation* for the Black-Crowned Night Heron

Chemical	Test species	Reference value (mg/kg-day)[a]	End point	Uncertainty factor[b] LOAEL	Duration	Species	Threshold criteria[c] (mg/kg-day)[a]	Reference
Inorganic Chemicals								
Lead	Mallard	6.2[d]	Altered blood chemistry	10	10	10	0.006	39
Mercury (organic)	Mallard	0.01[e]	Fish and wildlife criterion	1	1	10	0.001	19
Selenium	Black-crowned night heron	2.5[f]	Impaired chick skeletal development	10	10	1	0.025	41
Organic Compounds								
DDE	Black-crowned night heron	0.008[g]	Decreased eggshell thickness (NOAEL)	1	1	1	0.008	42, 43
PCBs	Black-crowned night heron	0.03[h]	Reduced embryo weights	10	1	1	0.003	43, 44
TCDD	Chicken	2.5 ng/kg[i]-day	Fish and wildlife criterion	10[j]	1	10	0.025 ng/kg-d	19

ᵃ Unless otherwise noted.

ᵇ Uncertainty factors applied on the following basis: factor of 10 for adjustment from LOAEL to NOAEL, factor of 10 for adjustment from subchronic exposure duration to chronic duration, factor of 10 for adjustment from species with markedly different dietary patterns.

ᶜ For the purpose of this analysis, defined as: daily dose that would not be likely to cause adverse effects in indicator species.

ᵈ Derived as follows: The 25 mg/kg diet concentration was converted to a dose by multiplying by a 0.212 kg/day ingestion rate and dividing by 0.85 kg/ body weight.

ᵉ Derived as follows: The 0.05 mg/kg diet concentration was converted to a dose as indicated in footnote (d).

ᶠ Derived as follows: The 10 mg/kg diet concentration was converted to a dose as indicated in footnote (d).

ᵍ Derived as follows: 1 mg/kg (=NOAEL in Black-Crowned Night Heron eggs) divided by diet-to-egg biotransfer factor of 34 to achieve a diet concentration of 0.03 mg/kg which was converted to a dose as indicated in footnote (d).

ʰ Derived as follows: 4 mg/kg egg concentration (=LOAEL in Black-Crowned Night Herons) divided by diet-to-egg biotransfer factor of 32 to achieve a dietary concentration of 0.125 mg/kg which was converted to a daily dose as indicated in footnote (d).

ⁱ Derived as follows: 120 ng/kg in diet which was converted to a dose as indicated in footnote (d).

ʲ Based on definition of the fish and wildlife criterion for dioxin (a diet concentration that *may* be nonhazardous to birds).

Table 8.8 Comparison of Estimated Concentrations of Chemicals of Interest in Soils at Zone 3 or 4 (Agricultural Fields) with TRVs for Corn

Indicator chemical	Concentration in soil[a] (mg/kg)	TRV (mg/kg)	HQ
Inorganic chemicals			
Arsenic	3.8E–06	2.0E+00	1.9E–06
Cadmium	3.2E–03	1.0E+00	3.2E–03
Chromium	2.6E–02	5.0E+01	5.3E–04
Copper	1.9E–04	5.0E+01	3.8E–06
Cyanide compounds	1.1E–03	1.0E+00	1.1E–03
Lead	1.2E–01	5.0E+01	2.4E–03
Mercury[b]	1.2E–03	5.0E–01	2.4E–03
Nickel	1.4E–02	3.0E+01	4.8E–04
Selenium	4.7E–05	1.6E+00	2.9E–05
Organic compounds			
Benzo(a)pyrene	3.8E–06	5.0E–05	7.6E–02
BEHP	5.5E–06	NA	NA
Captan	1.0E–07	4.4E–02	2.3E–06
Chlordane	8.6E–06	1.1E–02	7.8E–04
Dibenzofurans	3.0E–08	N/A	NA
Dichlorobenzene	1.1E–04	7.4E+00	1.5E–05
2,4-D[b]	1.5E–06	3.5E–02	4.3E–05
DDE	3.2E–05	NA	NA
Heptachlor	2.4E–06	4.5E–04	5.3E–03
Hexachlorobenzene	4.6E–05	N/A	NA
Hexachlorobutadiene	5.0E–06	N/A	NA
Lindane	2.0E–06	1.0E–01	2.0E–05
Methoxychlor	2.9E–06	1.0E–03	2.9E–03
N-Nitrosodimethylamine	1.5E–05	NA	NA
Parathion	8.4E–08	NA	NA
Pentachlorophenol	1.9E–05	1.0E–01	1.9E–04
PCB	3.3E–06	1.0E+00	3.3E–06
2,3,7,8-TCDD	3.0E–08	NA	NA
Toxaphene	2.1E–05	1.0E–03	2.1E–02
Trifluralin	4.2E–07	9.0E–03	4.7E–05
		HI	1.2E–01

Note: NA, not available; TRV, toxicity reference value; HQ, hazard quotient; and HI, hazard index.

[a] Derived as follows: ACE model predicted average soil concetration at zone 3 or 4 multiplied by a conversion factor of 0.001 mg/μg.

[b] Calculated values from zone 3; all other from zone 4 represent maximum concentrations predicted by ACE model.

Table 8.9 Comparison of Estimated Concentrations of Chemicals of
Interest in Soils at Zone 1 (Timber Wood Lot)
with TRVs for Loblolly Pine

Indicator chemical	Concentration in soil[a] (mg/kg)	TRV (mg/kg)	HQ
Inorganic chemicals			
Arsenic	1.1E–04	2.0E–01	5.5E–04
Cadmium	8.9E–05	1.0E–01	8.9E–04
Chromium	9.7E–04	5.0E+00	1.9E–04
Copper	5.8E–06	5.0E+00	1.2E–06
Cyanide compounds	2.7E–05	1.0E+01	2.7E–04
Lead	3.6E–03	5.0E+00	7.2E–04
Mercury	3.0E–04	5.0E–02	6.0E–03
Nickel	4.2E–04	3.0E+00	1.4E–04
Selenium	1.6E–06	1.6E–01	1.0E–05
Organic compounds			
Benzo(a)pyrene	1.6E–07	1.0E–02	1.6E–05
BEHP	1.7E–07	NA	NA
Captan	4.5E–09	4.4E–01	1.0E–08
Chlordane	3.7E–07	1.1E–02	3.4E–05
Dibenzofurans	7.2E–10	N/A	NA
Dichlorobenzene	4.9E–06	7.4E–01	6.6E–06
2,4-D	3.7E–07	8.0E–02	4.6E–06
DDE	1.2E–06	NA	NA
Heptachlor	1.0E–07	5.0E–03	2.0E–05
Hexachlorobenzene	1.1E–05	N/A	NA
Hexachlorobutadiene	9.1E–07	N/A	NA
Lindane	8.7E–08	5.0E–04	1.7E–04
Methoxychlor	1.3E–07	1.0E–03	1.3E–04
N-Nitrosodimethylamine	4.3E–07	NA	NA
Parathion	3.7E–09	NA	NA
Pentachlorophenol	4.8E–07	4.4E–02	NA
PCB	8.0E–08	1.0E–03	8.0E–05
2,3,7,8-TCDD	7.3E–10	NA	NA
Toxaphene	9.0E–07	1.0E–03	9.0E–04
Trifluralin	1.8E–08	4.9E–02	3.7E–07
		HI	1.0E–02

Note: NA, not available; TRV, toxicity reference value; HQ, hazard quotient; and
HI, hazard index.

[a] Derived as follows: ACE model predicted average soil concentration at zone 3
or 4 multiplied by a conversion factor of 0.001 mg/μg.

Fish species

The hazard quotient for the fish species was calculated using the following equation:

$$HQ_{fish} = \frac{Predicted\ Surface\ Water\ Concentration}{TRV_{fish}}$$

As noted above, the calculated water concentrations for the channel catfish were based on the modeled concentration for receptor location #311 multiplied by the chemical-specific partitioning factors. This is the location of the nearest catfish pond to the proposed waste treatment facility. The calculated water concentration for the fathead minnow was based on the maximum modeled concentration for either receptor location #269 or #470 (whichever location had the highest predicted water concentrations for a chemical), multiplied by the chemical-specific partitioning factors. The modeled chemical concentrations were assumed to be lower at all other surface ponds located farther from the proposed waste treatment facility.

As indicated in Table 8.10, an HI of 0.038 was calculated for the channel catfish. Similarly, an HI of 0.046 was calculated for the fathead minnow (Table 8.11). These results indicate that greater than a twenty-fold margin of protection would exist for fish species. Therefore, no adverse impacts for either the channel catfish or the fathead minnow are anticipated from chemical emissions from routine operation of the proposed waste treatment facility.

Mammalian species

The hazard quotient for the mammalian indicator species were calculated using the following equation:

$$HQ_{mammal} = \frac{Average\ Daily\ Dose}{TRV_{mammal}}$$

The average daily dose for the white-tail deer, cottontail rabbit, and eastern gray squirrel was the sum of the food and the water ingestion doses. The average daily dose based on food ingestion was calculated using the modeled chemical concentrations in plant tissue for zone 5, the area in closest proximity to the proposed waste treatment facility. For the seven COIs that do not significantly partition to sediment (arsenic, cyanide, selenium, dichlorobenzene, 2,4-D, lindane, and *n*-nitrosodimethylamine), the water ingestion dose was calculated based on modeled water concentrations for the nearest pond to zone 5.

Table 8.10 Comparison of Estimated Concentrations of Chemicals
of Interest in Water at Receptor Location #311
(Off-Site Catfish Pond) with TRVs for the Channel Catfish

Indicator chemical	Concentration in water[a] (μg/L)	TRV (μg/L)	HQ
Inorganic chemicals			
Arsenic	2.5E–03	4.8E+00	5.3E–04
Cadmium	2.1E–05	1.1E+00	1.9E–05
Chromium	1.0E–02	1.1E+01	9.1E–04
Copper	1.2E–06	1.2E+01	1.0E–07
Cyanide compounds	1.3E–02	5.2E+00	2.5E–03
Lead	9.2E–04	3.2E+00	2.9E–04
Mercury	3.6E–05	1.2E–02	3.0E–03
Nickel	1.4E–03	1.6E+02	8.6E–06
Selenium	4.0E–05	3.5E+01	1.1E–06
Organic compounds			
Benzo(a)pyrene	5.5E–07	1.0E–03	5.5E–04
BEHP	7.4E–06	3.0E+00	2.5E–06
Captan	5.5E–06	8.0E–03	6.9E–04
Chlordane	5.5E–07	4.3E–03	1.3E–04
Dibenzofurans	4.1E–10	4.0E–03	1.0E–07
Dichlorobenzene	3.8E–03	7.6E+01	5.0E–05
2,4-D	2.7E–03	9.7E+00	2.8E–04
DDE	6.4E–07	1.1E+03	6.1E–10
Heptachlor	5.5E–07	3.8E–03	1.4E–04
Hexachlorobenzene	1.0E–04	5.0E+01	2.1E–06
Hexachlorobutadiene	7.8E–06	9.3E+00	8.4E–07
Lindane	5.5E–05	8.0E–02	6.9E–04
Methoxychlor	5.5E–07	3.0E–02	1.8E–05
N-Nitrosodimethylamine (d)	6.7E–04	NA	NA
Parathion	5.5E–06	1.3E–02	4.2E–04
Pentachlorophenol (d)	3.2E–04	1.3E+01	2.5E–05
PCB	5.1E–08	1.4E–02	3.7E–06
2,3,7,8-TCDD	4.2E–10	1.0E–06	4.2E–04
Toxaphene	5.5E–06	2.0E–04	2.8E–02
Trifluralin	5.5E–07	2.1E–02	2.6E–05
		HI	3.8E–02

Note: NA, not available; TRV, toxicity reference value; HQ, hazard quotient; and
HI, hazard index.

[a] ACE model predicted water concentrations at receptor location #311 multiplied
by a partitioning factor.

Table 8.11 Comparison of Estimated Concentrations of Chemicals of Interest in Water at Receptor Location #269 or #470 (On-Site Stock Pond and other Off-Site Pond) with TRVs for the Fathead Minnow

Indicator chemical	Receptor location number[a]	Concentration in Water[b] (µg/L)	TRV (µg/L)	HQ
Inorganic chemicals				
Arsenic	470	4.4E–03	4.8E+00	9.1E–04
Cadmium	470	3.4E–05	1.1E+00	3.1E–05
Chromium	470	4.0E–04	1.1E+01	3.7E–05
Copper	470	2.0E–06	1.2E+01	1.7E–07
Cyanide compounds	470	2.2E–02	5.2E+00	4.2E–03
Lead	470	1.4E–03	3.2E+00	4.5E–04
Mercury	470	6.9E–05	1.2E–02	5.8E–03
Nickel	470	1.7E–04	1.6E+02	1.0E–06
Selenium	470	6.3E–05	3.5E+01	1.8E–06
Organic compounds				
Benzo(a)pyrene	269	6.2E–07	1.0E–03	6.2E–04
BEHP	470	8.2E–06	3.0E+00	2.7E–06
Captan	269	6.2E–06	8.0E–03	3.8E–07
Chlordane	269	6.2E–07	4.3E–03	1.4E–04
Dibenzofurans	470	2.9E–10	4.0E–03	7.4E–09
Dichlorobenzene	269	4.4E–03	7.6E+01	5.8E–05
2,4-D	269	5.2E–03	9.7E+00	7.4E–04
DDE	269	6.8E–07	1.1E+03	6.5E–10
Heptachlor	269	6.2E–07	3.8E–03	1.6E–04
Hexachlorobenzene	269	1.9E–04	5.0E+01	3.8E–06
Hexachlorobutadiene	269	1.4E–05	9.3E+00	1.5E–06
Lindane	269	6.3E–05	8.0E–02	7.9E–04
Methoxychlor	269	6.2E–07	3.0E–02	2.1E–05
N-Nitrosodimethylamine (d)	470	1.1E–03	NA	NA
Parathion	269	6.2E–06	1.3E–02	4.8E–04
Pentachlorophenol	470	2.9E–04	1.3E+01	2.3E–05
PCB	470	3.6E–08	1.4E–02	2.5E–06
2,3,7,8-TCDD	470	3.0E–10	1.0E–06	3.0E–04
Toxaphene	269	6.2E–06	2.0E–04	3.1E–02
Trifluralin	269	6.2E–07	2.1E–02	5.6E–06
			HI	4.6E–02

Note: NA, not available; TRV, toxicity reference value; HQ, hazard quotient; and HI, hazard index.

[a] Receptor location corresponds to the pond with the maximum predicted concentration from ACE modeling #260 and 420

[b] Ace model predicted water concentration at reception location multiplied by a partitioning factor.

As indicated in Table 8.12 the calculated HI for white-tail deer was 0.16. The predicted HI was 0.048 for the cottontail rabbit (Table 8.13). For the eastern gray squirrel, HI of 0.023 was calculated (Table 8.14). These results indicate that between a 20-fold and 50-fold margin of protection would exist for mammalian species. Therefore, no adverse effect on any of the mammalian indicator species are expected due to exposures to chemicals from routine operations at the proposed waste treatment facility.

Avian species

The hazard quotients for avian indicator species was calculated using the following equation:

$$HQ_{bird} = \frac{Average\ Daily\ Dose}{TRV_{bird}}$$

The average daily dose for the red-tailed hawk was calculated from estimated rodent tissue concentrations. These concentrations were estimated using the modeled chemical concentrations in plants for three distinct receptor locations in meadow habitat near on-site and off-site surface water ponds and multiplying by a bioaccumulation factor. These receptor locations were also used to predict impacts through the food web to the black-crowned night herons. These locations generally corresponded to points of maximum modeled media concentrations of COIs. The average daily dose for the black-crowned night heron was calculated from estimated fish tissue concentrations. Fish tissue concentrations were estimated using the modeled chemical concentrations in water (adjusted for partitioning) and multiplying by the chemical-specific bioconcentration factors.

As indicated in Table 8.15, HIs of 0.13, 0.031, and 0.1 were calculated for red-tailed hawks at receptor locations 470, 269, and 412, respectively. For the black-crowned night heron, HIs of 0.06, 0.63, and 0.03 were calculated for potential exposures at these same locations (Table 8.16). These results indicate that a margin of protection of between 1.5 and 30 times would exist for avian species. Therefore, no adverse impacts for the avian food web indicator species are expected due to potential exposures to chemical emissions from projected routine operations of the proposed treatment facility.

Uncertainty analysis for sensitive species

There are uncertainties inherently associated with any ERA. These uncertainties may be greater in a prospective assessment, such as this, where no chemical data can be obtained to calibrate model estimates.

Table 8.12   Comparison of Estimated Average Daily Doses of Chemicals
of Interest at Zone 5 (Proposed Facility)
with TRVs for the White-Tail Deer

Indicator chemical	Average daily dose (mg/kg-day)	TRV (mg/kg-day)	HQ
Inorganic chemicals			
Arsenic[a]	9.9E–08	1.0E–03	9.9E–05
Cadmium	1.8E–07	5.0E–03	3.6E–05
Chromium	42E–08	1.7E–04	2.4E–04
Copper	6.6E–09	8.0E–03	8.2E–07
Cyanide compounds[a]	4.5E–07	5.0E–01	9.0E–07
Lead	3.2E–07	4.0E–02	7.9E–06
Mercury	7.9E–08	1.8E–03	4.4E–05
Nickel	4.9E–07	2.5E–00	2.0E–07
Selenium[a]	6.3E–05	4.1E–04	1.5E–01
Organic compounds			
Benzo(a)pyrene	1.3E–08	5.0E–05	2.5E–04
BEHP	1.1E–07	1.0E–02	1.1E–05
Captan	4.7E–09	7.4E–03	6.4E–06
Chlordane	4.7E–09	2.5E–04	1.9E–05
Dibenzofurans	3.4E–09	1.0E–04	3.4E–04
Dichlorobenzene[a]	7.6E–07	1.9E–02	4.0E–05
2,4-D[a]	1.2E–07	4.0E–02	3.1E–06
DDE	2.1E–08	1.0E–03	2.1E–05
Heptachlor	8.4E–09	2.0E–03	4.2E–06
Hexachlorobenzene	1.0E–06	5.0E–04	2.0E–03
Hexachlorobutadiene	2.2E–08	9.0E–05	2.4E–04
Lindane	7.3E–09	2.5E–02	2.9E–07
Methoxychlor	5.5E–09	5.0E–02	1.1E–07
N-Nitrosodimethylamine[a]	4.4E–07	5.0E–04	8.8E–04
Parathion	4.7E–09	2.2E–03	2.1E–06
Pentachlorophenol[a]	8.7E–08	2.0E–04	4.4E–04
PCB	3.2E–09	5.0E–05	6.3E–05
2,3,7,8-TCDD	1.8E–11	1.0E–05	1.8E–06
Toxaphene	1.2E–07	1.3E–02	8.7E–06
Trifluralin	7.6E–09	3.7E–03	2.1E–06
		HI	1.6E–01

Note: TRV, toxicity reference value; HQ, hazard quotient; and HI, hazard index.

[a] Water ingestion was taken into account for these chemicals; for all other chemicals only food ingestion was a significant route of uptake.

Table 8.13 Comparison of Estimated Average Daily Doses of Chemicals of Interest at Zone 5 (Proposed Facility) with TRVs for the Eastern Cottontail

Indicator chemical	Average daily dose (mg/kg-day)	TRV (mg/kg-day)	HQ
Inorganic chemicals			
Arsenic[a]	5.0E–07	8.00E–05	6.3E–04
Cadmium	1.1E–06	5.00E–02	2.3E–05
Chromium	2.6E–07	1.70E–02	1.5E–05
Copper	4.2E–08	8.00E–03	5.2E–06
Cyanide compounds[a]	2.2E–06	1.70E–00	1.3E–06
Lead	2.0E–06	5.00E–05	4.0E–02
Mercury	5.0E–07	6.00E–03	8.3E–05
Nickel	3.1E–06	5.00E–01	6.3E–06
Selenium[a]	7.0E–09	2.00E–03	3.5E–06
Organic compounds			
Benzo(a)pyrene	8.0E–08	5.00E–04	1.6E–04
BEHP	7.2E–07	2.00E–00	3.6E–07
Captan	3.0E–08	7.40E–02	4.1E–07
Chlordane	3.0E–08	1.00E–01	3.0E–07
Dibenzofurans	2.2E–08	1.15E–05	1.9E–03
Dichlorobenzene[a]	4.7E–06	1.50E–00	3.1E–06
2,4,D[a]	6.9E–07	8.00E–02	8.6E–06
DDE	1.3E–07	1.80E–03	7.4E–05
Heptachlor	5.3E–08	2.50E–04	2.1E–04
Hexachlorobenzene	6.5E–06	2.60E–01	2.5E–05
Hexachlorobutadiene	1.4E–07	9.00E–04	1.5E–04
Lindane	4.5E–08	1.50E–02	3.0E–06
Methoxychlor	3.5E–08	5.00E–02	7.0E–07
N-Nitrosodimethylamine[a]	2.8E–06	5.00E–03	2.8E–03
Parathion	3.0E–08	1.00E–03	1.0E–07
Pentachlorophenol[a]	5.5E–07	3.00E–01	3.2E–04
PCB	2.0E–08	1.70E–03	1.7E–03
2,3,7,8-TCDD	1.1E–10	1.15E–05	3.4E–10
Toxaphene	7.7E–07	3.30E–01	1.5E–05
Trifluralin	4.8E–08	5.00E–02	9.7E–07
		HI	4.8E–02

Note: TRV, toxicity reference value; HQ, hazard quotient; and HI, hazard index.

[a] Water ingestion was taken into account for these chemicals; for all other chemicals only food ingestion was a significant route of uptake.

Table 8.14 Comparison of Estimated Average Daily Doses of Chemicals of Interest at Zone 5 (Proposed Facility) with TRVs for the Eastern Gray Squirrel

Indicator chemical	Average daily dose (mg/kg-day)	TRV (mg/kg-day)	HQ
Inorganic chemicals			
Arsenic [a]	4.0E–07	4.7E–03	8.4E–05
Cadmium	6.4E–07	5.0E–02	1.3E–05
Chromium	1.5E–07	1.7E–03	8.7E–05
Copper	2.4E–08	7.9E–02	3.0E–07
Cyanide compounds [a]	1.8E–06	9.3E–00	1.9E–07
Lead	1.1E–06	5.0E–05	2.3E–02
Mercury	2.8E–07	5.0E–02	5.6E–06
Nickel	1.8E–06	5.0E–00	3.5E–07
Selenium [a]	5.6E–09	2.5E–02	2.2E–07
Organic compounds			
Benzo(a)pyrene	4.5E–08	5.0E–04	9.0E–05
BEHP	4.0E–07	1.0E–00	4.0E–07
Captan	1.7E–08	9.0E–01	1.9E–08
Chlordane	1.7E–08	2.5E–02	6.8E–07
Dibenzofurans	1.2E–08	3.0E–04	4.1E–05
Dichlorobenzene [a]	2.7E–06	1.0E–01	2.7E–05
2,4-D a	4.8E–07	3.7E–02	1.3E–05
DDE	7.5E–08	1.0E–00	7.5E–08
Heptachlor	3.0E–08	2.5E–03	1.2E–05
Hexachlorobenzene	3.7E–06	2.0E–01	1.8E–05
Hexachlorobutadiene	7.8E–08	9.0E–03	8.7E–06
Lindane	2.7E–08	2.5E–00	1.1E–08
Methoxychlor	2.0E–08	5.0E–01	4.0E–08
N-Nitrosodimethylamine [a]	1.6E–06	5.0E–02	3.2E–05
Parathion	1.7E–08	1.3E–03	1.3E–05
Pentachlorophenol [a]	3.2E–07	3.0E–00	1.1E–07
PCB	1.1E–08	5.0E–04	2.3E–05
2,3,7,8-TCDD	6.3E–11	3.0E–04	2.1E–07
Toxaphene	4.3E–07	1.8E–00	2.4E–07
Trifluralin	2.7E–08	5.0E–02	5.5E–07
		Total Hazard Index	2.3E–02

Note: TRV, toxicity reference value; HQ, hazard quotient; and HI, hazard index.

[a] Water ingestion was taken into account for these chemicals; for all other chemicals only food ingestion was a significant route of uptake.

Table 8.15 Comparison of Estimated Average Daily Doses for the Chemicals of Interest with TRVs for the Red-Tailed Hawk

Indicator chemical	Receptor location number	Receptor location	Average daily dose[a] (mg/kg-day)	TRV (mg/kg-day)	HQ
Inorganic compounds					
Lead	470	On-site stock pond	4.2E-06	2.5E-01	1.7E-05
	269	On-site stock pond	7.9E-07	2.5E-01	3.2E-06
	412	Other off-site stock pond	1.2E-06	2.5E-01	4.6E-06
Mercury[a]	470	On-site stock pond	6.3E-04	1.9E-02	3.3E-02
	269	Off-site stock pond	5.1E-05	1.9E-02	2.7E-03
	412	Other off-site stock pond	1.8E-04	1.9E-02	9.5E-03
Selenium	470	On-site stock pond	1.0E-07	2.0E-03	5.0E-05
	269	Off-site stock pond	2.2E-07	2.0E-03	1.1E-04
	412	Other off-site stock pond	2.6E-08	2.0E-03	1.3E-05
Organic compounds					
DDE	470	On-site stock pond	7.9E-06	3.0E-02	2.6E-04
	269	Off-site stock pond	1.3E-06	3.0E-02	4.4E-05
	412	Other off-site stock pond	4.0E-06	3.0E-02	1.3E-04
PCB	470	On-site stock pond	4.9E-07	7.9E-01	6.2E-07
	269	Off-site stock pond	1.5E-07	7.9E-01	1.9E-07
	412	Other off-site stock pond	4.7E-07	7.9E-01	6.0E-07
2,3,7,8-TCDD	470	On-site stock pond	3.0E-09	3.0E-08	9.9E-02
	269	Off-site stock pond	8.6E-10	3.0E-08	2.9E-02
	412	Other off-site stock pond	2.7E-09	3.0E-08	9.0E-02
			HI	**Location**	
				470	1.3E-01
				269	3.1E-02
				412	9.0E-02

Note: TRV, toxicity reference quotient; HQ, hazard quotient; and HI, hazard index.

[a] Average daily dose = [fish tissue concentration] × [ingestion rate] / [body weight].

Table 8.16 Comparison of Estimated Average Daily Doses of Chemicals of Interest with TRV for the Black-Crowned Night Heron

Indicator chemical	Receptor location number	Receptor location	Average daily dose[a] (mg/kg-day)	TRV (mg/kg-day)	HQ
Inorganic compounds					
Lead	470	On-site stock pond	1.7E–05	6.0E–03	2.9E–03
	269	Off-site stock pond	1.2E–05	6.0E–03	2.0E–03
	412	Other off-site stock pond	4.5E–06	6.0E–03	7.5E–04
Mercury	470	On-site stock pond	4.0E–05	1.0E–03	4.0E–02
	269	Off-site stock pond	6.2E–04	1.0E–03	6.2E–01
	412	Other off-site stock pond	1.3E–05	1.0E–03	1.3E–02
Selenium	470	On-site stock pond	2.5E–07	2.5E–02	1.0E–05
	269	Off-site stock pond	1.8E–07	2.5E–02	7.0E–06
	412	Other off-site stock pond	6.5E–08	2.5E–02	2.6E–06
Organic compounds					
DDE	470	On-site stock pond	8.1E–06	8.0E–03	1.0E–03
	269	Off-site stock pond	8.6E–06	8.0E–03	1.1E–03
	412	Other off-site stock pond	4.1E–06	8.0E–03	5.1E–04
PCB	470	On-site stock pond	8.9E–07	3.0E–03	3.0E–04
	269	Off-site stock pond	5.4E–07	3.0E–03	1.8E–04
	412	Other off-site stock pond	8.0E–07	3.0E–03	2.7E–04
2,3,7,8-TCDD	470	On-site stock pond	3.7E–08	2.5E–08	1.5E–02
	269	Off-site stock pond	2.3E–08	2.5E–08	9.2E–03
	412	Other off-site stock pond	3.3E–08	2.5E–08	1.3E–02
			HI	**Location**	
				470	6.0E–02
				269	6.3E–01
				412	2.8E–02

Note: TRV, toxicity reference quotient; HQ, hazard quotient; and HI, hazard index.

[a] Average Daily Dose = [fish tissue concentration] × [ingestion rate] / [body weight].

To provide a quantitative analysis of some of the uncertainties associated with this assessment, predicted exposures to the indicator species with the lowest estimated margin of protection, the black-crowned night heron, were evaluated using probabilistic methods. The probabilistic analysis focused on estimating the distribution of potential doses of methylmercury, the compound which contributes most substantially to the HI calculated for this species. This analysis consisted of the following steps: (1) identifying potential exposure points for the herons in the vicinity of the proposed waste treatment facility, (2) estimating concentrations of mercury in pond and stream water, (3) estimating concentrations of methylmercury in fish in these water bodies, and (4) estimating the magnitude and distribution of potential methylmercury exposures to the black-crowned night heron using a Monte Carlo simulation of heron foraging behavior.

Identification of potential exposure points

Under the assumption that exposure can be adequately represented by ingestion of contaminated prey tissue, potential exposure points for the black-crowned night herons are locations in the vicinity of the proposed waste treatment facility site at which foraging could take place. Pearson[46] reported that the preferred foraging locations for black-crowned night herons included the edges of shallow tidal creeks, and the shorelines of shallow ponds, streams, and swamps. In addition, it has been observed that throughout the duration of a foraging event, herons use multiple foraging locations. Assuming that all shoreline areas of ponds and permanent streams within a five-mile radius of the proposed facility are accessible and equally desirable as feeding sites by herons (i.e., comparable prey abundance, etc.), potential exposure points were identified as discrete shoreline segments in ponds or along streams of 100 m in length. Ponds less than 100 m in diameter were assumed to be single locations. Streams less than 100 m in width were treated as a single shoreline. Both banks of rivers greater than 100 m in width were included as separate shorelines. Digitized U.S. Geological Survey (USGS) topographic maps (1:100,000 scale; 7.5-minute quadrangle) were used to locate shoreline segments and describe potential exposure point locations. Hydrographic features on these maps are presented in the AutoCAD DXF polyline format. A Fortran program was written to automatically locate all intermittent and permanent streams and ponds larger than 100 m in diameter within the study area and discretize their shorelines into 100-m segments. A total of 16,675 discrete segments were identified along stream shorelines within the study area. Of these, 5,031 occur on permanent streams and are thereby considered potential foraging locations/exposure points. Approximately 1,850 shoreline segments on larger ponds were also identified in the digitized maps. Due to the scale used in the digitized USGS maps, ponds smaller than 100 m in diameter are not shown. As such, these smaller ponds were manually located within the study area from 1:24,000 scale topographic maps. An additional 812 smaller ponds were located by this method. A total of 7,693 potential foraging locations were thus identified in proximity to the proposed facility (Figure 8.6).

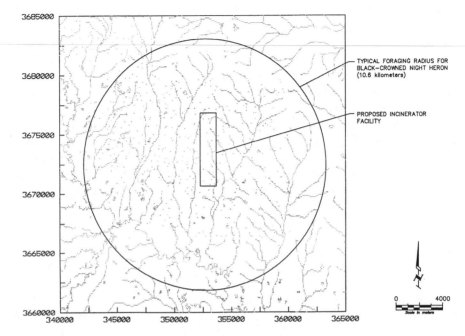

Figure 8.6 Plot of permanent ponds and streams surrounding the proposed waste treatment facility.

Estimation of mercury concentrations in ponds and streams

Methylmercury concentrations in pond water were estimated through the mass balance approach described previously, which considers the atmospheric deposition rate onto the pond surface and watershed area and the volumetric flow rate of water into each pond. Although intermittent streams may enter and/or drain some ponds, this interaction was not considered and ponds are considered separately from each other and from lotic systems that may be in the vicinity. The algorithm used in estimating pond water concentrations is assumed to be the mass rate of chemical contribution from runoff and direct deposition, divided by the volumetric flow rate of water from surface runoff, groundwater infiltration, and rainfall, minus evaporation.

Like the estimations of chemical concentration in pond water, a mass balance approach was used to estimate methylmercury concentrations at exposure points in streams. In utilizing this approach however, it was necessary to consider chemical mass contribution and volumetric water flow rates from both the immediate surroundings of a given stream segment as well as from all upstream segments. The result of such an approach is to obtain estimates of cumulative dissolved concentrations at each discrete location.

The initial step in this process was to develop a Fortran program that was capable of utilizing the location data (i.e., UTM coordinates) for each

100-m stream segment developed from the digitized USGS quadrangle maps to predict the routing and flow direction of all streams in the study area. As noted above, 16,675 individual 100-m stream segments were identified within the area on intermittent and permanent streams. These are distributed uniformly over the area due to the flat topography. Because of the flat topography and the vertical contour interval used on the USGS maps, it was not possible to use elevation data as the basis for modeling stream routing and flow direction.

This problem was addressed through the use of an algorithm which identifies "end point" segments of each stream and utilizes several assumptions to derive a "pointing" vector that stores all data regarding how stream flow is routed between segments. The routing model assumes that intermittent streams flow toward permanent streams and that junction points for two or more segments will have only one "exit" stream. For streams whose "origin" and/or "exit end" segments are outside the confines of the maps, the flow direction at the incoming segment was hard wired into the code. The model was run iteratively until the flow direction for all stream segments was solved.

After solving for flow direction, it is necessary to derive estimates of flow rates for each stream segment. To determine stream flow rates, the 57-year average volumetric rate of surface runoff for the watershed area of $1.1 \text{ ft}^3/\text{sec}/\text{mi}^2$ was obtained from USGS data for Mississippi. The total volumetric runoff rate represented by the study area was then derived and divided by the total number of stream segments identified in intermittent and permanent streams. The simplifying assumption was made that each segment contributes an equal volume of runoff to cumulative stream flow.

Estimation of fish tissue concentrations

The estimated fish tissue concentration is assumed to be a linear function of dissolved concentration in water. As such, the pond and stream models were set up to calculate the whole-body prey tissue concentration of methylmercury concurrently with the water concentration.

The calculations were performed for each potential exposure point, resulting in an estimate of whole body methylmercury concentration in tissue of hypothetical prey organisms, which remains at that point for a length of time sufficient to equilibrate with surrounding media.

Monte Carlo simulations using a random walk model

As noted above, the objective of the original screening evaluation was to use the black-crowned night heron as an indicator of potential food web impacts in aquatic systems in the vicinity of the proposed facility. The exposure scenario considered most pertinent to this objective was ingestion of impacted prey tissue. Although the results of the screening evaluation were presented in terms of a conservative point estimate, it is clear that the degree to which exposure could occur may vary over a wide range given the distribution of contaminant loadings to various locations throughout the study area, and the variation in the parameters affecting the average daily dose to individuals within a heron population. As such, the fundamental intent in applying Monte Carlo simulations to this analysis is to address the question, "Once exposure point concentrations are known at foraging sites throughout the study area, what is the distribution of possible daily doses that could occur to black-crowned night herons considering their normal foraging strategies and activity patterns?"

The conceptual approach for the random walk model is essentially a simulation of 3,000 foraging flights of a hypothetical black-crowned night heron having a home range centered on the proposed facility. For each Monte Carlo simulation performed, the model randomly chooses values for each parameter pertinent to the physical characteristics, life history, foraging behavior, and activity patterns of the black-crowned night heron. The ranges and distributions of values for each parameter utilized in the Monte Carlo simulations applied in the random walk model were obtained from published literature and from personal communication with individuals and organizations conducting contemporary research on the black-crowned night heron. The parameters, ranges and assumed distribution of values utilized in the Monte Carlo simulations are presented in Table 8.17.

Herons are assumed to forage from sundown to sunrise (12 hours). Equal likelihood of feeding in any permanent pond or stream section within 10.6-km radius of the roost and equal likelihood of feeding success is assumed. The first flight from the roost is random and varies from 100 m to 10.6 km in length, but based on energetic considerations most initial flights are assumed to be short (100 m is the most likely distance). Heron are assumed to forage between 10 and 39 minutes in each location. Secondary flights are assumed to be between 100 and 500 m of the last foraging location. The sum of the doses from food consumed at each foraging location for a day is the average daily dose. A plot of five hypothetical random foraging flights of a black-crowned night heron is presented in Figure 8.7. A plot of all 3,000 simulations of daily foraging events is provided in Figure 8.8.

The final output of the process is an estimated distribution of daily doses to the hypothetical exposed resident population. Once derived, this distribution can then be compared to the point estimate obtained under the screening analysis as a means for evaluating the degree of conservatism that may be inherent in that estimate.

The results of the 3,000 simulations are summarized in the cumulative distribution plot in Figure 8.9. As can be seen, the 95th percentile of estimated average daily dose distribution is below the worst-case point estimate for heron foraging solely in the pond with the high predicted concentration of mercury. Therefore, predicted exposures to the black-crown night heron population are substantially below toxic levels.

Table 8.17 Monte Carlo Model Exposure Parameters

Parameter	Point estimate	Reference	Probability distribution (values)	Reference
Body weight (kg)	0.850	34	Normal (mean = 0.820; standard deviation = 0.087)	47
Ingestion rate (kg/day)	0.212	29	Normal (mean = 0.183; standard deviation = 0.020)	29
Foraging time (hr/day)	12	50	Point estimate (12)	50
Foraging time per 100 m	Daily at pond with highest MeHg concentration	Assumed	Uniform (30.4–38.8)	29
Patch (min/patch)				
Flight speed (km/hr)	NC	N/A	Uniform (29.0–33.8)	49
Initial flight (m)	NC	N/A.	Logtriangular (most likely = 100; range = 100–10,600)	50
Secondary flight (m)	NC	N/A	Logtriangular (most likely = 100; range = 100–500)	51
Fraction of food that is fish	1	Assumed	Point estimate (1)	Assumed
MeHg concentration in fish (mg/kg)	2.5×10^{-3}	ACE model	Model predicted concentration in each water body used by the heron	ACE model

Note: NC, not considered and NA, not applied.

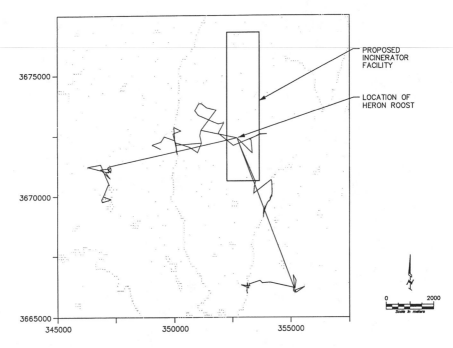

Figure 8.7 Plot of five random daily foraging flights for black-crowned night heron.

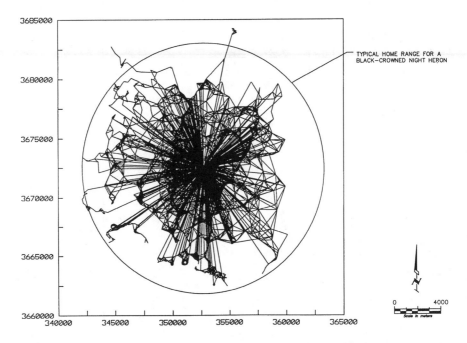

Figure 8.8 Plot of 3,000 random daily foraging events for the black-crowned night heron.

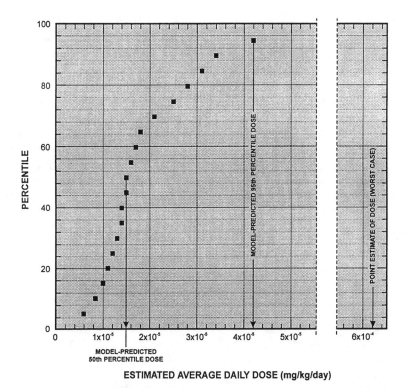

Figure 8.9 Comparison of cumulative percentile of hypothetical average daily dose to the worst case daily dose estimate for the black-crowned night heron exposed to mercury through consumption of fish.

Conclusions

This ERA is a prospective assessment of the potential for adverse impacts to agriculture, aquaculture, forestry, fisheries, and wildlife resources within a five-mile radius of a proposed waste treatment facility which may be associated with increased chemical concentrations in environmental media resulting from projected emissions from the facility. This assessment evaluated exposures to indicator species representing fish (fathead minnow and channel catfish), agricultural crops (corn, wheat, and soybeans), mammals (white-tail deer, cottontail rabbit, and gray squirrel), and birds (black-crowned night heron and red-tailed hawk) to 29 COIs having the greatest potential for ecological effects. Modeled concentrations of all COIs for surface water, soil plant tissue, and fish tissue were used as the basis for estimating exposure to indicator species. TRVs for indicator species and COIs were derived from established standards and/or published literature values, with the use of uncertainty factors for extrapolation. The calculated media concentrations or doses were compared to the TRVs for all indicators to characterize the potential for adverse affects as a result of potential exposures.

Based on these analyses, we conclude that exposures to hypothetical emissions of the 29 COIs from the proposed waste treatment facility are negligible for all indicator species evaluated. The margin of protection (i.e., the magnitude by which the TRV exceeds the predicted exposure) ranges from approximately two for mercury and the black-crowned night heron to 10 million for the fungicide captan and the loblolly pine. In general, predicted exposures of COIs to indicator species were at least 10 times less than applicable TRVs. As a highly conservative measure of cumulative hazard, the HI for all chemicals were summed for each indicator species. These cumulative hazard indices values are also less than 1.0, confirming that no adverse incremental effects are likely to occur in indicator species due to routine operation of the proposed waste treatment facility.

References

1. USEPA, *Risk Assessment Guidance for Superfund Volume Human Health evaluation Manual (Part A)*, EPA/540/1-89/002, U.S. Environmental Protection Agency, Washington DC, 1989.
2. Ney, R. E., *Where Did That Chemical Go? A Practical Guide to Chemical Fate and Transport in the Environment*, Van Nostrand Reinhold, New York, 1990, 192.
3. Conway, R. A., *Environmental Risk Analysis for Chemicals*, Van Nostrand Reinhold, New York, 1982.
4. Klaassen, C. D., M. O. Amdur, and J. Doull, *Casarett and Doull's Toxicology, the Basic Science of Poisons*, 3rd ed., Macmillan, New York, 1986.
5. USFW, U.S. Fish and Wildlife Service Research Information Bulletin, No. 84-78. 1984.
6. USEPA, *U.S.Environmental Protection Agency Policy and Guidance for Risk characterization*. Attachments to Memorandum for Carolyn Browner, Administrator of USEPA, 1995.
7. USEPA, *User's Guide for the Industrial Source Complex (ISC 3) Dispersion Models*, I, User Instructions, Research Triangle Park, September, EPA-454-/B-950039. 1995.
8. CAPCOA, *AB2588 air toxics hot spots risk assessment guidance*. California Air Pollution Control Officers Association. Office of Environmental Health Hazard Assessment, Sacramento, 1992.
9. Sehmel, G. Particle and gas dry deposition: a review. *Atmosph. Environ.* 14:983–1011. 1980.
10. CDHS. California Department of Health Services. California Site Mitigation Decision Tree Manual. Toxic Substances Control Division, Alternative Technology and Policy Development Section. May 1986.
11. Karickhoff, S. W., D. S. Brown, and T. A. Scott, Sorption of hydrophobic pollutants on natural sediments. *Water Res.* 13:241–248. 1978.
12. Dickson, K. L., A. W. Maki, and W. A. Brungs, *Fate and Effects of Sediment-Bound Chemicals in Aquatic Systems*, Pergamon Press, New York. 1987.
13. SCS, Soil survey of Noxubee County, Mississippi. U.S. Department of Agriculture Soil Conservation Service. 1983.
14. Akagi, H., D. C. Mortimer, and D. R. Miller, Mercury methylation and partitioning in aquatic systems. *Bull. Environm. Contam. Toxicol.* 28, 872, 1979.

15. Horowitz, A. J. *A Primer on Sediment-Trace Element Chemistry*. Lewis Publishers, Chelsea, MI, 1991.
16. Manahan, S. E., *Environmental Chemistry*, Lewis Publishers, Chelsea, MI, 1990.
17. USEPA, *The Health and Environmental Impacts of Lead and an Assessment of a Need for Limitations*. 560/2-79-001 U.S. Environmental Protection Agency Rep. 1979, 494.
18. Van Hattum, B., K. R. Timmermans, and H. A. Govers, Abiotic and biotic factors influencing in situ trace metal levels in macro invertebrates in freshwater ecosystems. *Environ. Toxicol. Chem.* 10, 275, 1991.
19. Aston, S. R. and I. Thornton, Regional geochemical data in relation to seasonal variations in water quality. *Sci. Total Environ.* 7, 247, 1977.
20. USEPA, *Water-related Environmental Fate of 129 Priority Pollutants*, I and II, Office of Water Planning and Standards Office of Water and Waste Management, USEPA, Washington, DC. 1979.
21. Eisler, R., *Selenium hazards to fish, wildlife, and invertebrates: a synoptic review*. 85(1.5), USFWLS Biol. Rep. 1985.
22. Eisler, R. *Arsenic hazards to fish, wildlife and invertebrates: a synoptic review*. 85, 1.12 USFWLS Biol. Rep., 1988.
23. Schintu, M., T. Kauri, and A. Kudo, Inorganic and methyl mercury in inland waters. *Water Res.* 23, 699, 1989.
24. National Research Council of Canada, *Distribution and Transport of Pollutants in Flowing Water Ecosystems*, 1, National Research Council of Ottawa, Canada, 1977.
25. Osborn, D., Personal communication. Biologist with the Arkansas Department of Fish and Game. Little Rock, 1992.
26. Genoways, H., *Current Mammalogy*. Vol. 2, Plenum Press, New York, 1990.
27. Raspopov, M. and Y. Isakov, *Biology of the Squirrel*. Amerind, New Delhi, 1980.
28. USEPA, *Interim Wildlife Criteria: Assessment of Screening Level Values*. U.S. Environmental Protection Agency. Office of Science and Technology, Duluth, 1991.
29. Kushlan, J. A., *Wading Birds*. National Audubon Society, New York, 1978.
30. Suter, G. W., *Ecological Risk Assessment*, Lewis, Chelsea, MI, 1993, 583.
31. Schmidly, D. J., *Texas Mammals East of the Valcones Fault Zone Texas*, ATM University Press, Texas, 1983.
32. Grzimek, B., W. Logiges, A. Portman, and E. Thenius, *Grizmeks's Animal Life Encyclopedia*, Vol. 13, Van Nostrand, New York, 1972.
33. Daniels, K. and G. E. Duke, Factors influencing respiratory rates in four species of raptors. *Comp. Biochem. Physiol.* 66A: 703, 1979.
34. Palmer, R. S., Loons through flamingos. *Handbook of North American Birds*. 1, London, Yale University Press, 1962.
35. USEPA, *Superfund Public Health Evaluation Manual*. Office of Emergency and Medical Response, Washington DC, 1986a.
36. USEPA, *Quality Criteria for Water 1986*. EPA/440/5-86-001, 1986.
37. PHYTOTOX, *PHYTOTOX Computerized Database*. University of Oklahoma Department of Botany and Microbiology. 1992.
38. Dragun, J. and A. Chiasson, Elements in North America Soils, *Hazardous material control resources institution*, Greenbelt, MD, 1991, 230.
39. Finley, M. T., M. P. Dieter, and L. N. Locke, Sublethal effects of chronic lead ingestion in mallard ducks, *J. Toxicol. Environ. Health* 1, 929, 1976.

40. Heinz, G.H., Methyl mercury: reproductive and behavioral effects on three generation of mallard ducks. *J. Wildl. Manage.* 43, 394, 1979.

41. Smith, G. J., G. H. Heinz, D. J. Hoffman, J. W. Spann, and A. J. Krynitsky, Reproduction in black-crowned night-herons fed selenium. *Lake Reservoir Manage.* 4, 2, 175, 1988.

42. Henny, C. J., L. J. Blus, and A. J. Krynitsky, Current Impact of DDE on Black-Crowned Night-Herons in the Intermountain West. *J. Wildl. Manage.* 48, 1, 1984.

43. Braune, B. M. and R. J. Nordstrom, Dynamics of organochlorine compounds in herring gulls. III. Tissue distribution and bioaccumulation in Lake Ontario gulls. *Environ. Toxicol. Chem.* 8, 957, 1989.

44. Hoffman, D. J., B. A. Rattner, C. M. Bunck, and A. Krynitsky, Association between PCBs and lower embryonic weight in black-crowned night herons in San Francisco Bay. *J. Toxicol. Environ. Health* 19, 383, 1986.

45. Eisler, R. Dioxin hazards to fish, wildlife and invertebrates: a synoptic review. *USFWLS Biol. Rep.* 85, 1.8, 1986.

46. Pearson, T. G., *Birds of America*, The University Society, New York, 1978

47. Span, J. Personal communication. Laurel, 1993.

48. Fasola, M. Acitivity rhythm and feeding success of nesting night-herons, *Nyciticorax Ardea* 72, 1984, 217.

49. Hollings, C. S., *Cross-scale morphology, geometry, and dynamics of ecosystems.* *Ecol. Monogr.* 62, 1992, 447.

50. Bateman, D., Movement-Behavior in Three species of Colonial-Nesting Wading Birds: A Radio Telemiting Study, unpublished Ph.D dissertation, Auburn University, Auburn, 1970.

chapter nine

Environmental monitoring for effects of hazardous waste incineration

Ann Bergquist Shortelle and William Thomas Marsh

Contents

Need for and purposes of environmental monitoring

Environmental monitoring can be defined as a long-term evaluation of the environment, including standardized measurement, in order to define its status and trends.[1] For hazardous waste incinerators and other waste treatment operations, environmental monitoring is generally conducted to determine trends in the quality of the environment which may be affected by the release of chemicals into the environment from the operation of the facility. The environmental monitoring program generally serves a variety of purposes:

- To determine baseline conditions (preoperational monitoring)
- To comply with permits (monitoring required by permit)
- To provide data sufficient to conduct ecological and human health risk assessments or natural resource damage assessments

Other objectives may also be met by some monitoring programs. Examples include providing an early warning system to avoid deleterious environmental effects and detecting episodic effects such as spills.[2]

Operating permits include source, discharge, and other monitoring to ensure compliance with applicable legislation such as the Clean Air Act, the Clean Water Act, and the Resource Conservation and Recovery Act. Although compliance monitoring can be a valuable core around which the environmental monitoring program for the facility may be built, the requirements are generally insufficient to ensure an adequate database to fulfill other objectives. On the other hand, variables to be monitored, sampling locations, and sampling frequencies must be carefully selected to provide maximum relevant data without a cost prohibitive study design.

Establishment of baseline conditions is a critical element of any monitoring program. Facilities are often located in industrialized areas with natural resources already significantly altered by ongoing anthropogenic activities. There is potentially significant liability associated with natural resource damages, which may be ameliorated by establishing a credible baseline[3] and conducting limited, focused sampling not required by compliance monitoring. The environmental manager should endeavor to define short-term and long-term data needs (and their potential benefits) in designing the monitoring program, and include as many long-term monitoring variables as feasible within the monitoring budget.

The conceptual site model

The site-specific conceptual site model is the initial framework around which the environmental monitoring plan will ultimately be designed. The conceptual site model defines the sources, environmental pathways of exposure, and the potential receptors and is a standard preliminary step in the human health and ecological risk assessment process.[4,5] For any hazardous waste incinerator, this process begins with characterization of potential emissions generated from the anticipated waste stream, along with ash residues, and wastewater compositions. A variety of organic and inorganic constituents can be anticipated in the waste streams,[6] and some portion of these will be routinely characterized at the source by compliance monitoring.

Environmental exposure pathways define the ways in which constituents from the facility waste streams find their way to potential receptors. This may be a single-step process (e.g., inhalation of an airborne constituent by a downwind receptor) but is frequently a multistep process involving the transport and fate of the constituents. For example, an airborne particulate may be deposited downwind on soils that are carried to a nearby water body via runoff and ultimately find its way into the aquatic food web. The rates at which such transport mechanisms are operational (or even likely) are governed by the physical and chemical characteristics of the constituents themselves[6] and the characteristics of the media within which the constituents are moving.

Potential ecological receptors are also identified. These may include terrestrial and aquatic habitats within the potential exposure area and the communities of organisms that live in these habitats. The list of potential receptors may be quite lengthy, but these species lists are usually readily available from existing sources. During the design of the monitoring program, it will be necessary to identify the key receptors, a small subset from the universe of potentially exposed biota.

Environmental monitoring program design

The environmental monitoring program should be devised based upon the specific program objectives, cost-effectiveness, patterns and variability of contamination, and other considerations (e.g. accessibility, security, etc.).[7] The most cost-effective plan in the short-term may be the one that meets the minimum compliance monitoring requirements, but consideration of longer-range objectives (e.g., risk assessment or natural resource damage assessment) may only require minor modifications to compliance monitoring with modest impact to costs and considerable savings over the long term.[8]

The environmental monitoring program may focus on the spatial distribution of contaminants or effects that requires a large number of sampling locations, environmental trends that require a high frequency of sampling, or on in-depth inventories of pollutants. A monitoring program that attempts to focus on all of these is likely to be cost prohibitive. The operational environmental monitoring program for the facility must carefully and fully define the study objectives and then prudently select key parameters, sampling locations, and sampling frequency to accommodate a long-term plan.

Existing data and ongoing monitoring programs

When the environmental monitoring program is being defined, it is helpful to gather relevant existing data from historical studies and ongoing monitoring efforts within the geographical area of investigation. A substantial database may be available with which to characterize the baseline conditions. Local colleges and universities may be sources of information. There are a number of large, comprehensive environmental monitoring programs conducted by federal agencies which may have relevant data. Selected examples include U.S. EPA's Environmental Monitoring and Assessment Program,[9] the National Status and Trends Program (overseen by the National Oceanic and Atmospheric Association), National Dry Deposition Network (overseen by USEPA), Chemical Contaminant Monitoring Program (Food and Drug Administration), and the National Contaminant Biomonitoring Program (U.S. Fish and Wildlife Service). State and local agencies and not-for-profit organizations may have similar programs. Finally, some industry groups may also have environmental monitoring ongoing within the area of interest. Local information is also useful in identifying components of the environment especially valued by the community. It is important to consider these values when designing the environmental monitoring program. Whether the valued resource is a sport fishery or a heron rookery, collecting data that address the status of the valued resource (as potentially impacted by contaminates from the hazardous waste incinerator) is important not only scientifically but also from a public relations standpoint.

Selection of analytical parameters and media

It is likely that the analytical parameters of primary interest for compliance monitoring will be a list of organic and inorganic constituents that may occur in the waste streams. From among these parameters, the study designer will want to select a subset of parameters around which the nonrequired portion of the environmental monitoring program can be designed. A number of factors should be considered when selecting parameters for long-term monitoring, and these are discussed by category of parameter (physical/chemical and biological).

Physical/chemical

The chemicals that may occur in the waste streams fall into a number of categories within the general categories of organic and inorganic chemicals. Examples of these classes or groups of organic chemicals include polycyclic aromatic hydrocarbons (PAHs), PCBs, dioxins, volatile organics, and others.[6] Within a group of chemicals, transport mechanisms and the ultimate fate of the chemical are dependent upon a variety of physical and chemical attributes affecting the behavior of the individual chemical in the environment.[10] Selection of the subset of constituents to be monitored should consider a number factors:

- The expected contribution of the constituent to the waste stream
- The environmental persistence of the constituent
- The bioaccumulation potential of the constituent
- The toxicity of the constituent, and the type(s) of environmental effects anticipated
- Analytical methodology and detection levels necessary to provide relevant monitoring data
- The constituent's ability to serve as a surrogate for other, similar constituents
- Public interest in or awareness of certain constituents, if known

By selecting a subset of chemicals to be sampled within the environment, monitoring costs can be reduced. The selected constituents should adequately represent the chemicals from the waste stream and their anticipated toxicities. For future evaluation of ecological effects, monitoring should be primarily concerned with chemicals that produce adverse ecological effects (e.g., reproductive failure) and/or that have a strong tendency to bioaccumulate, such as mercury, PCBs, or 2,3,7,8-TCDD. If possible, monitoring parameters should also be as unique to the facility as feasible for the site conditions. For example, although PAHs may constitute a large fraction of the semivolatile organics in the waste streams, PAHs are produced by any incomplete combustion and are ubiquitous in industrial areas, especially in soils and sediments, and thus may not represent the best selections for an environmental monitoring program specific to the hazardous waste incinerator. As another example, in some regions certain metals that may be present in the facilities' waste streams may occur naturally at elevated levels (e.g., potentially toxic to sensitive receptors) due to the surrounding geology. Native communities of organisms are already selected for or adapted to these conditions.

Some physical and chemical parameters should be a routine part of any sampling event regardless of the other chemical or biological elements being sampled.[11,12] Examples are meteorological conditions during air sampling, general weather conditions (temperature, cloud cover, precipitation), and some limnological parameters when sampling water bodies. Selection of other physical/chemical parameters should be directly associated with the intended use of the monitoring data and should be selected if the parameter has an important effect on other parameters being measured. In soils and sediments, the total organic carbon content can greatly influence the mobility and bioavailability of chemicals with an affinity for carbon. In stream sediments, if benthic invertebrate communities are being assessed and compared, it is necessary to determine grain size. Hardness, alkalinity, and pH influence the aquatic toxicity of a variety of dissolved metals, as well as controlling speciation. Failure to include such general parameters can significantly increase uncertainties when modeling chemical behavior in the environment and lead to unnecessarily conservative assumptions regarding bioavailability and the potential for adverse impacts.

Biological

Biological monitoring can be resource intensive, but rewarding, with proper selection and planning. Primary exposure media, such as air and water, associated with the hazardous waste incinerator are dynamic by nature, whereas the resident biota provide a sentinel that naturally integrates the exposure from all exposure pathways over time.[13] Bioaccumulation monitoring is the most common type of biomonitoring, although environmental monitoring of physiological or biochemical biomarkers is becoming more common as indicators of environmental change.[14] Histological and morphological methods have also been used. Biological monitoring is especially useful for chemicals that are present in abiotic media (soil, water, or air) at concentrations below analytical detection limits or that are short lived in the environment but highly toxic. Parameters to be monitored should be directly linked to the facility source chemicals, and this association should be as specific as possible. Elements to be included in a biological monitoring program should be linked to natural resources that are ecologically important and valued by the community. They should also be abundant, readily identified, and easily sampled.[15]

In addition, to provide maximum information, biological monitoring parameters should be interpretable at several trophic levels or provide a connection to other organisms not directly monitored.[16]

Additional attributes are sought for organisms that are to be monitored for tissue concentration. The organisms should be sedentary or of sufficiently small home range to ensure uniform exposure, and the individuals within the population should be relatively uniform. Selection of a long-lived species and a species that continuously accumulates over time may be desirable.[17]

The organisms selected for monitoring may also depend upon the primary abiotic receiving media (and thus, the primary transport mechanisms to the organisms) identified in the conceptual site model, and the availability of ecotoxicity data for interpretation. The transport medium also dictates the routes of exposure which predominate. For example, some airborne contaminants may produce foliar damage to plants by direct exposure, and be inhaled by animals. Selected terrestrial plants are commonly monitored for effects due to acid deposition, fluoride, or ozone exposure[18] and are the best choice if these parameters are prevalent. Plants, even aquatic macrophytes, are also sometimes used to monitor bioaccumulation of chemicals, especially metals, although the ecotoxicity database for interpretation of these monitoring results is weak in many areas, and should be carefully reviewed before selection.[19,20]

In terrestrial settings, small mammals have been used successfully to monitor trace elements, and bird feathers are considered excellent biomonitors for certain metals.[21] Terrestrial invertebrates have also been identified as very useful bioindicators in terrestrial ecosystems.[9]

In aquatic ecosystems, benthic macroinvertebrates,[22,23] plankton,[17] and fish[24] have all been commonly used in biomonitoring programs. Effects of numerous contaminants on benthic macroinvertebrates are available, but quantitative sampling is beyond the scope of many long-term, but single-site, studies.[25] A possible exception is the use of bivalves for monitoring contaminant concentrations in estuarine settings.[26] Plankton also present certain taxonomic and sampling difficulties.[27] On the other hand, fish communities are frequently identified as important natural resources by the public, and they also have a number of attributes that make them desirable as biomonitors.[28] Among these are that distributions and taxonomy are known, less frequent sampling is necessary than for smaller aquatic organisms, the communities tend to be persistent, and they recover rapidly from natural disturbance.[29]

Biomarkers have been found to be useful in environmental monitoring to identify organism stress and damage within organisms,[9,30] but these are frequently more useful as biological criteria, indicating overall status, rather than identifying chemical-specific responses. There are notable exceptions (e.g., cholinesterase inhibitors and metabolites of xenobiotic chemicals), but currently, these exceptions offer little benefit for routine sampling as part of the environmental monitoring for a hazardous waste incinerator. A variety of bioassays for aquatic and terrestrial environments are also available.[31]

Screening level bioassays can be very inexpensive; however, such screenings may generally be unnecessary at the point of exposure and thus are unlikely to be included in the environmental monitoring program for such a site except as compliance monitoring for permitted discharges.

Habitat assessment protocols have been employed in environmental monitoring,[32,33] but these methods are not designed to identify impacts associated with chemical-specific exposures. Communities and other associations of organisms are also sometimes monitored, and changes in community structure[34] or function[35] are analyzed to identify trends. Although such sampling and accompanying multivariate analysis is prevalent in community ecology, it is unlikely to be practical for environmental monitoring programs for single hazardous waste incinerators.

Selection of sampling locations

Because biological communities have distributions that tend to be clumped, not random, establishment of sampling locations must take into account the underlying distribution of the receptors of interest within the environment to be monitored. Sampling locations are frequently established in a haphazard fashion and may violate underlying assumptions of random sampling when data are analyzed. However, a statistically sound sampling design may allow for stratification or clustering of sites to suit the distribution of biological communities or sensitive habitat.[7] Strata may also include areas of high utilization by humans (e.g., recreational areas), which may also be included in risk assessments.[8] Within strata, samples are taken randomly. Samples taken from the various strata provide site-specific monitoring data to establish the current status of the environment, and trends, both spatial and temporal. These data will be instrumental in conducting future assessments without undue reliance on conservative assumptions, and will provide a framework within which modeling estimates (if needed) can be bounded in reality.

Environmental variability and sampling frequency

Ideally, environmental monitoring will identify any environmental alterations associated with the operation of the hazardous waste incinerator. However, habitats and environmental conditions are dynamic, and change, even deleterious change, can occur naturally or as a result of other, unrelated anthropogenic effects, or as a result of cumulative assaults from multiple sources.

Once monitoring parameters of interest are identified, the temporal and spatial scales of variability should be estimated to identify ideal and adequate sampling frequency. These sources of variability will also determine the length of monitoring necessary to adequately establish or supplement baseline conditions. Spatial and temporal variability also influence exposure of receptors to chemicals.[36] A variety of factors such as salinity, dissolved oxygen, food availability, seasonal weather patterns, and others, contribute to variability.

For most long-term environmental monitoring programs designed for hazardous waste incinerators with operational pollution control devices, relevant time frames are on the order of months to many years for parameters associated with biomass, reproductive success, population and community structure, and dynamics.[37] Seasonal variability is a natural part of temperate ecosystems,[38] and thus routine sampling conducted annually, but at the same time each year, will help to avoid the confounding effects of this variability. In some ecosystems, such as stratified lakes, sampling at turnover minimizes spatial and seasonal variability.

Established environmental monitoring programs in similar habitats may also contribute to an adequate estimate of variability. For some facilities with predictable winds and water flows, there may be an upgradient/downgradient aspect to sampling. In this case, it may be simpler to identify facility impacts from regional, cumulative, or natural impacts. At other sites, it may be advisable to establish one or more reference sites for long-term monitoring. To conserve monitoring resources, reference areas may be identified within existing governmental monitoring programs.

Statistical treatment and analysis of the environmental monitoring data depends upon sampling that is adequate. Water (especially flowing water) and air quality are dynamic and may require frequent sampling for some constituents to define trends and exposure. Spatial variability should also be evaluated within the context of the area to be monitored to determine adequate sample size within sampling areas.[1]

Data handling and quality assurance

Data handling, analysis, and interpretation are the last steps in the environmental monitoring program. To meet the objectives of the environmental monitoring program, data analysis and interpretation must be effectively understood by operators and managers at the incinerator facility, scientists conducting the study, regulators, and the public or other interested parties.[36] Quality assurance and use of standard analytical procedures are essential elements to ensure that all stakeholders have confidence in the environmental monitoring results. Any environmental monitoring program should include appropriate quality assurance and quality control,[30,39,40] and a quality assurance program is required for ecological assessments at hazardous waste sites under CERCLA.[41] Although this requirement does not apply to facilities undertaking a modest environmental monitoring effort outside the regulatory framework, a primary objective of the monitoring is to provide reliable, credible data for site-specific use as needed for risk assessment, defense in natural resource damage assessment, and the like. In such uses, the defensibility of the data is essential. The intended use or uses for the environmental monitoring data will identify the methods of data analysis, summary indices, and the like, appropriate for the parameters measured.[42]

References

1. Meybeck, M, V. Kimstach, and R. Helmer. Strategies for water quality assessment, in *Water Quality Assessments: A Guide to the Use of Biota, Sediments, and Water in Environmental Monitoring*, Chapman, D., ed., Chapman and Hall, New York, 1992, chap. 2.

2. Cairns, J. Jr. and E.P. Smith. The statistical validity of biomonitoring data, in *Biological Monitoring of Aquatic Systems*, Loeb, S.L. and A. Spacie, eds., Lewis Publishers, Boca Raton, FL, 1994, chap. 5.

3. Shortelle, A.B., J.L. Dudley, and T.M. Slocum. Natural resource damage assessment—a sleeping giant?, in *TAPPI Environmental Conference Proceedings*, TAPPI Press, 1993, 739.

4. U.S. EPA. *Risk Assessment Guidance for Superfund: Vol I. Human Health Evaluation Manual, Part A.*, Interim Final, Office of solid Waste and Emergency Response, OSWER Directive 9285.7-01, 1989.

5. U.S. EPA. *Proposed Guidelines for Ecological Risk Assessment*, USEPA Risk Assessment Forum, EPA630R-95002B, 1996 (Draft).

6. Teaf, C.M., I.K. Stabile, and P. Moffat. Characterizations of potential emissions from hazardous waste incinerators and related facilities, in *Hazardous Waste Incineration: Evaluating the Human Health and Environmental Risks*, Roberts, S., ed., CRC Press, Boca Raton, FL, 1997, chap. 4.

7. Gilbert, R. *Statistical Methods for Environmental Pollution Monitoring*, Van Nostrand Reinhold, New York, 1987, chap. 1.

8. Shortelle, A.B. Practical considerations for routine sampling plans to facilitate risk assessment, in *TAPPI Environmental Conference Proceedings*, TAPPI Press 1991, 155.

9. Hunsaker, C.T. and D.E. Carpenter, eds. *Ecological Indicators for the Environmental Monitoring and Assessment Program*, EPA 600/3-90/060. U.S. Environmental Protection Agency, office of Research and Development, Research Triangle Park, NC, 1990.

10. U.S. EPA. *Superfund Exposure Assessment Manual*, U.S. Environmental Protection Agency, EPA/540/1.88/001, Office of Remedial Response, Washington, DC, OSWER Directive 9285.5-1, 1988.

11. ASTM. *Data and Information Needs Checklist for Conducting an Ecological Risk Assessment at Contaminated Sites* (draft), ASTM Committee E-47 on Biological Effects and Environmental Fate, Subcommittee E 47.13, 1997, 257.

12. U.S. EPA. *Superfund Program Checklist for Ecological Assessment/Sampling* (draft), U.S. Environmental Protection Agency, Office of Emergency and Remedial Response and Office of Solid Waste and Emergency Response, Washington, DC, 1993.

13. Wiersma, G.B., R.C. Rogers, J.C. McFarlane, and D.V. Bradley, Jr. Biological monitoring techniques for assessing exposure, in *Biological Monitoring for Environmental Effects*, D.L. Worf, ed., Lexington Books, Lexington, MA, 1980, chap. 11.

14. Butterworth, F.M. Introduction to biomonitors and biomarkers as indicators of environmental change, in *Biomonitors and Biomarkers as Indicators of Environmental Change*, F.M. Butterworth, L.D. Corkum, and J. Guzman-Rincon, eds., Plenum Press, New York, 1995, chap. 1.

15. Hellawell, J.M. *Biological Indicators of Freshwater Pollution and Environmental Management*, Elsevier Applied Science, New York, 1986, chap. 3.
16. Yoder, C.O. and E.T. Rankin. Biological criteria program development and implementation in Ohio, in *Biological Assessment and Criteria: Tools for Water Resource Planning and Decision Making*, W. S. Davis, and T.P. Simon, eds., Lewis Publishers, Boca Raton, FL, 1994, chap. 9.
17. Patrick, R. What are the requirements for an effective biomonitor?, in *Biological Monitoring of Aquatic Systems*, Loeb, S.L. and A. Spacie, eds., Lewis Publishers, Boca Raton, FL, 1994, chap. 3.
18. Lange, C.R., S.R. Scott, and M. Tanner. Biomonitoring, *Water Environ. Res.*, 68(4), 801, 1996.
19. Kozuharov, S.I. Plants as bioindicators, in *Biological Monitoring of the State of the Environment: Bioindicators*, J. Salanki, ed., IRL Press, Oxford, 1986, chap. 3.
20. Jones, H.C. and W.W. Heck, Vegetation—biological indicators or monitors of air pollutants, in *Biological Monitoring for Environmental Effects*, D.L. Worf, ed., Lexington Books, Lexington, MA, 1980, chap. 10.
21. Jenkins, D.W., *Biological Monitoring of Toxic Trace Elements*, EPA-600/S3-80-090. U.S. Environmental Protection Agency, Environmental Monitoring Systems Laboratory, Las Vegas, NV, 1981.
22. Davis, W.S., Biological assessment and criteria: building on the past, in *Biological Assessment and Criteria: Tools for Water Resource Planning and Decision Making*, W. S. Davis, and T.P. Simon, eds., Lewis Publishers, Boca Raton, FL, 1994, chap. 3.
23. Resh, V.H. and E.P. McElravy, Contemporary quantitative approaches to biomonitoring using benthic macroinvertebrates, in *Freshwater Biomonitoring and Benthic Macroinvertebrates*, D.M. Rosenberg and V.H. Resh, eds., Chapman and Hall, New York, 1994, chap. 5.
24. USEPA. *National Study of Chemical Residues in Fish, Vol. I*, U.S. Environmental Protection Agency, EPA 823-R-92-008a, Office of Science and Technology, Standards and Applied Science Division, Washington, D.C., 1992.
25. Resh, V.H. Freshwater benthic macroinvertebrates and rapid assessment procedures for water quality monitoring in developing and newly industrialized countries, in *Biological Assessment and Criteria: Tools for Water Resource Planning and Decision Making*, W. S. Davis, and T.P. Simon, eds., Lewis Publishers, Boca Raton, FL, 1994, chap. 12.
26. O'Conner, T.P., A.Y. Cantillo, and G.G. Lauenstein. Monitoring of temporal trends in chemical contamination by the NOAA National Status and Trends Mussel Watch Project, in *Biomonitoring of Coastal Waters and Estuaries*, K.J.M. Kramer, ed., CRC Press, Boca Raton, FL, 1994, chap. 2.
27. Friedrich, G., D. Chapman, and A. Beim. The use of biological material, in *Water Quality Assessments: A Guide to the Use of Biota, Sediments, and Water in Environmental Monitoring*, Chapman, D., ed., Chapman and Hall, New York, 1992, chap. 5.
28. Simon, T.P. and J. Lyons. Application of the index of biotic integrity to evaluate water resource integrity in freshwater ecosystems, in *Biological Assessment and Criteria: Tools for Water Resource Planning and Decision Making*, W. S. Davis, and T.P. Simon, eds., Lewis Publishers, Boca Raton, FL, 1994, chap. 16.
29. Simon, T.P. *Development of Ecoregion Expectations for the index of Biotic Integrity. I. Central corn Belt Plain*. EPA 905-9-91-025. U.S. Environmental Protection Agency, Region 5, Chicago, IL, 1991.

30. U.S. EPA, *Ecological Assessment of Hazardous Waste Sites: A Field and Laboratory Reference*, U.S. Environmental Protection Agency, Environmental Research Laboratory, EPA 600/3-89-013, 1989.
31. Waters, M.D. and L.W. Little, Short-term bioassays of environmental samples, in *Biological Monitoring for Environmental Effects*, D.L. Worf, ed., Lexington Books, Lexington, MA, 1980, chap. 15.
32. Hayslip, G.A., *EPA Region 10 In-Stream Biological Monitoring Handbook (for Wadable Streams in the Pacific Northwest)*, EPA-910-9-92-013, U.S. Environmental Protection Agency, Region 10, Environmental Services Division, Seattle, WA, 1993.
33. Rankin, E.T., Habitat indices in water resource quality assessments, in *Biological Assessment and Criteria: Tools for Water Resource Planning and Decision Making*, W. S. Davis, and T.P. Simon, eds., Lewis Publishers, Boca Raton, FL, 1994, chap. 13.
34. Herricks, E.E. and J. Cairns, Jr. Biological monitoring, part III—receiving system methodology based on community structure, *Water Res.*, 16(2), 141, 1982.
35. Matthews, R.A., A.L. Buikema, Jr., J. Cairns, Jr., and J.H. Rodgers, Jr. Biological monitoring, part IIA—receiving system functional methods, relationships and indices, *Water Res.*, 16(2), 129, 1982.
36. Stewart, A.J. and J.M. Loar. Spatial and temporal variation in biomonitoring data, in *Biological Monitoring of Aquatic Systems*, Loeb, S.L. and A. Spacie, eds., Lewis Publishers, Boca Raton, FL, 1994, chap. 7.
37. Sastry, A.N. and D.C. Miller. Application of biochemical and physiological responses to water quality monitoring, in *Biological Monitoring of Marine Pollutants*, J. Vernberg, A. Calabrese, F.P. Thurberg, and W.B. Vernberg, eds., Academic Press, New York, 1981, chap. 11.
38. Livingston, R.J. Time as a factor in biomonitoring estuarine systems with reference to benthic macrophytes and epibenthic fishes and invertebrates, in *Biological Monitoring of Water and Effluent Quality*, J. Cairns, Jr., K.L. Dickson, and G.F. Westlake, eds., American Society for Testing and Materials, Philadelphia, 1976, chap. 18.
39. Aitio, A. and P. Apostoli. Quality assurance in biomarker measurement, *Toxicol. Lett.*, 77, 195, 1995.
40. U.S. EPA, *Environmental Monitoring Series: Quality Assurance Guidelines for Biological Testing*, U.S. Environmental Protection Agency, EPA-600/4-78-043, Environmental Monitoring and Support Laboratory, Las Vegas, NV, 1978.
41. U.S. EPA, *Quality Assurance Program Plan*, U.S. Environmental Protection Agency, Environmental Research Laboratory, Corvallis, OR, 1987.
42. Demayo, A. Data handling and presentation, in *Water Quality Assessments: A Guide to the use of Biota, Sediments, and Water in Environmental Monitoring*, D. Chapman, ed., Chapman and Hall, New York, 1992, chap. 9.

chapter ten

Evaluation of epidemiologic studies on the human health effects associated with hazardous waste incineration facilities

Lora E. Fleming, Judy A. Bean, Isabel K. Stabile, and Christopher M. Teaf

Contents

1-56670-250-X/99/$0.00+$.50
© 1999 by CRC Press LLC

Introduction

This chapter reviews the available scientific literature concerning human health effects associated with exposure hazardous waste incineration. As discussed in Chapter 11, there are various epidemiologic issues that impact any review of the current literature of the health effects of hazardous waste incineration. First, the existing database on human health effects from hazardous waste incineration is inadequate because this is a relatively new technology. Not only have few investigations in human populations been conducted but, those that have, studied a relatively small number of individuals. Furthermore, studies have not yet begun to evaluate possible chronic health effects, such as cancer, in human communities because there has not been a sufficient latency period from exposure to disease development. Finally, there has been inadequate assessment of exposure in affected human communities.

This chapter also reviews the scientific literature in other areas with possible relevance to hazardous waste incineration. The effects of air pollution on human health have been extensively studied over the last century and are relevant to hazardous waste incineration because inhalation could be a major route of exposure. An extensive literature also exists on the human health effects of the three major emission groups from hazardous waste incineration, i.e., the heavy metals, dioxins, and the volatile organic compounds, as well as on the possible waste stream contaminant, the PCBs; the known human health effects of these pollutants are briefly summarized.

Historical perspectives

Air pollution

The impact of airborne pollutants on human health has been recognized since the Roman Empire. Several large, acute episodes occurred in this century, clearly linking air pollution with human health effects. With regard to the infamous episodes of London fog from coal pollution, a review of death records by Goldsmith (1968) revealed many similar epidemics in that city's past, documented as far back as 1873 (Folinsbee, 1993; Gong, 1992; Momas et al., 1993). Since the 1960s, many of the industrialized nations have instituted relatively strict air standards. Unfortunately, recent air pollution conditions reported from Eastern European and former Soviet Union countries, as well as large urban areas in the developing nations, show that increased morbidity and mortality continue today secondary to air pollution exposures (Andrews, 1994; Jedrychowski, 1995; Moeller, 1992).

Characterization of air pollution

Air pollution has been defined as the presence in air of substances in concentrations sufficient to interfere with health, comfort, safety, or the full use and enjoyment of property (Moeller, 1992). The toxic components of various air pollutants depend on the contribution from local industries and other major sources such as transportation and the general burning of fossil fuels. The most commonly discussed and regulated pollutants are particulates, sulfur oxides, carbon monoxide, nitrogen oxides, ozone, metals such as lead, acid aerosols, and volatile organics (including hydrocarbons) (WHO, 1995).

One area of relatively new interest has been the effects of indoor air pollution on health. The average American worker spends up to 90% or more of her/his time indoors, while young children, the elderly, and the infirm may spend over 95%. In some urban areas, indoor concentrations of nitrogen oxides, carbon monoxide, airborne particulates, and other volatile organics can exceed the concentrations measured outside (Bachofen, 1993; Gong, 1992; Jaakkola et al., 1994; Moeller, 1992; Samet, 1987).

The major route of exposure from air pollution in humans is through the lungs; eye and skin irritation have been reported. However, through environmental deposition of certain persistent substances from air pollution, humans can be exposed through the food chain and through drinking water. Furthermore, children can be exposed to these same environmentally persistent pollutants *in utero* and through mother's milk.

Health effects

A large number of epidemiologic as well as animal and human laboratory studies have been performed to evaluate the effects of air pollution. In epidemiologic studies of large populations, it has been difficult to control for other individual factors affecting health such as infectious diseases and tobacco exposure. Some laboratory research does suggest that the toxicant constituents of air pollution may have synergistic effects on human health, such as ozone and sulfur dioxide or particulates and sulfur dioxide. More research is needed to evaluate the health effects and mechanisms of repeated pollutant mixtures (Fishbein, 1990; Folinsbee, 1993; Momas et al., 1993; Wanner, 1993).

The most frequent disorders associated with air pollution exposure are caused by the irritant gases and particulates on the mucous membranes and respiratory organs. Eye, nose, and throat inflammation as well as bronchoconstriction, diminished lung function, and increased susceptibility to respiratory infection have been reported in both epidemiologic and laboratory studies. On a molecular level, basic lung function activities and defenses are interfered with or destroyed, such as mucociliary clearance, changes in surfactant and gas exchange, release of proteolytic enzymes and inflammatory cells, decreased immune defense, etc. (Folinsbee, 1993; Sherwin, 1991; Wanner, 1993).

There is no specific "air pollution disease." Based on epidemiologic studies, air pollution, especially at high levels, is believed to cause and/or exacerbate the morbidity and mortality of the following lung conditions: asthma, chronic respiratory diseases [including chronic nonspecific respiratory disease, chronic obstructive pulmonary disease (COPD), chronic bronchitis, and emphysema], and, although controversial, possibly even lung cancer. There also appear to be interactions between air pollution and other carcinogenic exposures; for example, cigarette smoke increases the risk of lung cancer. In addition, if the effects of carbon monoxide are taken into account, air pollution can also have cardiovascular effects, especially in exacerbating existing cardiac conditions (Folinsbee, 1992; Gong, 1992; Momas et al., 1993; Schwartz and Dockery, 1992; WHO 1995).

Children, the sick and aged, and sensitive groups (such as asthmatics and persons with immune deficiencies) appear to be particularly effected by air pollution. Exercise, duration, concentration of a particular pollutant, and other exposure factors, such as tobacco smoke and concurrent occupational exposures, can all increase the effects of air pollution (Gong, 1992; Wanner, 1993; Schoni, 1993).

Solid and hazardous waste sites

A relatively small scientific literature exists concerning the health effects associated with solid and hazardous waste sites, especially with regard to chronic disease effects such as cancer since this is a relatively new area of epidemiologic research. In the communities surrounding hazardous waste sites, especially after the contamination of groundwater and the surrounding environment has been discovered, residents commonly report a myriad of symptoms including unpleasant odor and taste, skin and eye irritation, increased upper respiratory infections and ear infections, shortness of breath, and a variety of gastrointestinal symptoms including nausea and diarrhea.

Unfortunately, amidst the controversy, fear, and stress that inevitably surround the discovery of contamination, it is difficult to perform unbiased research, as is noted repeatedly in much of the existing literature on hazardous waste sites and human health effects. In particular, recall bias by those residents who live near the hazardous waste site and believe themselves to be harmed is particularly hard to eliminate in these studies, as is self-selection. Furthermore, small numbers, lack of individual exposure data, lack of objective disease data, and use of cross-sectional rather than longitudinal studies have added to rather than lessened the confusion surrounding this issue (Robinson, 1990; Sanjour, 1990; Shusterman et al., 1991; Travis, 1993). In addition, the issues of disease clusters and incomplete data collection can be very important problems in reports of human health effects near a hazardous waste facility. These make generalizability from such studies very difficult (Andrews 1995, Johnson 1995; Russell, 1993).

General health effects

Griffith et al. (1989) in an unmatched case control study examined cancer mortality rates in 3,065 U.S. counties for 1950 to 1979 for 13 anatomic sites, using 339 of the counties with proximity to hazardous waste sites as the cases. Although this study lacked individual exposure information, the fact that several cancers were increased in both males and females, predominantly gastrointestinal tract cancer, is suggestive of a possible association between the cancers and residence near a contaminating hazardous waste site.

Najem et al. (1983) reported that age-adjusted cancer rates of the gastrointestinal tract were higher in 20 of 21 New Jersey counties compared to national rates. In addition to population density and the degree of urbanization, the presence of toxic waste disposal sites was most frequently associated with increased gastrointestinal cancer risk. Cancer mortality from 1968 to 1977 in particular subgroups based on sex and race were elevated compared to national rates for cancers of the esophagus, stomach, colon, and rectum.

Logue et al. (1985) performed a community health survey in Pennsylvania in response to concerns about potential health effects secondary to residential exposure to chemical contaminants in the drinking water. Although the prevalence of a variety of reported symptoms was higher in the contaminated group, only for eye irritation was any dose response found based on the individual well water determinations of trichloroethylene. Baker et al. (1988) performed a cross-sectional health survey of 2,039 persons in 606 households located in two areas near the Stringfellow Hazardous Waste Disposal Site (California) and a nonexposed control community. Cancer incidence and pregnancy outcomes were similar between the different communities. Increased risks for a variety of reported diseases were noted in both hazardous waste communities. The investigators concluded that the study provided evidence against a strong association of the study area and significantly negative health outcome, but it could not rule out a slight or moderate association due to the relatively small population size, lack of biological monitoring and objective confirmation of disease, and issues of recall bias.

Budnick et al. (1984) performed a retrospective cohort study of cancer mortality in Clinton (Pennsylvania) and other surrounding counties near the Drake Superfund site. This Superfund site had stored many chemicals, including the human carcinogens benzidine dyes and benzene. An earlier study showed that bladder cancer was occurring at a relatively young age in the four surrounding counties. An increase in bladder cancer in white males was noted in Clinton during the 1970s; both males and females had significantly increased rates of lymphoma. The authors concluded that, although the bladder cancer was most likely due to occupational exposure, the increases in both sexes for other cancers were suggestive of a community effect from exposure to the hazardous waste site (Budnick, 1984; Lieben, 1963).

Reproductive effects
Reproductive effects have been studied in relation to hazardous waste and industrial sites for a variety of reasons. There is a relatively short latency period (i.e., conception to birth), human fetuses are relatively more sensitive to many chemicals than adults, and although 70% of birth defects are of unknown etiology, they are the leading cause of infant death in most

"developed" nations. In addition to the studies in contaminated communities such as Love Canal and Woburn, others showed an association between reproductive effects and exposure to hazardous waste (ATSDR, 1992a; Bove, 1995; Geschwind, 1992; Geschwind et al., 1992; Lie et al., 1994; Ozonoff and Aschengrau, 1993; Shaw et al., 1992).

Lie et al. (1994) used data from a population-based registry of birth defects that included information which allowed linkage of all records available on each child and both parents in Norway. Their results pointed to the possible strong contribution of environmental exposures as risks for birth defects rather than purely genetic factors (Lie et al., 1994; Cordero, 1994). Budnick et al. (1984) performed a retrospective cohort study of birth defect incidence in Clinton and other surrounding counties (Pennsylvania) near the Drake Superfund site. Of the 34 birth defects for the total of 3,098 live births from 1973 to 1978, no statistically significant clusters of any specific birth defect or of all birth defects were found.

In a case control study with extensive environmental monitoring, Swan et al. (1989) evaluated the incidence of congenital cardiac anomalies in Santa Clara (California) after a leak from an underground storage tank led to contamination of the community drinking water with the solvent 1,1,1 trichloroethene. There was a significant increase of major cardiac anomalies of 12 children with 6 expected (RR 2.2); no excess was seen subsequently when exposure ceased. However, ultimately the authors concluded that due to uncertainties about the timing and geographical distribution of the contamination, that the confirmed cluster of major cardiac anomalies could not be attributed to the solvent exposure. In a case control study of live birth deaths in the five-county San Francisco Bay area in 1983, Shaw et al. (1992) examined the risk of residence at the time of delivery in a census tract where environmental contamination had been documented, with the risk of congenital malformation and low birth weight. Heart/circulatory cardiac defects were significantly elevated (OR 1.5), especially conotruncal defects associated with exposures to direct exposure sites (OR 2.9), volatile organic compounds (OR 2.2), metals (OR 1.8), and cyanides (OR 4.7). No other associations were found between birth defects and environmentally contaminated census tracts (Shaw et al., 1992).

Geschwind et al. (1992) in a case control study in New York State evaluated the risk of congenital malformation with proximity of birth residence within 1 mile of 590 known hazardous waste sites in 1983 and 1984. Maternal residence near a known hazardous waste site was associated with a small but statistically significant increased risk (OR 1.12 with CI 1.06 to 1.18). In addition, higher malformation rates were associated with statistically increased risks of higher known exposure (Geschwind et al., 1992).

Using the Iowa Birth Defects Surveillance Project, ATSDR (1992a) examined the prevalence of birth defects in counties with and without NPL sites. Those counties with NPL sites containing heavy metals had higher prevalences of club foot and cleft lip, as well as cardiac defects, limb reductions, and urogenital defects; none of these associations was statistically significant. In a cross-sectional study, Bove et al. (1995) evaluated the effects of public drinking water contamination on birth outcomes in an area of northern New Jersey. Increased odds ratios for birth defects, small for gestational age, and low birth weight were associated with exposure during pregnancy to contaminated drinking water.

Changes in the sex ratio of births have been reported in areas of hazardous waste and industrial facilities. The hypothesis is that the Y-chromosome-bearing sperm are more sensitive to exposure to pollutants than the X-chromosome sperm; thus, the number of male infants is reduced. Of note, this has been associated with a number of paternal exposures including the pesticide DBCP, anesthesia gases and arsenic, as well as proximity to polluted air from smelters (Lyster, 1981; Lloyd et al., 1985).

Human health effects associated with hazardous waste incineration

Due to the relative newness of the technology, hazardous waste incineration and its possible effects on human health have been studied very little. The existing literature on human health effects concerns primarily other types of waste incineration (Johnson 1995, Andrews 1995).

Other incineration studies

Occupational
Bresnitz et al. (1992) performed a cross-sectional study of 86 of 105 actively employed male workers at a municipal waste incinerator in Philadelphia built in 1959. Five workers in the low exposure category had chest X-ray findings consistent with asbestos exposure; 8 workers had pulmonary interstitial opacifications, with 5 of the workers coming from the high exposure group; 29 (34%) of the workers were hypertensive, unrelated to exposure category; this prevalence was elevated compared to the U.S. population. Only 8 of 471 blood tests for lead and mercury in the 86 workers were abnormal, unrelated to exposure categories. As with all cross-sectional studies, no information was available for former workers.

In a retrospective cohort study, Gustavsson (1989) investigated the mortality of 176 male workers employed at a municipal waste incinerator between 1920 and 1985 in Stockholm. Compared to local and national rates, there were significant increases in all cancers (observed/expected = 22/16.27) and in circulatory diseases (obs/exp = 46/36.9). In particular, he found significant excesses of lung cancer deaths (obs/exp = 9/2.53) with 2 to 3.5 times excess risk and ischemic heart disease (obs/exp = 34/24.6); the latter risk increased after long follow-up which is consistent with ischemic heart disease associated with other occupational groups. Known but unquantified exposures were combustion products, nitrogen oxides, sulfur dioxide, carbon monoxide, dioxins, polycyclic aromatic hydrocarbons, and incineration-derived dusts. The incineration plant is noted by the author to be one of the oldest in Sweden with an increased tendency for fume leakage, open slag transport, little control over the composition of waste intake, and increased polycyclic aromatic hydrocarbon release due to poor control of the combustion process.

The association between lung cancer and polycyclic hydrocarbons has been established, with an excess of lung cancer found in several occupational groups exposed to combustion products including coke and gas production workers, chimney sweeps, certain tar distillation workers, and aluminum reduction plant workers. In addition, increased risks for ischemic heart disease and other cancers such as urinary bladder were found among chimney sweeps in Sweden and Denmark (Gustavsson et al., 1987; Gustavsson et al., 1988; Hansen, 1983; Hansen et al., 1982).

Community

Zmirou et al. (1984) performed a cross-sectional study in 1981 of the reported health effects of air pollution caused by incineration of industrial and household waste in a village of France's Isere Department. Respiratory exposure was considered the predominant exposure to this air pollution. The investigators evaluated the use of medicines over the prior two years by reviewing Social Security forms filed in three very similar groups of people living at differing distances from the incinerator. For all types of medication, the number and variety of medications was highest in persons living closest to the incinerator and lowest in those living 2 kilometers from the incinerator, especially antibiotics, expectorants, and bronchodilators. Allegedly due to these results which were attributed to the exposure to the incinerator, the public health authorities imposed more stringent regulations on the incinerator.

A disease cluster of acute lymphoblastic leukemia (ALL) was reported among children living near the Wrekenton incinerator in Gateshead, England, between 1968 and 1977. No exposure details were given except that heavy metal emissions were allegedly within control limits, but grit and dust emissions exceeded the standards. A follow-up standardized mortality study of all cases of ALL in Gateshead between 1974 and 1983 reportedly showed no increased rate of ALL compared to national rates. No individual confounding data were presented. Of note, the disease cluster started before the initiation of the incinerator and disappeared while it was still functioning (Gateshead, 1991).

Hazardous waste incineration studies

Occupational

Decker et al. (1983) monitored employee exposure to organic vapors at a liquid/fluid chemical waste incinerator facility in Cincinnati, Ohio. During the routine operation, worker exposure was less than the action level for all the compounds tested except benzene. The highest exposure operations were cleaning pump strainers after receiving waste, and benzene distillations in the laboratory; nonroutine storage tank entry resulted in the highest exposure of any operation. In a biological monitoring study of 45 workers in a modern waste combustion plant over a 2-year period, Bloedner et al. (1986) found no increase in the concentration of cadmium, lead, or mercury in the blood or urine, except for increases of cadmium in the blood of seven workers who smoked.

The National Institute of Occupational Safety and Health (NIOSH, 1992) attempted to perform two studies of former employees of the Caldwell Incinerator (described above) in a somewhat tense legal–political atmosphere. Fifteen former employees identified by a local physician as having neurologic problems had been evaluated. Among the subjects, there was a myriad of neurologic complaints and some objective neurologic abnormalities. The abnormalities included abnormal gait and myoclonus with tremor (2), possible Parkinson disease (1), possible peripheral sensory neuropathy (2), and tremor or station abnormalities (2). Twelve of the 15 people had diminished concentration by digit span. Alcohol abuse or dependence was the only consistent psychiatric diagnosis. In November 1991, NIOSH attempted to examine all current and former employees of the incinerator; 54 (21%) of the 328 eligible participants responded. Thirty-six (67%) of the participants had no abnormalities on neurologic examination. The 18 persons with multiple neurologic complaints had very little objective evidence on examination by a neurologist. Eight of the participants did have mild postural tremor.

Community

A cross-sectional study was performed by ATSDR (1993) of the prevalence of reported symptoms and health effects in persons living within a 1.5 mile radius of the former Caldwell Systems Incinerator (CSI) site (North Carolina). The study took place two years prior to closure. There were 713 exposed residents; these were compared to 588 residents of an unexposed but similar community. Respiratory symptoms, but not diseases, were significantly increased among exposed residents (OR = 9.0) even when adjusted for multiple confounders. Furthermore, those who lived closest to the site (within 0.9 miles) had a significant increase in reported respiratory symptoms compared to those who lived farther away (OR = 1.8). Exposed residents were also significantly more likely to report neurological symptoms (OR = 1.4) and neurologic diseases (OR = 2.4). There was also a significant increase in the reports of irritative symptoms from residents north and south of the incinerator compared with east and west; irritative symptoms were also more prevalent in persons living closer to the incinerator. There was no significant difference in the prevalence of overall physician diagnosed illnesses nor hospital admissions between the exposed and unexposed group, nor were there significant differences for cancer or adverse reproductive outcomes. No exposure information was available. As with all cross-sectional studies, no information was available for former residents. A similar study by Feigley et al. (1993) of community residents living near the ThermalKEM hazardous waste incinerator reportedly found significantly higher prevalence of reported respiratory symptoms (morning cough and phlegm) in residents living near the incinerator compared with an unexposed comparison community. As with the previous study, there was no increase in the prevalence of physician-confirmed diagnoses of respiratory disease.

Rothenbacher et al. (1994) evaluated three communities for respiratory symptom prevalence. The communities were situated close to, respectively, a biomedical waste incinerator, a municipal waste incinerator, and a liquid hazardous waste industrial furnace; three other communities were chosen as control populations, located upwind and at least five miles from the incinerator sites. There was no increased risk of reported symptoms in the biomedical waste community; in the municipal waste community, there was an increased risk of chronic morning cough (OR = 1.3) and sore throat within the past month (OR = 1.3); in the hazardous waste community, there was an increased risk of reported chronic morning cough (OR = 1.5), being awakened by a cough (OR = 1.5), as well as for "doctor diagnosed" emphysema (OR = 2.7) and sinus trouble (OR = 1.3). Of note, the reported smoking prevalence (active and passive) was less in all the exposed communities compared to their control communities; analysis of perception of outdoor air quality did not vary between the exposed and unexposed communities. The differences in reported symptoms for the municipal and hazardous waste-exposed communities from their control communities were not explained by age, race, education, tobacco smoke, occupation, use of unvented heaters, or indoor mold problems.

In Bonnybridge, Scotland, an increased number of cases of a rare congenital eye deformity, microphthalmos, was noted from 1976 to 1985 in the area of one of the two incineration plants licensed to burn polychlorinated biphenyls (PCBs). Local residents also noted increased cancer rates (especially leukemia) and bronchitis death rates. In addition, there were complaints of multiple diseases among local cattle on two nearby farms including deaths, birth deformities, and even a wasting illness allegedly similar to Michigan cattle exposed to PCBs; dioxins and furans, products of incomplete PCB incineration were measured in the soil and fat and milk samples of cattle from these farms. An official but controversial investigation determined that the increased cancer rates were due to detection bias and chance, and that the incidence of microphthalmos was increased but unrelated to the incinerator. Further investigation confirmed the lack of association between the risk of the birth defect and the incinerator. There still remains considerable controversy with respect to the possible association with animal illness (Handysides, 1993; Lenihan, 1985; Lloyd et al., 1987; Lloyd et al., 1988; Russell, 1993; Scottish Home and Health Dept, 1988).

Increased twinning was noted in both human and cattle populations in the Bonnybridge area of Scotland by residents and farmers. Lloyd et al. (1988)

reviewed data on human births in hospitals in central Scotland for 1975 to 1983. The twinning rates in areas exposed to airborne pollution from the PCB hazardous waste incinerators were compared to background rates from neighboring areas and found to be elevated especially after continuous PCB incineration was instituted. A similar increase in cattle twinning was reported for the two farms near the incinerator, although recall bias may have been a major factor. The authors hypothesized that the increase in twinning in humans and animals might be due to exposure to the known estrogenic effects of PCBs and their metabolites. No individual human, animal, or environmental exposure data were available (Lloyd, 1988). Reexamination of the human data by Jones (1989) failed to show any consistent trend over time, suggesting that the increase among humans may have been a random occurrence.

Gatrell and Lovett (1992) in an investigation of a disease cluster of laryngeal cancer around an industrial incinerator in Lancashire (England) performed a standardized mortality study of cancer mortality from 1974 to 1983 among Lancashire residents. The incinerator actively burned industrial liquid wastes (e.g., solvents and oils) from 1972 to 1980. Fifty-eight cases of laryngeal cancer were found. Although the number of total cases was not elevated compared to national rates, four of the cases were located within 2 km of the incinerator. However, the latency period was inadequate to associate the incinerator exposure from 1972 to 1980 with laryngeal cancers during the same time period (Diggle et al., 1990; Gatrell and Lovett, 1992).

Psychological effects

In a variety of psychological studies of communities around hazardous waste and incineration sites, a continuous finding has been the apparent discrepancy between low-level airborne chemical exposures and prominent reported symptoms. In many cases, the perception of "chemical odors" by community members was important in the identification of hazardous waste sites as environmental problems. Although worry and perception of harm seem to be universal symptoms in communities surrounding hazardous waste sites, these studies show that community participation and individual involvement can be important factors in decreasing the anxiety and even decreasing reported health effects in communities around hazardous waste sites (Andrews 1994; Bachrach and Zautra, 1985; Baker et al., 1988; Elliot et al., 1993; Eyles et al., 1993; Faust and Brilliant, 1981; Horowitz and Stefanko, 1989; Johnson and Covello, 1987; Kleindorfer, 1986; Ozonoff et al., 1987; Robinson, 1990; Roht et al., 1985; Shusterman et al., 1991; Taylor et al., 1991; Travis, 1993; 1991; Vyner, 1988).

Hazardous waste sites

Bachrach and Zautra (1985) in a cross-sectional survey to evaluate coping mechanisms for stress, interviewed 99 residents of a rural community near

Phoenix where a hazardous waste facility was planned. The authors distinguished between individual and community stressors. Increased ability to regulate one's emotional response to stress was associated with decreased community involvement. However, increased self-efficacy and sense of community were associated with increased community involvement.

Horowitz and Stefanko (1989) performed a telephone survey of 426 persons living at different distances from a toxic landfill in Southern California to evaluate the stress-related behavioral effects. There was no association between the different effects and the distance of residence from the site. The younger the person, the more reported total effects, anger-hostility, and demoralization. Those who owned their dwellings were more likely to express anger-hostility, demoralization, and total effects. Finally, those with less education were more likely to express demoralization and total effects. Residents were reportedly unaware of the purpose of the questionnaire with relation to the hazardous waste site.

Shusterman et al. (1991) interviewed more than 2,000 adults living near three different hazardous waste sites in California for a variety of symptoms. Of particular interest to the researchers were self-reported "environmental worry" and the frequency of perceiving environmental odors. Both these issues correlated significantly with the prevalence of headache (OR 5.0 for odors; OR 10.8 for worried), nausea (OR 5.2 odor; OR 11.9 worry), eye (OR 4.6 odor; OR 5.4 worry), and throat irritation (OR 4.3 odor; OR 9.3 worry). There was also strong correlation between the perception of odor, environmental worry and symptom prevalence. In comparison neighborhoods with no local hazardous waste sites, environmental worry also correlated with symptom prevalence. These associations were not affected by a variety of confounders.

ATSDR (1992b) identified the health concerns of communities living near hazardous waste sites; in 233 public health assessment reports, ATSDR listed 755 health concerns expressed by individuals living near the sites where the assessments were performed. Of these health concerns, 167 (22%) were considered cancer health end points, 267 (35%) noncancer end points, 126 (17%) trauma such as lacerations, and 195 (26%) nonspecific complaints such as headache, nausea, and dizziness (Andrews 1994).

Hazardous waste incineration

Eyles et al. (1993), Elliot et al. (1993), and Taylor et al. (1991) reported an investigation of psychosocial effects in populations (n = 696) living near three solid waste disposal facilities including incineration in southern Ontario. There was significant economic and environmental concern among the residents of all three sites; in particular, residents worried about depressed property values, traffic volume, pests, and air and water quality. There was also considerable distrust of government, which was mitigated by the presence of community surveillance committees and participation by local residents in decision making. General concern about the sites was found among the residents in a dose-response fashion; reported health concerns increased as the distance from the site decreased. Furthermore, reported health concerns decreased if the resident had taken part in any site-specific actions. The authors concluded that distance from the waste site was an important variable. The authors recommend that strategies aimed to address and alleviate psychosocial effects need to be specific to the characteristics of the populations in particular settings.

Human health effects associated with potential emission and waste stream substances

The following summaries of the large scientific literature on the human health effects of the heavy metals, the dioxins, PCBs, and the volatile organic compounds are included because these are believed to be the primary pollutant groups in the emissions and waste streams of hazardous waste incineration. As discussed before, people living in surrounding communities could be exposed to these pollutants through inhalation of the contaminated air and/or ingestion of contaminated water and contaminated food chain.

The major issues relevant to human health effects are the efficiency of combustion and the emissions. In particular, heavy metals and dioxins are groups of substances which have been produced in hazardous waste incineration and in other contexts and cause significant human health effects. An additional group of chemicals of possible concern are those created as byproducts of the incineration process, i.e., the products of incomplete combustion (PICs). In addition, respirable particulate emissions may preferentially carry many carcinogens (especially certain metals and polycyclic organic compounds), thus enabling them to come into close contact with lung tissues when inhaled.

Of interest, studies have shown that the emissions and fly ash of coal and oil-fired boilers can be mutagenic in *in vitro* systems such as the Ames test and sister chromatid exchange in human macrophages. The degree of mutagenicity is highly dependent on the operation efficiency of the boilers. More recent studies of incineration of solid wastes, including plastics, have found mutagenicity especially high in the particulate emissions when operated at incineration temperatures between 700°C and 900°C. The implications of these findings for human health effects are unknown (Chiang et al., 1992; Costner and Thornton, 1990; Oppelt, 1990; Peters et al., 1990; Sedman and Esparza, 1991; Young and Vorhees, 1992).

Heavy metals

The heavy metals may be the most important emission pollutant group from hazardous waste incineration with respect to possible human health effects.

The scientific literature concerning these pollutants is enormous because heavy metals are extremely well studied in both occupational and nonoccupational human communities over many hundreds of years. This literature is briefly summarized below.

Heavy metals, unlike organics, are not thermally destroyed during incineration; instead they are partially or completely vaporized during the combustion process. Although air pollution devices such as scrubbers at the incinerator should remove some portion of the particulates with metals, a significant fraction can be released to the outside air. In addition, those heavy metals which are captured in the bottom ash and by the scrubbers must be disposed of carefully as a new form of hazardous waste (ASME, 1988; Costner and Thornton, 1990; Oppelt, 1990; Travis and Cook, 1989; Travis et al., 1987).

Metals are naturally occurring elements ubiquitous in the environment. Metals tend to accumulate in living systems and up the food chain, especially mercury and lead. Once metals have contaminated a food chain, they are very difficult, if not impossible, to remove. Furthermore, in general, once a metal has been absorbed into the human system and has caused physiologic effects, these effects are irreversible, especially in fetuses and developing children.

Arsenic, barium, beryllium, chromium, cadmium, lead, mercury, nickel, and zinc are all metals which have been detected in the bottom ash and stack emissions (particulate and/or vapor) from hazardous waste incineration (Costner and Thornton, 1990; Lisk, 1988; Oppelt, 1990; Travis and Cook, 1989; Travis et al., 1987). The human health effects of these metals depend significantly on both their route of exposure and the chemical formulation of the particular metal. Many of the metals are neurotoxicants and reproductive toxicants since they are able to pass through both the blood brain barrier and the placenta with relative ease. Their effects on the developing fetus and young child, predominantly neurologic, can be devastating, much more so than on the adult.

The majority of the metals, especially with chronic ingestion exposures, can cause systemic diseases with permanent damage. In addition to chronic neurologic disease, cardiac and circulatory disease are also seen even with contaminated water ingestion, and respiratory disease has been reported with chronic cadmium inhalation. Finally, several are known or suspected carcinogens (Clarkson, 1992; Klassen et al., 1991; Landrigan, 1989; Needleman, 1989; Needleman and Gatsonis, 1990; Paul, 1993; Roelevald et al., 1990; Rom, 1992; Rosenstock and Cullen, 1994; Tamaddon and Hogland, 1993).

Dioxins and furans

Although present in smaller amounts in hazardous waste incineration emissions compared with heavy metals, the dioxins as a group have caused considerable controversy due to their extreme toxicity in animals and their environmental persistence. In recent years, new evidence as to the possible human health effects of dioxins has accumulated.

Polychlorinated dibenzo-*p*-dioxins (PCDDs) and polychlorinated dibenzofurans (PCDFs) are products of incomplete combustion (PICs) as well as contaminants of certain chemicals. They have no known use and are not manufactured intentionally. As a chemical group, dioxins have been detected in the emissions of both municipal and hazardous waste incineration facilities. Dioxins have also been found as ubiquitous trace contaminants in air, water, and soil, where they are quite environmentally persistent and bioaccumulate in biological systems. Trace amounts of dioxins have been detected in breast milk and fat tissue samples of humans throughout the world due to their ubiquitous contamination of the environment and aquatic organisms. Although readily absorbed through the respiratory system, the human food chain, especially the consumption of contaminated agricultural and dairy products, is believed to be the major source (98%) of this human contamination by dioxins (Costner and Thornton, 1990; Dickson and Buzik, 1993; Fries and Paustenbach, 1990; Kimbrough et al., 1984; Masuda, 1991; Orban et al., 1994; Paustenbach et al., 1992; Sedman and Esparza, 1991; Travis and Cook, 1989; Travis et al., 1989; Travis and Hattemer Frey, 1991; Vainio et al., 1989).

In many nonprimate animal species in laboratory experiments, the dioxins are highly toxic compounds in minute amounts, both acutely and chronically. Dioxins are immunotoxic, teratogenic, embryotoxic, and carcinogenic in animals even at very low doses, especially by the oral route (Dickson and Buzik, 1993; Holsapple et al., 1991; Johnson, 1992, 1993; Neubert et al., 1993; Travis and Hattemer Frey, 1991; Vanden Heuvel and Lucier, 1993). Based on epidemiologic studies, humans do not appear to be as susceptible to any of the toxic effects of dioxins compared with many of the other animal species. However, there have been several consistent issues with these epidemiologic studies. In general, humans have been exposed to chemical mixtures in which dioxins are present in very small amounts as a contaminant. In addition, possible interviewer bias and recall bias are major problems with the small numbers of humans, especially industrial worker populations, studied epidemiologically (Johnson, 1992, 1993).

Through studies of workers exposed in industry and various accidents, the following health effects have been noted in humans. Chloracne is the classic skin lesion of dioxin exposure, although *Porphyria cutanea tarda* and hyperpigmentation have also been reported. Liver damage has been reported, ranging from mild abnormalities of the liver enzymes to mild fibrosis and fatty changes, and disorders of fat and carbohydrate metabolism. Polyneuropathies with sensory impairments and lower extremity weakness have also been reported. In addition, sleep disturbance, depression, and personality disorders have been documented (Dickson and Buzik, 1993; WHO, 1989).

In epidemiologic reports concerning possible teratogenic effects in humans after exposures to dioxins and/or herbicides from Vietnam and elsewhere, there has not been any consistent pattern of birth defect increase across studies. In laboratory studies in primates, following maternal exposure to dioxins, it has not been possible to produce teratogenic effects. Among monkey offspring in experimental studies by Bowman et al. (1989), alterations in social behavior were noted, and performance on learning tasks was correlated with body fat concentrations (Stellman et al., 1988; Mastroiacova et al., 1988; Pearn, 1985; Lilienfeld and Gallo, 1989).

Recent research has concentrated on the effects of dioxins on the human immune system. In animals, effects on the immune system are among the earliest and most sensitive indicators of TCDD toxicity. TCDD appears to have effects on both cell-mediated and humoral immunity. In one recent study, workers exposed to TCDD over 17 years earlier had more natural killer cells, antinuclear antibodies, and immune complexes in their peripheral blood than unexposed controls. In Times Beach, Missouri, exposure to TCDD in 1971 resulted in depression of the delayed-type hypersensitivity reaction when tested in 1984. In children at Seveso, Italy, 6 years after exposure, increased levels of serum complement protein correlated with the incidence of chloracne (Hoffman et al., 1986; Holsapple et al., 1991; Neubert et al., 1993).

Cancer risk in humans from exposure to dioxins, specifically soft tissue sarcomas, non-Hodgkin lymphomas, and liver cancer as well as possibly cancers of the stomach, thyroid, and lung, is controversial. This issue has been studied in a number of occupational groups including accidentally exposed groups, herbicide applicators, chemical manufacturers, and Vietnam veterans. It appears that the risk of soft tissue sarcomas is not consistently associated with dioxin exposure, although an increase was found in the heaviest exposed group in the largest occupational epidemiologic study to date by Fingerhut et al. (1991). Among Agent Orange-exposed Vietnam veterans, the increase in lymphoma was seen only in the least exposed individuals. Liver cancer was found to be increased in a study performed in Vietnam where Agent Orange was sprayed heavily as a herbicide; however, information concerning confounding variables was not available. Stomach cancer was found to be elevated among herbicide applicators, but not in more heavily exposed chemical plant workers.

One problem with the majority of existing studies on cancer risk has been the small sample sizes of the various occupational cohorts, as well as the multiple concomitant carcinogenic exposures. Some groups, such as the population of Seveso, have not been followed for significant time periods to allow for the development of cancers such as sarcomas. Furthermore, individual dose estimation has been very difficult, and some exposures are one-time accidental, while others are continuous but variable. In some cases, especially with the soft tissue sarcomas, the cancer diagnoses were incorrect due to incorrect pathologic classification. Again, issues of recall bias have also plagued these studies. There are several ongoing epidemiologic studies which will hopefully clarify the cancer risk of dioxin exposure in humans in the future (CDC, 1988; Eriksson et al., 1990; Fingerhut et al., 1991; Gough,

1991a, 1991b; Hardell and Eriksson, 1988; Johnson, 1992, 1993; The Selected Cancers Cooperative Study Group, 1990; Tollefson, 1991).

Volatile organic compounds

The final group of pollutant emissions from hazardous waste incineration has been largely created within the last century with technological advances. However, there exists an extensive occupational literature concerning human health effects which is briefly summarized below.

Volatile organic compounds are a diverse group of chemicals that are liquid at room temperature, relatively nonreactive, and characterized by their ability to dissolve oils, fats, resins, rubber, and plastics. Most of these organic compounds are also quite volatile even at room temperature. Volatile organic compounds have been ubiquitously used in industry and now are found in the environment and in the food chain in measurable amounts throughout the world, even in supposedly pristine areas. In particular, chlorine and the other halogens have been incorporated into many commonly used volatile organic compounds, resulting in significant human health impacts.

Volatile organic compounds have been measured in significant amounts as volatile and semivolatile emissions from incinerators, boilers, kilns, and even in test-run data from hazardous waste incinerators. The most commonly measured compounds in terms of presence and amount tend to be the most volatile. Benzene, toluene, carbon tetrachloride, chloroform, methylene chloride, trichloroethylene, tetrachloroethylene, 1,1,1-trichloroethane, and chlorobenzene are the most frequently found volatile organic compounds; naphthalene, phenol, bis(2-ethylhexy)phthalate, diethylphthalate, butylbenzylphthalate, and dibutylphthalate are the most frequently measured semivolatile compounds.

The more volatile organic compounds are absorbed through the respiratory tract, while the less volatile can be absorbed through the skin and/or by ingestion. In nonoccupational settings, the most likely routes of exposure to volatile organic compounds will be through ingestion (especially through the water supply) and to a lesser extent, through inhalation. Most volatile organic compounds are metabolized by the liver. This biotransformation is intended to produce less biologically active compounds, but it can lead to more toxic metabolites. The amount of biotransformation is dependent on individual variability in terms of the presence of specific metabolizing enzymes, as well as the presence of other toxicants such as ethanol, certain medications, or even other volatile organic compounds.

Although usually rapidly distributed and metabolized in the body, volatile organic compounds are lipophilic, and as such can be stored in the fat of living systems. Although the majority of volatile organic compounds are easily absorbed through inhalation, some volatile organic compounds, such as formaldehyde, are highly irritating to the respiratory tract. Many volatile organic compounds pass easily across the skin barrier, although skin exposure to most volatile organic compounds can lead to skin irritation and dermatitis. They pass easily across the blood brain barrier and across the placenta. In humans and other animals, volatile organic compounds are neurotoxicants and possibly fetotoxicants (especially if ethanol is considered as a model). Volatile organic compounds may be teratogenic, especially to the cardiac system of the developing fetus.

As neurotoxicants, acutely volatile organic compounds can cause intoxication, psychosis, and even anesthesia through inhalation, ingestion, and even skin absorption; chronically, an entity known as "chronic solvent dementia" has been associated with prolonged occupational exposure. Many volatile organic compounds and/or their metabolites are toxic to the liver, ranging from increased mild fat storage to complete liver necrosis; the classic hepatotoxic solvent is carbon tetrachloride. Volatile organic compounds are also believed to be toxic to the kidney, especially with chronic occupational exposure. Some volatile organic compounds cause extensive harm to the bone marrow and blood cells; in the case of benzene, this can lead to complete elimination of the bone marrow blood cells as with aplastic anemia. Some volatile organic compounds have been shown to have specific effects on the reproductive system; for example, the glycol ethers are spermatotoxicants.

Many volatile organic compounds have toxic side effects on the cardiovascular system including atherosclerosis and cardiac arrythmias. Finally, some volatile organic compounds, especially the chlorinated volatile organic compounds, are carcinogenic in humans, in particular cancers of the hematopoietic, hepatic, and nervous systems (Klassen et al., 1991; LaDou, 1990; Oppelt, 1990; Paul, 1993; Roelevald et al., 1990; Rom, 1992; Rosenstock and Cullen, 1994).

Polychlorinated biphenyls and polychlorinated dibenzofurans

Polychlorinated biphenyls (PCBs)

Although burned at another temperature and in other incineration facilities, and therefore not intentionally associated with hazardous waste incineration, polychlorinated biphenyls (PCBs) are included because they may be a waste stream contaminant in relatively small amounts. Similar to dioxins, they are manmade and environmentally persistent. In the past, significant human disease was attributed to PCBs, in part due to two large community contamination episodes in Taiwan and Japan; recent evidence has shown that most of the human health effects are attributable to the more toxic polychlorinated dibenzofurans (PCDFs), not to the concomitant PCB exposure, as detailed below.

PCBs are mixtures of aromatic chemicals which were used extensively since the 1930s for their fire retardant and dielectric properties, especially in transformers and large capacitors, as well as plasticizers and hydraulic fluids. Humans are primarily exposed to PCBs through the food chain, especially fish consumption, and in certain specific cases, accidental food contamination. Occupational exposures have resulted from skin and inhalation exposures of PCBs, but these are much less common. Since PCBs are lipophilic, they are stored in fat tissues and are present in concentrated amounts in human and animal breast milk. Variable but detectable amounts of PCBs have been measured in human samples throughout the world (Masuda, 1991; Fitzgerald et al., 1991; Safe, 1994).

In laboratory animals, including nonhuman primates, PCBs affect reproduction and the immune response, and, in rodents, they cause liver tumors (hepatocellular carcinomas) and possibly stomach cancers. The susceptibility, especially with respect to cancer, is highly dependent on both the particular PCB mixture and the particular animal species (Kimbrough, 1987; Safe, 1989). Recent studies in alligators and turtles have raised the possibility that PCBs are hormone-disrupting chemicals that may have significant effects on male reproduction (Toppari 1996).

Skin irritation and chloracne are seen with all routes of PCB exposure in humans. In addition, liver abnormalities, pulmonary defects, decreased birth weight, sensory neuropathies, possibly immune dysfunction, and consistent neuropsychological complaints of weakness and fatigue have been reported in the past. Two reviews by James et al. (1993) and Safe (1994) have extensively summarized the evidence of human health effects secondary to PCB exposure; these reviews conclude that only the skin effects and possibly cancer can legitimately be attributed to the PCB exposure. Other reported effects are secondary to contaminants, are not found consistently in heavily exposed populations, and/or were derived from significantly flawed scientific studies (James, 1993; Safe, 1994). Of note, recent work has indicated that *in utero* exposure to PCBs (through consumption of contaminated fish) may be associated with decreased intelligence, even as long as 11 years from exposure (Jacobson 1990a, 1990b, 1996).

The risk of cancer in humans with exposure to PCBs has not been established. Some occupational groups with chronic exposure to PCBs have been studied although often the numbers are relatively small; there have been concurrent carcinogenic exposures (including PCDFs and chlorinated benzenes), and characterization of exposure has been difficult. In addition, the existing studies have shown increases in a variety of cancers, without any apparent consistent pattern. Nevertheless, given the carcinogenicity of PCBs in animals, their persistence both in the environment and in organisms, and the reported existence of cancer increases in various exposed worker populations, continued monitoring of worker-exposed populations is warranted (Bertazzi et al., 1987; Brown, 1987; Chase, 1991; Gustavsson 1986; Gustavsson et al., 1986; James et al., 1993, Safe, 1993, 1994; Sinks 1992).

Polychlorinated dibenzofurans (PCDFs)

In 1968 and 1979, two very similar outbreaks of poisoning occurred in Japan (Yusho Disease) and in Taiwan (Yu-Cheng Disease) when rice oil was contaminated with PCBs and widely dispersed, affecting several thousand people. Subsequent investigations have shown that the PCB levels of the two exposed populations were not greater than those reported in other PCB-exposed populations. However, the levels of the more toxic polychlorinated dibenzofurans (PCDFs) were much higher. In addition, studies in monkeys were able to duplicate Yusho-like signs only with a mixture of PCBs and PCDFs (similar to the contaminated rice oil) or with the PCDFs alone.

In both epidemics, the latency period from the beginning of ingestion to the onset of symptoms was about 2 to 3 months. Besides chloracne (the presenting complaint in both epidemics), various other clinical effects were noted. These included skin and nail pigmentation, eye discharge, and complaints of headache, nausea, and numbness in the extremities. In addition, neuropsychological complaints such as dizziness, depression, sleep and memory disturbances, nervousness, fatigue, and impotence were noted. Sensory neuropathies confirmed by nerve conduction testing were found in a number of patients. In addition, persistent chronic bronchitis developed in 20% of the Taiwanese cases (Ahlborg and Victorin, 1987; Chen et al., 1992; Holsapple et al., 1991; Hsu et al., 1985; Reggiani and Bruppacher, 1985).

Increased spontaneous abortion, infant mortality, and increased intrauterine growth retardation were documented at the time of the Yu-Cheng episode. In both epidemics, infants born to mothers exposed in these epidemics typically had hyperpigmentation of the skin and nails, as well as abnormal teeth, eye discharge, and abnormal liver function tests. In older exposed children, there were significant decreases in height and weight growth after exposure (Ahlborg and Victorin, 1987; Chen et al., 1992; Holsapple et al., 1991; Hsu et al., 1985; Rogan et al., 1985; Yen, 1989).

Recent studies in Taiwan of Yu-Cheng children show significant developmental delay which persists over time, correlating with smaller body size and a positive history of neonatal symptoms. Similar deficits in cognitive development have been noted at different ages in children born to women who consumed PCB-contaminated fish from Lake Michigan (USA), as prolonged as 11 years from exposure (Chen et al., 1992, 1994; Gladden and Rogan, 1991; Jacobson et al., 1990a, 1990b, 1996; Yen et al., 1989; Yu et al., 1991).

Yusho disease has been associated with immunosuppression. A study of 30 exposed cases and 23 unexposed controls of the Taiwan episode showed decreased IgA and IgM levels, as well as decreased total T cells. In a related study, the number of Fc receptors on monocytes and polymorphonuclear cells were coordinately reduced. In another study of 30 exposed and 50 unexposed persons, the delayed hypersensitivity was reduced significantly and correlated with the degree of chloracne (Holsapple et al., 1991; Chang et al., 1982).

Summary

From the above review of the existing literature of the human health effects of hazardous waste incineration, several aspects emerge (Florida, 1995). First, the existing database on human health effects from hazardous waste incineration is inadequate because this is a relatively new technology. Not only have few investigations in human populations been conducted, but those that have, have studied a relatively small number of individuals. Furthermore, studies have not yet begun to evaluate possible chronic health effects, such as cancer, in human communities because there has not been a sufficient latency period from exposure to disease development.

The few available studies of occupational groups show that hazardous exposures to waste incineration workers are highly variable by facility. In addition, these studies were performed in older facilities with minimal environmental controls or personal protection. Therefore, it is difficult to draw conclusions as to their relevance to current incineration occupational exposure.

A few cross-sectional studies have been performed in communities surrounding hazardous incineration plants. Although again very dependent on the operation and exposures of a particular plant, there is some suggestion that persons who reside closer to incineration plants are more likely to complain of respiratory problems (such as cough and phlegm production). This elevation continued even after controlling for factors such as age, gender, cigarette smoking, and environmental concern. In no study was the prevalence of physician-diagnosed respiratory disease increased in residents compared to unexposed comparison communities. However, in one study, the number of medications specific for respiratory ailments was significantly increased in those persons living closest to an incinerator.

There have also been several investigations of disease clusters of cancer and birth defects around various waste incinerators in Europe. However, none of these disease clusters has ultimately been shown to be definitively associated with proximity to the incinerator. In at least one of the cancer studies, the time from exposure to the onset of the disease (i.e., the latency) was completely inadequate. One area of possible concern identified from the hazardous waste literature is reproductive effects in children born to mothers living near these facilities, especially if they are consuming contaminated water during the pregnancy.

This chapter also reviewed the scientific literature in other areas of possible relevance to hazardous waste incineration. The effects of air pollution on human health have been extensively studied over the last century and are relevant to hazardous waste incineration because inhalation of polluted air could be a major route of exposure. An extensive literature also exists on the human health effects of the three major emission groups from hazardous waste incineration, i.e., the heavy metals, dioxins, and the volatile organic compounds, as well as the PCBs as a possible waste stream contaminant.

With the exception of some research in air pollution, there is almost no research on the effects of mixed low-level exposures over long periods of time in normal human communities which consist of young and old, healthy and sick people. This would be the appropriate epidemiologic scenario for human health effects from hazardous waste incineration. Furthermore, the exposures from hazardous waste incineration will not only vary over time, but also by geographic area, even by individual hazardous waste incinerator.

A cause for concern with respect to all of the above-mentioned emissions is that, in general, once these substances enter the food chain, and even more importantly, the human body, it is very difficult to eliminate them. The literature shows that all these substances can cause human health effects (predominantly respiratory, neurologic, and in some cases, cancer) and that these effects can be irreversible, especially in fetuses and growing children.

Finally, in the little literature available concerning human health effects, incineration, and hazardous waste, it appears that persons living in communities surrounding such sites often suffer from substantial anxiety concerning the existence of possible health risk, as well as substantial economic concerns such as decreased property values. It appears that much of this anxiety can be alleviated if the local community is educated and actively involved in the overseeing of the waste site from the beginning.

References

Agency for Toxic Substances and Disease Registry (ATSDR), *Iowa Birth Defects Registry*, U.S. Department of Health and Human Services, Atlanta, 1992a.

Agency for Toxic Substances and Disease Registry (ATSDR), *Study of Symptom and Disease Prevalence at the Caldwell Systems Inc. Hazardous Waste Incinerator, Caldwell County, North Carolina*, U.S. Department of Health and Human Services, Atlanta, February 1993.

Agency for Toxic Substances and Disease Registry (ATSDR), Hazardous waste sites: priority health conditions and research strategies, *Morb Mort Week Rep*, 41, 72, 1992b.

Ahlborg, U. G., Victorin, K., Impact on health of chlorinated dioxins and other trace organic emissions, *Waste Manage Res*, 5, 203, 1987.

American Society for Mechanical Engineers, *Hazardous Waste Incineration*, ASME, New York, January 1988.

Andrews, J. S., Frumkin, H., Johnson, B. L., Mehlman, M. A., Xintaras, C., Bucsela, J. A., eds., *Hazardous Waste and Public Health: International Congress on the Health Effects of Hazardous Waste*, Princeton Scientific, Princeton, NJ, 1994.

Bachofen, H., Air pollution, *Schweiz Med. Wochenschr*, 123, 183, 1993.

Bachrach, K. M., Zautra, A. J., Coping with a community stressor: the threat of a hazardous waste facility, *J Health Soc Behav*, 26, 127, 1985.

Baker, D. B., Greenland, S., Mendlein, J., Harmon, P., A health study of two communities near the Stringfellow waste disposal site, *Arch Environ Health*, 43, 325, 1988.

Bertazzi, P. A., Riboldi, L., Pesatori, A., Radice, L., Zochetti, C., Cancer mortality of capacitor manufacturing workers, *Am J Ind Med*, 1, 165, 1987.

Bloedner, C. D., Reimann, E. O., Schaller, K. H., Weltle, D. Evaluation of internal cadmium, lead, and mercury exposure in workers of a modern waste combustion plant — a comparison of the results o two years of surveillance, *Zentralbl Arbitsmed*, 36, 322, 1986.

Bove, F. J., Fulcomer, M. C., Klotz, J., Esmart, J., Dufficy, E. M., Savrin, J. E., Public drinking water contamination and birth outcomes, *Am J Epidemiol*, 141, 850, 1995.

Bresnitz, E. A., Roseman, J., Becker, D., Gracely, E., Morbidity among municipal waste incinerator workers, *Am J Ind Med*, 2, 363, 1992.

Brown, D. P., Mortality of workers exposed to PCBs, *Arch Environ Health*, 42, 333, 1987.

Budnick, L. D., Sokal, D. C., Falk, H., Logue, J. N., Fox, J. M., Cancer and birth defects near the Drake Superfund site, Pennsylvania, *Arch Environ Health*, 39, 409, 1984.

Centers for Disease Control, Veterans Health Study: Serum 2,3,7,8 tetrachloro-dibenzo-p-dioxin levels in US Army Vietnam-era veterans, *JAMA*, 260, 1249, 1988.

Chang, K.J., Hsieh, K. H., Tang, S. Y., Tung, T. C., Immunologic evaluation of patients with polychlorinated biphenyl poisoning: evaluation of delayed-type hypersensitive response and its relation to clinical studies, *J Toxicol Environ Health*, 9, 217, 1982.

Chase, K. H., The epidemiological basis for assessing human health risk from PCBs, *J Occup Med*, 33, 538, 1991.

Chen, Y.-C. J., Yu, M.-L. M., Rogan, W. J., Gladen, B. C., Hsu, C.-C., A 6 year follow-up of behavioral and activity disorders in the Taiwan Yu-Cheng children, *Am J Public Health*, 84, 415, 1994.

Chen, Y.-C. J., Guo, Y.-L. L., Hsu, C.-C., Cognitive development of children prenatally exposed to polychlorinated biphenyls (Yu-Cheng children) and their siblings, *J. Formosan Med Assoc*, 91, 704, 1992.

Chiang, P.-C., You, J.-H., Chang, S.-C., Wei, Y.-H., Identification of toxic PAH compounds in emitted particulates from incineration of urban solid wastes, *J Hazard Mater*, 31, 29, 1992.

Clarkson, T. W., Mercury: major issues in environmental health, *Environ Health Perspect*, 100, 31, 1992.

Cordero, J. D., Finding the causes of birth defects [editorial], *N Engl J Med*, 331, 48, 1994.

Costner, P., Thornton, J., *Playing with Fire: Hazardous Waste Incineration*, Greenpeace, Washington, DC, 1990.

Decker, D. W., Clark, C. S., Elia, V. J., Kominsky, J. R., Trapp, J. H., Worker exposure to organic vapors at a liquid chemical waste incinerator, *Am Ind Hyg Assoc J*, 44, 296, 1983.

Dickson, L. C., Buzik, S. C., Health risks of dioxins: a review of environmental and toxicological considerations, *Vet Hum Toxicol*, 35, 68, 1993.

Diggle P. J., Modelling the prevalence of cancer of the larynx in part of Lancashire: a new methodology for spatial epidemiology, in *Spatial Epidemiology*, Thomas, R. W., ed., Pion Limited, 1990, 34.

Elliot, S. J., Taylor, S. M., Walter, S., Stieb, D., Frank, J., Eyles, J., Modelling psychosocial effects of exposure to solid waste facilities, *Soc Sci Med*, 37, 791, 1993.

Eriksson, M., Hardell, L., Adami, H., Exposure to dioxins as a risk factor for soft tissue sarcoma: a population based case control study, *J Natl Cancer Inst*, 82, 486, 1990.

Eyles, J., Taylor, S. M., Johnson, N., Baxter, J., Worrying about waste: living close to solid waste disposal facilities in southern Ontario, *Soc Sci Med*, 6, 805, 1993.

Faust, H. S., Brilliant, L. B., Is the diagnosis of "Mass Hysteria" an excuse for incomplete investigation of low-level environmental contamination?, *J Occup Med*, 23, 22, 1981.

Feigley, C. E., Hornung, C. A., Macera, C. A., Draheim, L. A., Wei, M., Oldendick, R., Community study of health effects of hazardous waste incineration: preliminary results, in *Hazardous Waste and Public Health: International Congress on the Health Effects of Hazardous Waste*, Andrews, J. S., Frumkin, H., Johnson, B. L., Mehlman, M. A., Xintaras, C., Bucsela, J. A., eds., Princeton Scientific, Princeton, NJ, 1994, 765.

Fingerhut, M., Halperin, W., Marlow D., et al., Cancer mortality in workers exposed to 2,3,7,8 tetrachlorodibenzo-*p*-dioxin, *N Engl J Med*, 324, 212, 1991.

Fishbein, L., Sources, nature and levels of air pollutants, *Air pollution and Human Cancer European School of Oncology Monographs*, Tomatis, L., ed., Springer-Verlag, Berlin, 1990, 9.

Fitzgerald, E., Bush, B., Hwang, S., Brix, K., Quinn, J., Cook, K., Dietary exposure to polychlorinated biphenyls (PCBs) from hazardous waste, *Am J Epidemiol*, 134, 784, 1991.

Florida Center for Solid and Hazardous Waste Management, *Evaluation of the Human Health Impacts Associated with Commercial Hazardous Waste Incinerators, Report #95-4B)*, Florida Dept. of Environmental Protection, Gainesville, July 1995.

Folinsbee, L. J., Human health effects of air pollution, *Environ Health Perspect*, 100, 45, 1993.

Fries, G. F., Paustenbach, D., Evaluation of potential transmission of 2,3,7,8-tetrachlorodibenzo-*p*-dioxin-contaminated incinerator emissions to humans via foods, *J Toxicol Environ Health*, 29, 1, 1990.

Gateshead Metropolitan Borough Council, Incidence of Acute Lymphoblastic Leukemia in Gateshead, in *Hazardous Waste and Human Health*, British Medical Association, Oxford University Press, 1991.

Gatrell, A. C., Lovett, A. A., Burning questions: incineration of wastes and implications for human health, in *Waste Location: Spatial Aspects of Waste Management, Hazards and Disposal*, Clark, M., Smith, D., Blowers, A., eds, Routledge, 1992, 142.

Geschwind, S. A., Stolwijk, J., Bracken, M., Fitzgerald, E., Stark, A., Olsen, C., Melius, J., Risk of congenital malformation associated with proximity to hazardous waste sites, *Am J Epidemiol*, 135, 1197, 1992.

Geschwind, S. A., Should pregnant women move? Linking risks for birth defects with proximity to toxic waste sites, *New Direct Stat Comput*, 5, 40, 1992.

Gladen, B. C., Rogan, W. J., Effects of perinatal polychlorinated biphenyls and dichlorodiphenyl dichloroethene on later development, *J Pediatr*, 119, 58, 1991.

Goldsmith, J. R., Effects of Air Pollution on Human Health, *Air Pollution and its Effects*, AC Stern, ed., Academic Press, New York, 1968.

Gong, H. Jr., Health effects of air pollution. A review of clinical studies, *Clin Chest Med*, 13, 201, 1992.

Gough, M., Human health effects: what the data indicate, *Sci Total Environ*, 104, 129, 1991b.

Gough, M., Agent orange: exposure and policy, *Am J Public Health*, 81, 289, 1991a.

Griffith, J., Riggan, W. B., Duncan, R., Pellom, A., Cancer mortality in U.S. counties with hazardous waste sites and ground water pollution, *Arch Environ Health*, 44, 69, 1989.

Gustavsson, P., Mortality among workers at a municipal waste incinerator, *Am J Ind Med*, 15, 245, 1989.

Gustavsson, P., Hogstedt, C., Rappe, C., Short term mortality and cancer incidence in capacitor manufacturing workers exposed to PCBs, *Am J Ind Med*, 10, 341, 1986.

Gustavsson, P., Gustavsson, A., Hogstedt, C., Excess mortality among Swedish chimney sweeps, *Br J Ind Med*, 44, 738, 1987.

Handysides, S., Reports of anophthalmia under scrutiny, *Br Med J*, 306, 416, 1993.

Hansen, E., Mortality from cancer and ischemic heart disease in Danish chimney sweeps: a five-year follow-up, *Am J Epidemiol*, 117, 160, 1983.

Hansen, E., Olsen, J., Tilt, B., Cancer and non-cancer mortality of chimney sweeps in Copenhagen, *Int J Epidemiol*, 11, 356, 1982.

Hardell, L., Eriksson, M., The association between soft tissue sarcomas and exposure to phenoxyacetic acids. A new case referent study, *Cancer*, 62, 6562, 1988.

Hoffman, R. E., Stehr-Green, P. A., Webb, K. W., Health effects of long-term exposure to 2,3,7,8-tetrachloro-dibenzo-p-dioxin, *JAMA*, 255, 2031, 1986.

Holsapple, M. P., Snyder, N. K., Wood, S. C., Morris, D. L., A review of 2,3,7,8-tetrachlorodibenzo-p-dioxin-induced changes in immunocompetence, *Toxicology*, 69, 219, 1991.

Horowitz, J., Stefanko, M., Toxic waste: behavioral effects of an environmental stressor, *Behav Med*, 15, 22, 1989.

Hsu, S. T., Ma, C. I., Hsu, S. K., Wu, S. S., Hsu, N. H., Yey, C.-C., Wu, S. B., Discovery and epidemiology of PCB poisoning in Taiwan: a four year follow-up, *Environ Health Perspect*, 59, 5, 1985.

Jaakkola, J. J., Tuomaala, P., Seppanen, O., Air recirculation and sick building syndrome: a blinded crossover trial, *Am J Public Health*, 84, 422, 1994.

Jacobson, J. L., Jacobson, S. W., Intellectual impairment in children exposed to polychlorinated biphenyls in utero, *N Engl J Med*, 335, 783, 1996.

Jacobson, J. L., Jacobson, S. W., Humphrey, H. E., Effects of exposure to PCBs and related compounds on growth and activity, *Neurotoxicol Teratol*, 12, 319, 1990b.

Jacobson, J. L., Jacobson, S. W., Humphrey, H., Effects of in utero exposure to polychlorinated biphenyls and related contaminants on cognitive functioning in young children, *J Pediatr*, 116, 38, 1990a.

James, R. C., Busch, H., Tamburro, C. H., Roberts, S. M., Harbison, R. D., Polychlorinated biphenyl exposure and human disease, *J Occup Med*, 35, 136, 1993.

Jedrychowski, W., Review of recent studies from Centeral and Eastern Europe associating respiratory health effects with high levels of exposure to "traditional" air pollutants, *Environ Health Persp*, 103 (suppl. 2), 15, 1995.

Johnson, E. S., Important aspects of the evidence for TCDD carcinogenicity in man, *Environ Health Perspect*, 99, 383, 9, 1993.

Johnson, E. S., Human exposure to 2,3,7,8-TCDD and risk of cancer, *Crit Rev Toxicol*, 21, 451, 1992.

Johnson, B. B., Covello, V. T., *The Social and Cultural Construction of Risk*, D. Reidel, Dordrecht, 1987.

Johnson, B. L., DeRosa, C. T., Chemical mixtures released from hazardous waste sites: implications for health risk assessment, *Toxicology*, 105, 145, 1995.

Jones, P. W., Twinning in the human population and in the cattle exposed to air pollution from incinerators [letter], *Brit J Ind Med*, 46, 215, 1989.

Kimbrough, R. D., Human heath effects of polychlorinated biphenyls (PCBs) and polybrominated biphenyls (PBBs), *Annu Rev Pharm Toxicol*, 27, 87, 1987.

Kimbrough, R. D., Falk, H., Stehr, P., Fries, G., Health implications of 2,3,7,8-tetrachlorodibenzodioxin (TCDD) contamination of residential soil, *J Toxicol Environ Health*, 14, 1, 47, 1984.

Klassen, C. D., Amdur, M. O., Doull, J., eds., *Cassarett and Doull's Toxicology: the Basic Science of Poisons*, MacMillan, New York, 1991.

Kleindorfer, P. R., Compensation and negotiation in the siting of hazardous-waste facilities, *Sci. Total Environ*, 51, 197, 1986.

LaDou, J., *Occupational Medicine*, Appleton & Lange, Norwalk, CT, 1990.

Landrigan, P. J., Toxicity of lead at low dose, *Br J Ind Med*, 46, 593, 1989.

Lenihan, J., *Bonnybridge/Denny Morbidity Review: Report of Independent Review Group*, Scottish Home and Health Dept, Edinburgh, Scotland, 1985.

Lie, R. T., Wilcox, A. J., Skaerven, R., A population based study of the risk of recurrence of birth defects, *N Engl J Med*, 331, 1, 1994.

Lieben, A., An epidemiologic study of occupational bladder cancer, *Acta Union Int Cancer*, 19, 749, 1963.

Lilienfeld, D. E., Gallo, M. A., 2,4-D, 2,4,5-T, and 2,3,7,8-TCDD: an overview, *Epidemiol Rev*, 11, 28, 1989.

Lisk, D. J., Environmental implications of incineration of municipal solid waste and ash disposal, *Sci Total Environ*, 74, 39, 1988.

Lloyd, O. L., Smith, G. A., Lloyd, M. M., Williams, F., Hopwood, D., Health hazards from chemicals in the environment; problems of evaluation, *Chem Br*, 23, 31, 1987.

Lloyd, O. L., Smith, G., Lloyd, M. M., Holland Y., Gailey, G. A., Raised mortality from lung cancer and high sex ratios of births associated with industrial pollution, *Br J Ind Med*, 42, 475, 1985.

Lloyd, O. L., Lloyd, M. M., Williams, F., Lawson, A., Twinning in human populations and in cattle exposed to air pollution from incinerators, *Br J Ind Med*, 45, 556, 1988.

Logue, J. N., Stroman, R. M., Reif, D., Hayes, C. W., Sivarajah, K., Investigation of potential health effects associated with well water chemical contamination in Londonderry township, Pennsylvania, U.S.A., *Arch Environ Health*, 40, 155, 1985.

Lyster, W. R., Pollution and sex ratio of births [letter], *Med J Aust*, 2, 151, 1981.

Mastriacova, P., Spagnolo, A., Marni E., Birth defects in the Seveso area after TCDD contamination, *JAMA*, 259, 1668, 1988.

Masuda, Y., Toxic evaluation of chlorinated aromatic hydrocarbons in human environments, *Toxicol Ind Health*, 7, 137, 141, 1991.

Moeller, D. W., *Environmental Health*, Harvard University Press, Cambridge, MA, 1992.

Momas, I., Pirard, P., Quenel, P., Medina, S., Moullec, Y., Ferry, R., Dab, W., Festy, B., Urban air pollution and mortality: contribution of epidemiologic studies published from 1980–1991, *Rev Epidemiol Sante Publique*, 41, 30, 1993.

Najem, G., Thind, I., Lavenhar, M., Louria, D., Gastrointestinal cancer mortality in New Jersey counties and the relationship with environmental variables, *Int J Epidemiol*, 12, 276, 1983.

National Institute of Occupational Safety and Health (NIOSH), *Health Hazard Evaluation Report: The Caldwell Group, North Carolina*, HETA 90-240-259, Cincinnati, Ohio, 1992.

Needleman, H., The persistent threat of lead: a singular opportunity, *Am J Public Health*, 75, 643, 1989.

Needleman, H., Gatsonis, G. A., Low level lead exposure and the IQ of children, *JAMA*, 263, 673, 1990.

Neubert, R., Stahlmann, R., Korte, M., van Loveren, H., Vos, J. G., Golor, G., Webb, J. R., Helge, H., Neubert, D., Effects of small doses of dioxins on the immune system of marmosets and rats, *Ann N Y Acad Sci*, 685, 662, 1993.

Oppelt, E., Air emissions from the incineration of hazardous waste, *Tox Ind Health*, 6, 23, 1990.

Orban, J. E., Stanley, J. S. Schwemberger, J. G., Remmers, J. C., Dioxins and dibenzo-furans in adipose tissue of the general US population and selected subpopula-tions, *Am J Public Health*, 84, 439, 1994.

Ozonoff, D., Colten, M. E., Cupples, A., Heeren, T., Schatzkin, A., Mangione, T. Dresner, M., Colton, T., Health problems reported by residents of a neighborhood contaminated by a hazardous waste facility, *Am J Ind Med*, II, 581, 1987.

Ozonoff, D., Aschengrau, A., Community exposure to toxic substances, *Occupational and Environmental Reproductive Hazards: A Guide For Clinicians*, Paul, M., ed., Williams & Wilkins, Baltimore, 1993, 379.

Paustenbach, D. J., Wenning, R. J., Lau, V., Harrington, N. W., Rennix, D. K., Parsons, A. H., Recent developments on the hazards posed by 2,3,7,8-tetrachlorodibenzo-*p*-dioxin in soil: implications for setting risk-based cleanup levels at residential and industrial sites, *J Toxicol Environ Health*, 36, 103, 1992.

Pearn, J. H., Herbicides and congenital malformations: a review for the pediatrician, *Aust Pediatr J*, 21, 237, 1985.

Peters, W. A., Darivakis, D. S., Howard, J. B., Solids pyrolysis and volatiles secondary reactions in hazardous waste incineration: implications for toxicants destruction and PIC's generation, *Haz Waste Haz Mater*, 7, 89, 1990.

Reggiani, G., Bruppacher, R., Symptoms, signs and findings in humans exposed to PCBs and their derivatives, *Environ Health Perspect*, 60, 225, 1985.

Robinson, F., Rollins hazardous waste incinerator at Alsen, Louisiana, *Waste Not*, 113, July 26, 1990.

Roeleveld, N., Zielhuis, G. A., Gabreels, F., Occupational exposure and defects of the central nervous system in offspring, *Br J Ind Med*, 47, 580, 1990.

Rogan, W. J., Gladen, B. C., Wilcox, A. J., Potential reproductive and postnatal mor-bidity from exposure to polychlorinated biphenyls: epidemiological consider-ations, *Environ Health Perspect*, 60, 233, 1985.

Roht, L., Vernon, S. W., Weir, F. W., Pier, S. M., Sullivan, P., Reed, L. J., Community exposure to hazardous waste disposal sites: assessing reporting bias, *Am J Epi-demiol*, 122, 418, 1985.

Rom, W. N., ed., *Environmental and Occupational Medicine*, Little Brown & Company, Boston, 1992.

Rosenstock, L., Cullen, M. R., eds, *Textbook of Clinical Occupational and Environmental Medicine*, W.B. Saunders Company, Philadelphia, 1994.

Rothenbacher, D., Dgenan, D., Rhodes, V., Shy, C., Respiratory symptom prevalence in three North Carolina incinerator communities, in *Hazardous Waste and Public Health: International Congress on the Health Effects of Hazardous Waste*, Andrews, J. S., Frumkin, H., Johnson, B. L., Mehlman, M. A., Xintaras, C., Bucsela, J. A., eds., Princeton Scientific, Princeton, NJ, 1994, 757.

Russell, E. B., Microphthalmos and anophthalmos and environmental pollutants, *Br Med J*, 306, 790, 1993.

Safe, S., Polychlorinated biphenyls (PCBs): environmental impact, biochemical and toxic responses, and implications for risk assessment, *Crit Rev Toxicol*, 24, 87, 149, 1994.

Safe, S., Polychlorinated biphenyls (PCBs): mutagenicity and carcinogenicity, *Mutat Res*, 22, 31, 1989.

Safe, S., Toxicology, structure-function relationship, and human and environmental health impacts of polychlorinated biphenyls: progress and problems, *Environ Health Perspect*, 100, 259, 68, 1993.

Samet, J. M., Marbury, M. C., Spengler, J. D., Health effects and sources of indoor air pollution, *Am Rev Resp Dis*, 136, 1486, 1987.

Sanjour, W., Hazardous waste incinerators, *Waste Not*, 112, July 19, 1990.

Schoni, M. H., Air pollution and the juvenile lungs, *Schweiz Med Wochenschr*, 123, 188, 1993.

Schwartz, J., Dockery, D. W., Increased mortality in Philadelphia associated with daily air pollution concentrations, *Am Rev Resp Dis*, 145, 600, 1992.

Scottish Home and Health Dept., *Report of Working Party on Microphthalmos in the Forth Valley Health Board Area*, Scottish Home and Health Dept., Edinburgh, Scotland, 1988.

Sedman, R., M., Esparza, J., Evaluation of volatile organic emissions from hazardous waste incinerators, *Environ Health Perspect*, 94, 169, 180, 1991.

Shaw, G. M., Shulman, J., Frisch, J. D., Cummins, S. K., Harris, J. A., Congenital malformations and birthweight in areas with potential environmental contamination, *Arch Environ Health*, 47, 147, 1992.

Sherwin, R. P., Air pollution: the pathobiologic issues, *J Toxicol Clin Toxicol*, 29, 385, 1991.

Shusterman, D., Lipscomb, J., Neutra, R., Satin, K., Symptom prevalence and odor-worry interaction near hazardous waste sites, *Environ Health Perspect*, 94, 25, 1991.

Sinks, T., Steele, G., Smith, A. B., Watkins, K., Shults, R. A., Mortality among workers exposed to polychlorinated biphenyls, *Am J Epidemiol*, 136, 389, 1992.

Stellman, S. D., Stellman, M. J., Sommer, J. F., Health and reproductive outcomes among American Legionnaires in relation to combat and herbicide exposure in Vietnam, *Environ Res*, 47, 701, 1988.

Swan, S., Shaw, S., Harris, J. A., Neutra, R. R., Congenital cardiac anomalies in relation to water contamination, Santa Clara, California, 1981–1983, *Am J Epidemiol*, 129, 885, 1989.

Tamaddon, F., Hogland, W., Review of cadmium in plastic waste in Sweden, *Waste Manage Res*, 11, 287, 1993.

Taylor, S., Elliott, S., Eyles, J., Frank, J., Haight, M., Streiner, D., Walter, S., White, N., Willms, D., Psychosocial impacts in populations exposed to solid waste facilities, *Soc Sci Med*, 33, 441, 1991.

The Selected Cancers Cooperative Study Group, The association of selected cancers with service in the US military in Vietnam, *Arch Intern Med*, 150, 2473, 1990.

Tollefson, L., Use of epidemiology data to assess the cancer risk of 2,3,7,8-tetrachlorodibenzo-*p*-dioxin, *Rev Toxicol Pharm*, 13, 150, 1991.

Toppari, J., Male reproductive health and environmental xenoestrogens, *Environ Health Persp*, 104 (suppl. 4), 741, 1996.

Travis, C. C., ed., *Use of Biomarkers in Assessing Health and Environmental Impacts of Chemical Pollutants*, Plenum Press, New York, 1993.

Travis, C. C., Cook, S. C., eds., *Hazardous Waste Incineration and Human Health*, CRC Press, Boca Raton, FL, 1989.

Travis, C. C., Hattemer-Frey, H. A., Human exposure to dioxin, *Sci Total Environ*, 104, 97, 1991.

Travis, C. C., Holton, G. A., Etnier, E. L., Cook, S. C., O'Donnell, F. R., Hetrick, D. M., Dixon, E., Potential health risk of hazardous waste incineration, *J Haz Mater*, 14, 309, 1987.

Vainio, H., Hesso, A., Jappinen, P., Chlorinated dioxins and dibenzofurans in the environment — a hazard to public health?, *Scand J Work Environ Health*, 15, 377, 1989.

Vanden Heuvel, J. P., Lucier, G., Environmental toxicology of polychlorinated diben-zo-p-dioxins and polychlorinated dibenzofurans, *Environ Health Perspect*, 100, 189, 1993.

Vyner, H. M., *Invisible Trauma: Psychosocial Effects of the Invisible Environmental Contaminants*, DC Health Toronto, Toronto, Canada, 1988.

Wanner, H., Effects of atmospheric pollution on human health, *Experientia*, 49, 754, 1993.

World Health Organization (WHO), *Polychlorinated Dibenzo-para-dioxins and Dibenzofurans*, WHO Environmental Criteria Document #88, Geneva, 1989.

World Health Organization (WHO). *Update and Revision of the Air Quality Guidelines for Europe*. WHO, Copenhagen, 1995.

Yen, Y. Y., Lan, S. J., Ko, Y. C., Chen, C. J., Follow-up study of reproductive hazards of multiparous women consuming PCBs-contaminated rice oil in Taiwan, *Bull Environ Contam Toxicol*, 43, 647, 1989.

Young, C. M., Voorhees, K. J., Thermal decomposition of 1, 2-dichlorobenzene. II. Effect of feed mixtures, *Chemosphere*, 24, 681, 1992.

Yu, M. L., Hsu, C. C., Gladen, B. C., Rogan, W. J., In utero PCB/PCDD exposure: relation of developmental delay to dysmorphology and dose, *Neurotoxicol Teratol*, 13, 195, 1991.

Zmirou, D., Parent, B., Potelon, J. L., Epidemiologic study of the health effects of atmospheric waste from an industrial and household refuse incineration plant, *Rev Epidemiol Sante Publique*, 32, 391, 1984.

chapter eleven

The role of epidemiology in evaluating potential public health consequences of hazardous waste incineration

Lora E. Fleming and Judy A. Bean

Contents

Introduction

In this chapter, a brief summary of the scientific discipline of epidemiology — its uses and limitations, and its relation to the study of human health effects from hazardous waste incineration — is presented.

Principles of epidemiology

Epidemiology is the study of the distribution and determinants of diseases in human populations. In the case of hazardous waste incineration, environmental epidemiology attempts to determine associations between pollutants

1-56670-250-X/99/$0.00+$.50
© 1999 by CRC Press LLC

and particular human health effects, and how to prevent these effects in exposed populations.

Since epidemiology is predominantly an observational rather than experimental science, it relies heavily on data collected in records, question-naires, and elsewhere. Using measures of risk (i.e., the relative risk and the odds ratio), comparing unexposed and exposed populations for disease risk or comparing diseased and well populations for exposure risk, epidemio-logic studies can associate a disease risk with a particular exposure (Andrews, 1994; Checkoway et al., 1989; Florida, 1995; Goldsmith, 1986; Grandjean, 1993, 1994; Hernberg, 1992; Johnson, 1995; Leaverton, 1982; Lil-ienfeld and Stolley, 1994; Monson, 1990; Talbott, 1995).

In epidemiology, proof of causality (or the association of a particular pollutant exposure with a particular disease) is based on a variety of criteria. The association must be shown repeatedly in different studies of different populations. Furthermore, the association should be preferably strong, as determined by one of the above-mentioned measures of risk. The exposure to the particular pollutant exposure must precede the onset of disease, and the time period from first exposure to the onset of the disease (i.e., the latency) must be sufficient. There should be some evidence of a dose-response relationship: with an increased dose of the pollutant, there is an increased risk of disease. The association between the pollutant exposure and the disease must make scientific sense (i.e., biological plausibility). If possible, it should be reproducible in toxicologic studies with laboratory animals and other systems, such as *in vitro*. Finally, ideally, if the pollutant exposure is removed, the amount of disease (i.e., the incidence) should decrease.

In science, proof that a given pollutant causes human health effects is established by a hierarchy of evidence. First, there is often the existence of a medical literature with multiple individual case reports which associate human disease with exposure to a particular pollutant. There can be toxico-logic evidence in experimental animals in which exposure to the pollutant causes diseases in animals similar to those seen in humans. Finally, epide-miologic studies are considered to be the highest level of scientific evidence for proving an association between a particular toxic exposure and human health effects.

In addition to issues of whether a study was conducted properly, among epidemiologic studies there is an additional hierarchy of evidence dependent on the type of study performed. The simplest type of epidemiologic studies are cross-sectional studies; these are performed at a single point in time. Cross-sectional studies can suggest an association. However, because cross-sectional studies are performed at a single point in time, it is impossible to distinguish whether the pollutant exposure came before and thus caused the disease, or came after and is not truly associated. Therefore, in general, cross-sectional studies are considered "hypothesis generating" but do not provide proof of causation.

A more complicated epidemiologic study that can suggest an association more strongly is the case control study. In case control studies, persons with the disease (the cases) and without the disease (controls) are investigated for their different exposures; if the pollutant is associated with the disease, then the cases will report a much greater exposure to the pollutant in the past than the controls. In certain cases, such as extremely rare diseases with a very strong disease-toxin association, such as maternal exposure to the hormone DES and vaginal carcinoma in the female offspring, case control studies can provide proof of causation.

The experimental equivalent of a toxicologic study (i.e., controlled laboratory studies in animals and humans) in epidemiology is the clinical trial in which a group of people are given a particular exposure (usually a medication) or not (the placebo), and then followed to observe the effect of the medication or the lack of the medication. Obviously, this type of epidemiologic study is rarely performed in the study of toxic exposures in humans for ethical reasons. Therefore, although clinical trials obviously provide proof of causation, they would probably not be relevant in the study of possible human health effects from hazardous waste incineration.

Finally, cohort studies are considered the *sine qua non* of epidemiologic studies. These studies follow people over time so that the exposure is known to precede the health effect. Furthermore, cohort studies produce population-based rates of disease. In cohort studies, a group of people (the cohort) is distinguished by their exposures (usually exposed and nonexposed) and examined for the diseases which have developed over time. If a particular pollutant is associated with a particular disease, then the pollutant-exposed group would be expected to have many more cases of the particular disease than the nonexposed group.

As stated above, a disease-pollutant association would be considered established if there are repeated similar findings in both toxicologic studies in animals and in multiple epidemiologic studies. Further proof would be toxicologic and epidemiologic studies that show that, when the toxic exposure is removed, the amount of the particular disease decreases or disappears.

Obviously disease-pollutant connections are much easier to establish in the case of acute, as opposed to chronic, health effects in both humans and laboratory animals. The acute effects of carbon monoxide, i.e., death by asphyxiation, have been easy to establish. However, the long-term effects of asbestos exposure associated with lung disease and cancer have been much more difficult to prove because the animals or people must be followed for long time periods and may be affected by many other concurrent exposures during that time. In addition, humans have much longer life spans than many other animals as well as subtle differences in enzymatic systems and often different routes of exposure. Therefore, the extrapolation between diseases found in laboratory animals from toxicologic studies to human diseases in the general population associated with a pollutant exposure is problematic, especially for chronic diseases such as cancer (Checkoway et al., 1989;

Grandjean, 1993, 1994; Hernberg, 1992; Leaverton, 1982; Lilienfeld and Stolley, 1994; Monson, 1990; Talbott 1995).

Pertinent epidemiologic issues

Exposure and population

Key components in establishing the relationship between an environmental pollutant and a health effect are the type of exposure and the selection of the population. Traditionally, many disease-pollutant connections in humans have been established by the evaluation of exposed workers and their occupational diseases. This is because, with rare exceptions (such as methylmercury exposure in Minamata, Japan), workers tend to have much higher exposures in the workplace than the general public to pollutants (Buffler, 1985; Checkoway et al., 1989; Grandjean, 1993, 1994; Hernberg, 1992; Leaverton, 1982; Lilienfeld and Stolley, 1994; Monson, 1990; Talbott 1995).

However, for issues such as community exposure to hazardous waste incineration, occupational exposure information may not be appropriate for extrapolation to the community exposure and their chronic health effects. There is a variety of reasons. Community exposures tend to be much lower and may occur over the entire lifetime of a person, not just during a 40-hour work week. In addition, communities are composed of young and old people, healthy and sick, while working populations tend to be young and healthy. Another important issue is that the effects of pollutants on fetuses and growing children can be devastating at levels which are relatively tolerated by adults (as with the example of the neurologic effects of lead and mercury) (NRC, 1993). To determine low-level exposures in communities in epidemiologic studies, it is necessary to follow large populations of people for long periods of time to see any disease effects; these studies are exceedingly expensive to perform and difficult to interpret.

Mixed low-level exposures with multiple different pollutants, as would be expected from the incineration of hazardous waste, is another issue which has been very difficult for both toxicologic and epidemiologic investigations in the area of hazardous waste. Mixed exposures are difficult to classify, quantify, and even to measure. Furthermore, the particular mixture of exposures from one hazardous waste incineration site may not be the same mixture found in another site. In addition, it is possible that mixed exposures may cause more or less health effects than exposures to single pollutants, since there is the possibility of synergism and/or antagonism of pollutants within the organism (Andrews, 1994; Buffler, 1985; Jain, 1993; Landrigan, 1983; Vainio, 1990).

Another exposure issue relevant to hazardous waste incineration and human health effects is the effect of brief and/or intermittent exposures. For example, much of the existing hazardous waste literature is based on brief "accidental" exposures such as that seen in Seveso, Italy; however, the relevance of the health outcomes seen with these single exposures to the more likely scenario of consistent, chronic, low-level exposures is unknown. Finally, recent interest in the scientific community has focused on the issue of indoor as well as outdoor exposures, since most persons in the "developed" nations spend the majority of their time (over 90%) indoors; in some studies, indoor exposures to various pollutants exceeded outdoor exposures by 10- to 100-fold (Jaakkola et al., 1994; Samet, 1987).

The routes of exposure can determine the presence and/or type of human health effect. Toxic exposures in the occupational setting are traditionally through inhalation and/or through the skin, whereas community exposures to hazardous waste have been predominantly through contaminated water (such as at Woburn, Massachusetts), and/or the food chain (such as fish consumption at Minamata, Japan). With the incineration of hazardous waste, the possibility of both inhalation and ingestion as routes of exposure must be considered. Inhalation of the airborne pollutant emissions and ingestion of both foods from the food chain and water contaminated by either the emissions and/or inadequately disposed ash could be expected.

Particular to the food chain are issues of bioaccumulation and bioconcentration (as are seen with mercury and lead as well as with many lipophilic pollutants). This means that as a pollutant goes up the food chain, the concentrations of the pollutant increase due to increased storage; therefore, humans, often at the top of the food chain, will receive the highest doses. Furthermore, within human beings, the ability of certain pollutants to concentrate in fat (i.e., lipophilic) means that increased doses of a pollutant can be delivered to the fetus and to the nursing child from the mother, as has been shown with DDT and other organochlorine chemicals (NRC, 1993). In addition, low dose lifetime exposures can be provided with slow but continual release from fat-stored lipophilic pollutants, even when external exposure has ceased.

Finally, individuals even in the same species can process exogenous chemicals differently, depending on particular enzyme systems, baseline health and nutrition, and concomitant exposures. Therefore, not just children and infants, but adults who are sick and/or malnourished can be at greater risk from exposure to pollutants than the rest of the population (Grandjean, 1993, 1994; NRC, 1993).

Human health effects

In addition to the issues of exposure, human health effects must be defined. In the past, end-stage diseases have been established as the appropriate human health effects to be studied. But due to issues of prevention, as well as the ability to detect much subtler physiologic changes which are possibly reversible before the development of fully developed diseases, human health effects can no longer be defined in this fashion. For example, in the past, lead toxicity was considered to be frank encephalopathy and even death;

now lead levels are being set based on the prevention of subtle neurobehavioral effects on fetuses and young children (Andrews, 1994; Buffler, 1985; Landrigan, 1983).

The new area of biomarkers of effect and/or exposure is being explored in environmental epidemiology. Biomarkers are indicators of specific pollutants and their metabolites within the organism, and thus represent true individualized measures of exposure and/or effect. For example, DNA adducts can be formed with exposure to certain carcinogenic pollutants (such as benzo(a)pyrene); these biomarkers can denote both exposure to the carcinogen and possibly a marker of effect. Extensive epidemiologic studies in multiple populations are needed to establish the prognostic implications of these biomarkers. Nevertheless, biomarkers represent a fruitful area of epidemiologic research for communities with chronic low level exposure to hazardous waste incineration (Hulka and Wilcosky, 1990; Hulka et al., 1992).

Furthermore, epidemiologic studies relevant to the possible human health effects of exposure to hazardous waste incineration must focus on more acute diseases such as reproductive or respiratory complaints. Furthermore, these studies must monitor apparently trivial complaints (such as mucous membrane irritation), or if available, biomarkers, rather than waiting for the establishment of definitive physiologic damage. Even psychological effects should be more fully studied, not only because they are acute and valid markers of exposure effects, but because psychological effects can have such an important impact on the overall community attitude toward the hazardous waste incineration facility.

Other concerns

Epidemiologic studies have their own intrinsic problems in the study of possible human health effects from hazardous waste exposure, besides those already discussed — disease, exposure, and population definitions. There is the reliance on already collected data, such as records and questionnaires, which allows for inaccuracy of individual information. Competing or confounding exposures, such as cigarette smoking, can be very difficult to determine accurately, especially for individuals, and can affect the results of a study. In the hazardous waste literature, the majority of epidemiologic studies have not collected individual confounding information from their subjects (Andrews, 1994; Checkoway et al., 1989; Florida, 1995; Goldsmith, 1986; Grandjean, 1993, 1994; Hernberg, 1992; Johnson, 1995; Lilienfeld and Stolley, 1994; Monson, 1990; Talbott, 1995).

Other biases, especially so-called reporting bias, can make interpretation of results difficult. Reporting bias is especially common when a community which believes itself to be ill, based on known proximity to a hazardous waste site, is examined for complaints of disease and exposures. Often, their heightened awareness and fear leads to the reporting of falsely elevated human health effects (Shusterman et al., 1991; Elliot et al., 1993; Taylor et al., 1991; Andrews, 1994).

A further epidemiologic problem, especially in the area of hazardous waste research, is the inevitable existence of disease clusters. Clusters are increases of disease in time and space. Sometimes these disease increases are due to the same and/or shared toxic exposure in a given community, but more often they are due to chance; in other words due to randomness, a group of individuals who live in the same geographic area has the same disease for different reasons. Therefore, it is always difficult to draw conclusions from a single epidemiologic study which indicates a positive disease–pollutant connection (Fleming et al., 1991; Aldrich, 1993).

A further problem plaguing the epidemiologic research of the human health effects of hazardous waste is the relatively small populations studied. The small numbers makes valid statistical analysis and generalizability of the study conclusions questionable. Not only are the numbers of people small, but again these populations are often studied in crisis (such as the Love Canal studies) with considerable anxiety as well as legal involvement; this atmosphere makes objective scientific investigation very difficult (Andrews, 1994; Elliot et al., 1993; Shusterman et al., 1991; Taylor et al., 1991).

The final issue with epidemiologic studies of hazardous waste, in particular concerning hazardous waste incineration, is the relatively short latency time. In other words, large-scale, organized hazardous waste incineration is a relatively new technology so that populations have only been exposed for a few years. Chronic diseases, such as cancer, have not had sufficient time to develop. Thus, conclusions concerning the health effects in humans of the incineration of hazardous waste must be drawn from a frankly inadequate database.

Surveillance

As discussed above, in epidemiologic studies to determine the human health effects of low level exposures, it is necessary to follow large populations of people for long periods of time to observe any chronic human health effects. These studies are exceedingly difficult to perform and interpret. Due to the dearth of information concerning hazardous waste incineration and human health effects, a basic form of epidemiologic study known as surveillance is recommended for populations exposed to hazardous waste incineration (Kipen, 1992; Stallones, 1992; Schaub, 1994; Grandjean, 1993, 1994; Thacker, 1996; Herz-Picciotto, 1996; Reilly, 1995; Stebbings, 1981).

Ideally, surveillance of populations, both community and worker, with proposed hazardous waste incineration exposure should begin with a pre-exposure baseline of health status. Surveillance can be used for relatively acute disease end points such as respiratory illness or reproductive changes, both of which have been linked with exposure to hazardous waste. Surveillance should target not only fully developed disease (as has been the case in the past), but rather it should try to detect subtle physiologic changes which are ideally reversible, or even biomarkers of exposure and/or effect. Tests can range from neurobehavioral testing to DNA adduct monitoring for specific carcinogens, depending on the exposures as well as the wishes of the local community and its scientific advisors. Surveillance can also be age and community specific; for example, surveillance of growing children might focus on subtle developmental changes. After a hazardous waste incinerator is built, there should be monitoring of the community at appropriate intervals to ascertain the risk of disease and/or physiologic change.

Ideally, there should be a similar but unexposed community studied at the same time as a control population. This would allow for an appropriate comparison and interpretation of changes in health effects of the exposed community over time against the "norm" or control of the unexposed community. Collection of confounding information (such as tobacco use) in both populations would be important to assure appropriate comparison. In addition, it has been suggested that health promotion education concerning existing competing risks (such as tobacco use) in communities exposed to hazardous waste incineration may help to decrease these risks (Andrews 1994).

There can also be surveillance of key species of flora and fauna in the local environment, again monitoring for subtle, reversible physiologic changes rather than obvious abnormalities or destruction. These creatures are often more susceptible to pollutant exposure, and can be used as sentinel species to detect environmental impact long before physiologic or disease effects are noted in humans.

In addition to population and environmental surveillance, accurate exposure monitoring of all routes of hazardous waste incineration emissions is crucial. This should not only include air monitoring, but also water, soil and food chain measurement. Occupational exposures both inside the plant and among the surrounding community would need to be assessed as well. Again, the issues of multiple exposures with possible synergism and/or antagonism must be taken into account.

All the information collected above could be integrated using geographic information systems (GIS). GIS has data management, data analysis, and data presentation mapping functions which make it uniquely applicable to single point pollution sources with extensive data collection (Stallones, 1992; Herz-Picciotto, 1996). Although more work has to be done by epidemiologists, risk assessors, and toxicologists to integrate existing epidemiologic data into the risk assessment, as well as guide data collection in future epidemiologic studies, the data can be continually integrated into mathematical models of risk assessment to predict and to prevent any irreversible human and/or environmental damage before the damage occurs. In addition, even though these surveillance studies would consist of fairly small communities with relatively few individuals, if properly set up, results from multiple studies could be pooled for meaningful analysis.

Finally, as mentioned above, communication, education, and community involvement in both surveillance and monitoring should be integral parts of any proposed study. This would be important not only for cooperation by the community in such a venture but also for the community to protect itself, since this is an area in which scientific knowledge is lacking (Neutra, 1985; Shusterman et al., 1991; Elliot et al., 1993; Taylor et al., 1991; Johnson, 1995; Andrews, 1994; Florida, 1995; Sims, 1983).

Summary

Epidemiology studies human populations in their natural environments. Epidemiologic evidence provides the *sine qua non* for evidence of human health effects from particular exposures. Nevertheless, certain epidemiologic issues are associated with the determination of whether or not hazardous waste incineration can cause human health effects.

Intrinsic to epidemiology are the issues of epidemiologic causation, bias, (especially confounding), and exposure and disease measurement. A further epidemiologic issue is that these are costly studies to perform, both in terms of time and person power. The *sine qua non* epidemiologic study is the prospective cohort study; this requires following a population over time into the future.

With regard to the study of possible human health effects from hazardous waste incineration, a number of epidemiologic issues must be considered. Community low-level exposures in mobile populations of the whole spectrum of society must be studied, not just more homogeneous and relatively manageable working populations. The time period from exposure to the development of disease (i.e., the latency), especially for chronic diseases, is still too short for this relatively new technology for definitive epidemiologic study. Furthermore, the existing studies suffer from lack of precise exposure information (including confounders) and from recall bias, since these are often communities in crisis.

Some conclusions can be made about the possible human health effects of hazardous waste incineration from the study of the available literature of hazardous waste incineration and related issues such as air pollution. However, proactive surveillance of current and future exposed populations and their environment must be performed to truly understand the human health ramifications of hazardous waste incineration. This surveillance of the population and environment, as well as exposure monitoring, can serve the dual functions of health effect prevention in a given community, as well as the gathering of essential scientific data.

References

Aldrich, T. E., Leaverton, P.E., Sentinel event strategies in environmental health, *Ann Rev Pub Health*, 14, 205, 1993.

Andrews, J. S., Frumkin, H., Johnson, B. L., Mehlman, M. A., Xintaras, C., Bucsela, J. A., eds., *Hazardous Waste and Public Health: International Congress on the Health Effects of Hazardous Waste*, Princeton Scientific, Princeton, NJ, 1994.

Buffler, P. A., Crane, M., Key, M. M., Possibilities of detecting health effects by studies of populations exposed to chemicals from waste disposal sites, *Env Health Perspectives*, 62, 423, 1985.

Checkoway, H., Pearce, N. E., Crawford-Brown, D. J., *Research Methods in Occupational Epidemiology*, Oxford University Press, New York, 1989.

Elliot, S. J., Taylor, S. M., Walter, S., Stieb, D., Frank, J. Eyles, J., Modelling psychosocial effects of exposure to solid waste facilities, *Soc Sci Med*, 37, 791, 1993.

Fleming, L. E., Ducatman, A. M., Shalat, S. L., Disease clusters: a central and ongoing role in occupational medicine, *J Occup Med*, 33, 818, 1991.

Florida Center for Solid and Hazardous Waste Management., *Evaluation of the Human Health Impacts Associated with Commercial Hazardous Waste Incinerators, Report #95-4B*, Florida Dept of Environmental Protection, Gainesville (FL), July 1995.

Goldsmith, J. R., *Environmental Epidemiology: Epidemiological Investigation of Community Environmental Health Problems*, CRC Press, Boca Raton, FL, 1986.

Grandjean, P., Epidemiology of environmental hazard, *Public Health Rev*, 21, 255, 1993/1994.

Hernberg, S., *Introduction to Occupational Epidemiology*, CRC Press, Boca Raton, FL, 1992.

Herz-Picciotto, I., Comment: toward a coordinated system for the surveillance of environmental health hazards. *Am J Public Health*, 86, 638, 1996.

Hulka, B. S., Wilcosky, T., eds, *Biological Markers in Epidemiology*, Oxford University Press, New York, 1990.

Hulka, B. S., Wilcosky, T., Griffith, J. D., Methodological issues in epidemiologic studies using biological markers, *Am J Epidemiol*, 136, 200, 1992.

Jaakkola, J. J. K., Tuomaala, P., Seppanen, O., Air recirculation and sick building syndrome: a blinded crossover trial, *Am J Public Health*, 84, 422, 1994.

Jain, R. K., Urban, L. V., Stacey, G. S., Balbach, H. E., *Environmental Assessment*, McGraw-Hill, New York, 1993.

Johnson, B. L., DeRosa, C. T., Chemical mixtures released from hazardous waste sites: implications for health risk assessment, *Toxicology*, 105, 145, 1995.

Kipen, H. M., Craner, J., Sentinel pathologic conditions: an adjunct to teaching occupational and environmental disease recognition and history taking, *Environ Res*, 59, 93, 1992.

Landrigan, P. J. Epidemiologic approaches to persons with exposures to waste chemicals. *Environ Health Perspect*, 48, 93, 1983.

Leaverton, P. E., *Environmental Epidemiology*, Praeger, New York, 1982.

Lilienfeld, D. E., Stolley, P. D., *Foundations of Epidemiology*, Oxford University Press, Oxford, 1994.

Monson, R. R., *Occupational Epidemiology*, CRC Press, Boca Raton, FL, 1990.

National Research Council, *Pesticides in the Diets of Infants and Children*, National Academy Press, Washington, DC, 1993.

Neutra, R. R., Epidemiology for and with a distrustful community, *Environ Health Perspect*, 62, 393, 1985.

Neutra, R., Roles for epidemiology: the impact of environmental chemicals, *Environ Health Perspect*, 48, 99, 1983.

Reilly, M. J., Rosenman, K. D., Use of hospital discharge data for surveillance of chemically-related respiratory disease, *Arch Environ Health*, 50, 26, 1995.

Samet, J. M., Marbury, M. C., Spengler, J. D., Health effects and sources of indoor air pollution, *Am Rev Resp Dis* 136, 1486, 1987.

Schaub, E. A., Bisesi, M. S., Medical and environmental surveillance, *N J Med*, 91, 715, 1994.

Shusterman, D., Lipscomb, J., Neutra, R., Satin, K., Symptom prevalence and odor-worry interaction near hazardous waste sites, *Environ Health Perspect*, 94, 25, 1991.

Sims, J. H., Baumann, D. D., Educational programs and human response to natural hazards, *Environ Behav*, 2, 165, 1983.

Stallones, L., Nuckols, J. R., Berry, J. K., Surveillance around hazardous waste sites: geographic information systems and reproductive outcomes, *Environ Res*, 59, 81, 1992.

Stebbings, J. H., Epidemiology, public health, and health surveillance around point sources of pollution, *Environ Res*, 25, 1, 1981.

Talbott, E. O., Craun, G. F., *Introduction to Environmental Epidemiology*, CRC Press, Boca Raton, FL, 1995.

Taylor, S. M., Elliott, S., Eyles, J., Frank, J., Haight, M., Streiner, D., Walter, S., White, N., Willms, D., Psychosocial impacts in populations exposed to solid waste facilities, *Soc Sci Med*, 33, 441, 1991.

Thacker, S. B., Stroup, D. F., Parrish, G., Anderson, H. A., Surveillance in environmental public health: issues, systems and sources, *Am J Public Health*, 86, 633, 1996.

Vainio, H., Sorsa, M., McMichael A. J., et al., *Complex Mixtures and Cancer Risk*, International Agency for Research on Cancer [#104], Lyon, France, 1990.

chapter twelve

Medical monitoring for human health impacts from hazardous waste incineration

Isabel Stabile and Stuart Brooks

Contents

1-56670-250-X/99/$0.00+$.50
© 1999 by CRC Press LLC

Introduction

Incineration is only one of the many different methods that can be used to eliminate, reduce, and dispose of hazardous waste. Strictly speaking, incineration does not actually eliminate waste, but rather reduces its volume and changes its form. Some potentially hazardous compounds are destroyed by incineration, while others are created or concentrated into other residues such as ash, gases, or vapors.[1] Prior to the Clean Air Act of 1970, incinerators were poorly controlled and often polluted the air. When these antiquated incinerator systems were replaced by unlined landfills, contamination of groundwater and soil led to an EPA ban on land burial of volatile organic hazardous waste. Although the industry maintains that modern, properly designed, and regulated incineration systems have received worldwide scientific acceptance as the best available method for destroying and detoxifying hazardous waste,[2] these facilities have received much public opposition due to the perception that the old incinerators are typical of all incinerators.

Incineration has several advantages over other disposal methods. Controlled hazardous waste incinerators utilize extremely high temperatures to ensure nearly complete combustion. These incinerators can handle solid, liquid, and gaseous toxic wastes. When burned at optimal temperatures, these compounds are converted to less harmful ones, and the waste volume is greatly reduced.[1] However, there are many public health and environmental concerns about some of the solid and gaseous residual products of hazardous waste combustion, such as dioxins, polycyclic aromatic hydrocarbons, and heavy metals such as lead and arsenic.[3-5] The EPA has set standards for acceptable environmental concentrations, but little is known about the health effects of cumulative chronic exposure to these low-level pollutants.

The by-products of hazardous waste incineration can potentially contaminate air, water, and soil. Although exposure to airborne pollutants is perceived to be the most important public health threat, air concentrations of the most feared pollutants, such as dioxins, in the vicinity of incinerators are not increased above background levels.[6] Moreover, because of efficient air quality control technology, levels of other airborne pollutants, such as lead, carbon monoxide, nitrogen oxides, and total suspended particulates, are typically well below national air quality standard levels.[7] More significant health hazards stem potentially from the leaching of toxic substances from improperly disposed ash in landfills. This potential is of significant concern in the state of Florida where most of the drinking water comes directly from the underground aquifer system.

Relatively few studies concerning the health effects associated with hazardous waste sites have been published, and even fewer exist on human health effects of hazardous waste incineration. Unfortunately amid the controversy, fear, and stress which inevitably surround the possible location of a new incinerator facility, it is difficult to perform unbiased research. Most of the studies that have been published suffer from recall bias, self selection, small numbers, lack of individual exposure data, and lack of objective disease data. All of these factors make it very difficult to accurately generalize from such studies.

The existing data concerning human health effects of hazardous waste incineration are inadequate. Not only have few investigations been conducted, but existing studies have included relatively small numbers of individuals. The few available studies of occupational groups show highly variable effects on workers depending on which facility they worked in.[8-10] Most of these occupational studies (which, in principle, should give a worst-case scenario) were performed in older facilities with few environmental controls, if any. In one study of populations surrounding *municipal* solid waste incineration facilities, there was a negligible theoretical excess in cases of cancer from inhalation of metal emissions.[11] Studies of the health impacts on *municipal* (as compared to hazardous) waste incineration workers have concluded that there was an excess of deaths due to ischemic heart disease and lung cancer,[12] increased risk of esophageal cancer,[13] increase in mutagens in the urine of incineration workers,[14,15] and an increase in blood lead levels.[5] In contrast, the National Institute for Occupational Safety and Health (NIOSH) conducted five health hazard evaluations of waste incineration workers in the 1980s and identified very few health problems.[16]

There have been a few cross-sectional studies in communities surrounding hazardous waste incineration plants. Bearing in mind the limitations of this type of exercise (as listed above), there is some suggestion in the literature that those individuals residing closer to the plants are more likely to complain of respiratory problems, even after controlling for age, gender, and cigarette smoking.[17-19] However, these were all self-reported complaints, and the prevalence of physician-diagnosed respiratory disorders in residents was no different from that in unexposed comparison communities. However, in one study, the number of medications prescribed for respiratory disorders was significantly higher in those living closest to an incinerator.[20]

There have also been several investigations of clusters of cancer and birth defects around various incinerators, primarily in Europe. However, none of these disease clusters has been unequivocally shown to be associated with living near an incinerator. There is, however, evidence that persons living in communities surrounding incinerators and hazardous waste sites may suffer anxiety concerning the possible health risk related to the proximity of their residence to the presumed source of pollution.[21-23] In many cases there is also concern about decreased property values.

This chapter introduces the basic principles and approaches used in medical monitoring and will describe the principal methods for exposure assessment and population surveillance of health outcomes. The roles of risk assessment, risk communication, and community involvement associated with hazardous waste incineration facilities will also be described.

Principles of medical monitoring

The major goal of medical monitoring is to accurately collect and fully analyze clinical, epidemiologic, and other biomedical data for any evidence that the public and the environment are being, or could be, harmed by an existing or potential exposure scenario, e.g., a hazardous waste incineration facility. Surveillance strategies are implemented to serially monitor the health of the public and the environment so that any alterations are discovered and investigated at the earliest opportunity. The two components of a comprehensive surveillance strategy are measurement of exposures and health outcomes in potentially exposed individuals.

In order to comprehensively evaluate the cumulative health and environmental impacts of hazardous waste incineration, it is important to follow the fundamental principles of population epidemiologic research. Specifically, the complete cohort of persons who have the potential for exposure must be identified and characterized. This process includes the collection of baseline information through individual and family health status questionnaires, often using validated instruments such as those developed by the National Center for Health Statistics (NCHS) and the Agency for Toxic Substances and Disease Registry (ATSDR). Informed consent for the collection and banking of biologic materials must also be obtained. Well-defined rules for stratified sampling of populations must be developed and followed for the subsequent collection of data and/or specimens.

Medical monitoring or surveillance of a potentially exposed population may take many forms. At its simplest, a cross-sectional study of a typically self-selected population is performed utilizing history, physical examination, and investigations to screen for diseases likely to result from a specific exposure scenario. This will help to determine whether, and to what extent, any population members have been harmed by the potential exposure. Once initially screened, the population can then be subjected to periodic medical examination of exposed or affected individuals in an attempt to detect whether disease has occurred and to provide appropriate treatment.

Screening

Since the basis for medical surveillance is in fact screening of potentially exposed individuals, some or all of whom are asymptomatic, it is important for the reader to appreciate the principles behind screening for disease. The goal of any screening test (be it clinical examination or investigation) is to detect diseases before signs or symptoms develop. Several general principles (White criteria) must be considered before embarking on a screening program for one or more specific diseases. First, the disease must have recognizable characteristics; second, a cost-effective screening test must be available; third, an effective treatment for the disease must exist; fourth, the amount of disease in a preclinically detectable stage must be high. The ideal screening test is inexpensive, accurate, and easy to administer, with a consistent level of accuracy and without risks to the screened population. Examples of cancer-related screening tests endorsed by the American Cancer Society include the Pap test, breast self examination, mammography, and digital rectal examination.

Medical surveillance systems

Most medical surveillance systems typically consist at a minimum of a questionnaire survey of individuals residing in geographic rings around the hazardous waste incinerator site. The questionnaire will typically include a wide range of acute and chronic health conditions which might be related to chemical exposures to agents likely to be found in close proximity to hazardous waste incinerators. Most surveys initially have a broad scope, but, once specific problems are identified, the questionnaire may be focused on certain body systems, e.g., reproductive health or certain categories of disease, e.g., cancer.

In the assessment of health risks, such questionnaire data may be combined with information obtained from studies of birth or death certificates or data from statewide disease registries, such as tumor or birth defects registries. If available, these data facilitate the process of health risk assessment by enabling site-specific incidence in the affected area to be compared with incidence elsewhere in the state. Despite the benefits, there are some well-described methodological problems with this type of epidemiologic approach to medical surveillance.[24] First, the population being studied is usually relatively small, considering the relative rarity of cancer or birth defects at particular sites over short periods of time. Second, since in most cases, the census tract population being studied is relatively large (compared to the area of the incinerator and the first two or three rings of potentially exposed homes), a large proportion of individuals in the census tract have in fact relatively low exposure, thus diluting rates of disease in those who are exposed. Third, since the latency period for cancer development after chemical exposure is typically assumed to be at least 10 to 15 years, and often more if dose levels are low, it is unlikely that increased incidence would appear at the time that the exposure is of concern to current residents. Even assuming that exposure can be defined, and that a sufficient number of persons would be exposed to allow an increased risk to be detected, the increase in tumor incidence is unlikely to be detectable by tumor registries until a decade or two after exposure begins.

Another approach that is sometimes used to detect possible health effects around contaminated sites, including incinerator sites, is the measurement of certain subclinical abnormalities such as chromosomal aberrations or nerve conduction velocity studies. Interpreting the results of these measurements is hampered by uncertainty about the actual exposure status of those persons tested, limitations in sample size, and the question of latency. Moreover, it is important to evaluate whether and for how long subclinical changes may persist after the alleged exposure and putative biological damage. Finally, the investigator must attempt to determine to what extent these subclinical abnormalities predict the development of clinical disease in the future.

Whether the type of medical monitoring being proposed around a hazardous waste site is based on detection of clinical or subclinical abnormalities, the overriding problem is that of interpreting these data without precise evidence for exposure in specific individuals. This is particularly problematic when the likely chemicals to which residents might have been exposed do not persist in tissue or fat storage. In this case, the only way to determine what past levels of chemicals in the area might have been is to extrapolate from current findings, subject to obvious limitations. Thus, past levels of exposure and cumulative dose usually can only be inferred when no biologically persistent chemical markers are available.

Association or causation?

Contrary to common belief, most of the symptoms, signs, or diseases that can be caused by environmental factors are often nonspecific. So, for example, while headache or angina can be caused or precipitated by carbon monoxide poisoning, there are many other explanations that need to be considered before concluding that a certain disease has resulted from a particular exposure. This leads to the question of "association versus causation." Little evidence is needed to invoke an association between exposure to some chemical and a new disease, but a great deal of effort is required to investigate it. The first step is to consider the range of possible explanations for the disease. An association may then be suggested, and, if it appears to have merit, scientists attempt to investigate it. However, with any apparent risk factor, the relationship needs to be challenged. Are we seeing cause and effect or is a confounding factor exerting its influence? Once the confounders have been examined and it can be concluded that an exposure factor is directly and independently linked to a disease, the next step is to decide whether the association is strong enough to be considered causative. To clarify the distinction between association and causation with respect to environmental or occupational diseases careful observation and accurate enumeration of defined events are necessary. The difficulty lies in passing from observed association to a clear verdict of causation.

In his seminal contribution to the Proceedings of the Royal Society of Medicine in 1965, Sir Austin Bradford Hill[25] described nine separate aspects of association that should be considered before deciding that causation is the most likely interpretation of the facts of the case. However, apart from temporality none of these criteria individually can bring indisputable evidence for or against the cause and effect hypothesis and none can be required as a *sine qua non*. The first of these is the *strength of the association*, e.g., a heavy cigarette smoker is 30 times more likely to develop lung cancer than is a nonsmoker. The second criterion for causation is that of *consistency of the association* by different persons, in different places, circumstances, and times. In other words, the evidence should hang together in a consistent manner, both within the confines of a study and from one study to another. If an association between a substance and a disease is demonstrated time and time again through a number of studies that utilize different populations and different study techniques, the argument for causation is improved.

The third factor to be considered is the *specificity of the association*, i.e., if the association is limited to a particular group of individuals and to particular sites and types of diseases, it increases the strength of the association. It is important to note, however, that diseases may have more than one cause; indeed multicausation is generally more likely than single causation. Moreover, a particular risk factor may be only one of several possible causes of the disease. Problems may then occur when one weak but definite cause-and-effect link operates among other much stronger ones.

Fourth, the *temporal relationship* of the association is clearly important, i.e., the cause must precede the effect. Fifth, a *biological gradient* in the form of a dose-response curve adds to the suggestion of causation. In simplest terms this means that greater exposures (and presumably greater absorbed doses) should produce more pronounced effects than lower exposures. Thus, in the case of residents around a hazardous waste incinerator site, one would expect that if the primary exposure route was the air and groundwater, that those who lived closest to the site would have had the greatest exposure and therefore the greatest burden of disease.

At this point it is worth noting the distinction between dose and exposure from the toxicological point of view. While exposure simply refers to opportunity for contact with a substance in a medium, the actual dose absorbed depends on the substance itself, the medium, the organism, and the magnitude of the exposure. This distinction is important because even if substance A is known to cause disease B in humans, simply being exposed to substance A does not necessarily mean that a sufficient dose was absorbed to have caused the disease in question. Moreover, determining that a particular individual was exposed to a single substance is difficult, even if she or he were exposed on the job, as workplaces are filled with a multitude of chemicals, making it difficult to ascertain which chemical is likely to be responsible for the disease. This often overlooked point is particularly relevant in the case of hazardous waste incineration sites. Not only is it certain that some of the persons who may be exposed may also have used other chemicals in the course of their work, home life, or hobbies, but the chemicals present in hazardous waste sites are almost always found in mixtures. It is rarely possible to identify the precise chemical content of these mixtures. Even if such a complete analysis were possible, mixtures of chemicals can interact in unpredictable ways, including additivity, synergism and antagonism.

The sixth point is that of *biological plausibility*, i.e., that the chemical in question is known to produce the effect in man via a well-defined biological mechanism. In other words, the association should make biological sense and be consistent with available anatomical, physiological, pharmacological, and pathological information. Biological plausibility is closely linked to the next point, that of *coherence*, where the cause-and-effect interpretation of the data should be consistent with the known natural history and biology of the disease. However, because what is known today differs both qualitatively and quantitatively from that which was known yesterday or will be discovered tomorrow, biological plausibility and coherence are not essential to the argument for causation. Nevertheless, if present, they add to the strength of the association and may contribute to a conclusion of causation.

Strong support for the cause and effect hypothesis may come from *experimental evidence* (the eighth point). Does taking preventive action alter the observed association with respect to the frequency of associated events? This emphasizes the importance of correct design in the experiments from which such conclusions will be drawn. If for example, the usual occurrence of disease is one case in 100, and if a study examines only 25 people, the rate would have to be approximately four times normal to be detectable by that study.

In conclusion, the search for causation is always conducted under the shadow of confounding factors. The purpose of each of these Hill criteria is not to exclude the possibility of causation when they are absent, rather it is to help reach a conclusion on the fundamental question of whether there is any other way of explaining the available facts concerning a particular cause-and-effect hypothesis.

Methods for exposure assessment

The public health approach to the effects of hazardous waste incineration focuses on the recognition, control, and prevention of adverse health outcomes caused by human exposure to pollutants. A complex chain of events begins when pollutants are released into the environment and are distributed in one or more environmental media. The chain may extend to actual human exposure, to internal or delivered dose, and finally to environmentally induced disease or injury. Documentation of these links is important because, although contaminants may be present in a variety of environmental media surrounding hazardous waste sites, the chemicals must enter the body in order to cause disease. The following is an analysis of methods for exposure measurement that may be useful in establishing a health surveillance strategy.

Direct and indirect measurement

Environmental exposure can be measured directly in a variety of body fluids and tissues (i.e., blood, urine, adipose tissue, or bronchial washings), or indirectly in environmental media such as the air, water, soil, or even plants and animals with which humans may come in contact. Exposure measurements should take into account the four basic characteristics of an exposure: route (inhalation, ingestion, or dermal exposure), magnitude (pollutant concentration), duration (minutes, years, lifetime), and frequency (daily, weekly, seasonally). A variety of techniques are available to do this. For example, microenvironmental samplers and personal monitors worn by individuals, can measure concentrations of chemicals such as airborne lead in the home or workplace. This information can be combined with biological measurements (e.g., blood lead levels) in human tissues.

When increased levels of chemicals of concern are detected in hazardous waste incinerator stack emissions, the surrounding air, water, and soil should be monitored. Ideally, monitoring stations would be established in the vicinity to continuously monitor ambient air, soil, and vegetation and establish background conditions *before* the incinerator begins operation.[26]

When elevated levels of chemicals are found in the ambient air, water, or soil, measurement of the deposition and uptake of these chemicals in plants can be used as one indicator of potential exposure.[27] For example, experimental plots of kale, lettuce, carrots, and potatoes were used to assess plant uptake of chemicals surrounding wood plants.[4]

Sentinel animal species may also be used in the early detection of an environmental hazard.[28] Because most animals have an accelerated life span compared to humans, they may exhibit adverse health effects due to hazardous chemical exposures more quickly than humans.

In the event that monitoring techniques detect increased levels of chemicals in stack emissions, ambient air, soil, or plants, the next step may be to perform biological measurements in humans. Most investigations using biologic measurements as indicators of exposure to incinerator emissions have studied incinerator worker populations. Blood lead levels were found to be elevated in a study of municipal solid waste incinerator workers in New York,[5] but when workers wore their personal protective devices, abstained from smoking, and rotated responsibility for cleaning the precipitators, lead levels fell. Another study found slightly elevated levels of organic substances, such as benzene and polychlorinated biphenyls (PCBs), in the blood and urine of incinerator workers.[29]

Measurement of biologic markers

In the context of environmental health, biologic (or subclinical) markers may be indicators of events in biologic systems or samples. Once an exposure has occurred, a continuum of biologic events can be detected. These events may serve as markers of the initial exposure, administered dose, biologically effective dose, altered structure or function with no ensuing pathological effect, or potential or actual health impairment.[30] Biologic markers can represent a marker of effect (e.g., blood glucose), a marker of susceptibility (e.g., presence of atopy), or a marker of exposure (e.g., the presence of a metabolite of the product of interaction between a hazardous agent and a target molecule).

Some indicators of exposure could have the potential to be categorized as predictive outcomes (e.g., altered enzyme levels or chromosomal mutations).

A study in Poland of biologic markers of damage due to environmental pollutants from combustion, such as polycyclic aromatic hydrocarbons,[31] found that aromatic adducts on DNA were significantly correlated with chromosomal changes. Cytogenic monitoring of populations potentially exposed to low levels of environmental pollutants from chemical waste sites examined the frequency of sister-chromatid exchanges in peripheral blood lymphocytes. In one such study, there were significant differences in exposed and nonexposed adults and children and, as expected, between smokers and nonsmokers.[32] As described earlier, interpretion of these types of studies is directly and often fatally limited by uncertainty about the actual exposure status of those being tested and often small sample sizes. Another uncertainty is the extent to which these subclinical abnormalities actually predict the subsequent development of clinical disease or are actually indicative of genetic changes.

Use of mathematical models

Since it may not be feasible to obtain exposure measurements on all members of a community, an alternative approach is to extrapolate from analyses of more focused measurements to fit the larger population of interest using mathematical models.[33] These include the concentration or fate and transport models that estimate the concentration of a pollutant in a particular environmental medium; the contact or exposure models that estimate the contact between pollutant and persons (exposure); and dose models that estimate the amount of pollutant that enters the body (the internal dose).

One of the problems with this approach is that errors in the assumptions may be greatly magnified in the final exposure estimates. For instance, in his report for Greenpeace, Costner[34] assumed that the average hazardous waste incinerator burns 70 million pounds of waste per year and that 100% of the waste is hazardous. Therefore, when 99.99% destruction and removal efficiency (DRE) is calculated, Costner reports that 7,000 pounds of unburned hydrocarbon wastes will be released in the stack emissions per year. However, as pointed out by Santoleri et al.,[35] toxic organics entering incinerators average 25 to 30% of the total waste stream, the remainder being inorganic (e.g., soil or clay) and water. Using these assumptions, the unburned hydrocarbon wastes would be reduced from 7,000 to 2,100 pounds per year. Variations in other assumptions may have similar or greater effects.

Other investigators feel that the hazards posed by chemicals are often exaggerated when "conservative" estimates are used in the exposure calculations.[36] Exposure estimates commonly use a maximally exposed individual (MEI): a person living for 70 years at the fence line of the incinerator facility breathing the outdoor air, which represents a highly unlikely occurrence in real-life situations. When these conservative estimates are reported, many in the public interpret this as the average exposure levels expected, not the estimate of a worst-case scenario. The impact of reducing such exposure assumptions to realistic values may be dramatic, and may mean the difference between predicting an effect and assuming no such effect.

Use of environmental exposure databases

Many available databases are appropriate for environmental exposure estimation.[33, 37] However, database collection systems currently exist for regulatory rather than for public health purposes.[38] There is a tremendous need to develop exposure databases that can be used for the identification of potential risks and to provide surveillance for the protection of public health. Adapting existing and future databases to be more sensitive to public health needs will require documentation of special populations at risk, provisions for early warning of new problems, methods for monitoring changes over time, and approaches for enhancement of documentation.[39-41]

Population surveillance of health outcomes

To be effective, population surveillance strategies to assess health outcomes associated with hazardous waste incineration must be based on sound epidemiologic principles. A study population must be defined and characterized before subsequent steps in the assessment process can be undertaken. Baseline characteristics such as demographics, physical and mental health status and prevailing community attitudes can be collected by questionnaire or phone survey. Changes in these characteristics after a facility is operational cannot be reliably detected without accurate and relevant baseline data. Measured outcomes should include indicators of exposure, such as decreased pulmonary function, as well as cancer and death.

To establish a link between health outcomes and exposure to hazardous waste incinerator emissions, factors such as the location of the source as well as climatic factors must be considered. The types of hazardous waste incineration emissions may differ depending on the way they are generated (e.g., heat of combustion), by the type of pollutant group present (e.g., metals or dioxins), by location, or by the rate and pattern of emissions. A number of human factors may influence an individual's response to emissions. Genetic factors may increase the risk for developing a response, although there is very little knowledge concerning this issue.[42] Personal habits or lifestyles, including tobacco smoking, may influence the risk for certain diseases.[43] The following are examples of outcome measurements that could be used in a public health surveillance program surrounding a hazardous waste incineration facility.

Sentinel event strategies

Sentinel event strategies follow the philosophy that the occurrence of some events, such as rare cancers or congenital anomalies, is suggestive of exposure to environmental hazards.[44] A system must be in place to monitor sentinel events around hazardous waste incineration facilities so that patterns of unusual health events will be detected. Identifying clusters of disease events may also uncover a hazardous substance point source that was not previously recognized.

The occurence of rare health events in a small aggregate or in unusual populations might represent a sentinel event, as might changes in the expected pattern of disease on a population scale. For example, an increase in the distribution of bladder cancer in younger individuals may reflect chemical contamination of a water supply. Another example is the finding of a small cluster of rare cancers, which has marginal statistical significance, but a concordant increase in birth defects can provide additional evidence for a possible common environmental source.

Geographic information system mapping

Geographic information systems (GIS) are tools which can integrate diverse multimedia data, geographic features, census information, information about ecological regions, and chemical toxicity information into a common database with spatial characteristics.[37] This approach may provide the initial step in defining the geographic distribution of environmental releases of chemicals and potential exposure zones.

Investigations of spatial patterns and geographical comparisons of rates of diseases across county regions in states can be accomplished using these mapping procedures.[45, 48] Analyses of clusters of diseases have utilized algorithms coupled with population density equalized maps (e.g., cartograms).[46] Williams[45] used three-dimensional mapping techniques to investigate the sex ratio of births in an area in Scotland containing two incineration plants. This study showed a statistically significant excess of female births in residential areas at risk from airborne pollution from the incinerators. Spatial clustering of diseases around geographic formations such as rivers and hazardous waste sites has been evaluated using proximity analysis.[47]

GIS and other mapping systems have tremendous value for generating future hypotheses. For example, if there is a match between a carcinogen emission region and a county with a high cancer rate, then more definitive hypothesis-testing methodology (such as a case control investigation) is warranted.[49] When monitoring excess cancer rates in the vicinity of a hazardous waste incinerator facility, it must be stressed that there may be other factors in the area that may also contribute to this outcome.

Reporting databases and vital statistics

A number of local, state, and national databases are now available both in digital and hard copy format and may be helpful in answering specific community concerns about adverse health effects. Rarely, if ever, can this information directly identify causal relationships between exposures and outcomes, but the information they contain may assist in addressing community concerns about increased occurrence of disease.[50]

At the national level, data sources [e.g., the National Center for Health Statistics, the National Cancer Institute (NCI), the National Institute for Occupational Safety and Health (NIOSH)] may not be site specific, but they may be used for comparison purposes. The U.S. Bureau of Census collects geographic-specific demographic data, while state health departments maintain census data for their areas.

Some states have registries for diseases, cancers, and congenital malformations. State health departments collect vital statistics, such as birth and death records. Local sources of health outcome data include health records from local health departments (including complaint logs and health studies), hospital discharge records, hospital emergency room logs, private physicians' records, school records, and records from occupational or free-standing health clinics.[50]

The information obtained from analysis of reporting databases and vital statistics may help in the identification of disease patterns in geographic areas, although limitations such as completeness and accuracy must be addressed when utilizing secondary health outcome data in a public health surveillance strategy.

Risk communication

An important component of the public health approach to community exposure to hazardous waste incineration emissions is communicating accurate and relevant information about the potential risks and benefits of proposed facilities, so individuals can make informed decisions.[51, 52] In this context, the perception of risk is important because it adds to the stress, fear, uncertainty, and misconceptions experienced by those living in the vicinity of hazardous waste incineration facilities.

The siting of incinerators, particularly those that burn hazardous waste, often generates unbalanced fears. Individuals may worry about incinerator residues in their food, but continue to smoke or eat an unhealthy diet. Others may worry about the siting of an incinerator in their community while neglecting to test their homes for radon. Risk communication and health education offer the opportunity to put these risks into perspective. For example, it is likely that a product such as air bags in motor vehicles, the manufacture of which generates hazardous waste, will save more lives than exposure to combustion products of the waste will potentially endanger.

Ostry[53] conducted an environmental risk perception survey in a community with a municipal solid waste incinerator. The aims of the survey were to study attitudes about waste management techniques including incineration and to identify sociologic attributes that helped shape attitudes about the incinerator. A third of the community was unaware that there was an incinerator facility in the area, and those people were more opposed to incinerator technology. The investigators suggest that informing the public about technology perceived as hazardous may not lead to alarm, but may increase acceptance. The usefulness of scientific and technical material ultimately relies upon how effectively that information is communicated.[54]

These complex issues about risk must be a part of the communication between all stakeholders in the proposed facility: the public, industry, environmentalists, regulators, and public health officials. Typical questions that need to be explored at a community level include: Will a centralized facility disproportionately burden one community with possible environmental contamination? Are "acceptable" cancer and respiratory outcomes acceptable to the community? Will a hazardous waste incineration facility provide jobs to a community that will increase the standard of living and access to medical care and subsequently protect the public's health and decrease other risks in their lives?

Community involvement

Community participation must be an integral part of communication efforts and decision-making processes about risks and benefits associated with hazardous waste incineration. Indeed, opposition to any project is encountered when a community feels that they have no control over the proceedings. Thus, according to federal and state regulations, community participation is required in the permitting process of hazardous waste incinerator facilities. The public can also participate in the enforcement process.[55]

The community includes (1) community groups (i.e., religious leaders, civic leaders, educators, social organizations, cultural groups, recreational organizations, and political party representatives), (2) immediate neighbors of the facility site, (3) local activist groups, (4) local and state government representatives, (5) the medical community, and (6) the media. Accurate and timely information must be provided to them on a continuous basis in order to prevent inaccurate messages from other sources.

Minorities and other sensitive population groups, who are likely to experience increased susceptibility to potential adverse effects, must be included in the discussions. Studies have shown that most of the communities surrounding incinerator facilities were not predominantly minority or poor when the facilities were sited, but that demographics shifted subsequently.[56] An executive order was signed in February 1994 stating that federal agencies have one year to institute policies that ensure their environmental actions do not disproportionately burden America's poor and minority communities (environmental equity).

Public relations is a planned effort designed to influence public opinion using responsible techniques and good intentions through mutually satisfactory two-way communication. This means communicating accurate and relevant information, allowing the community to express its concerns, and responding appropriately. Information must be highly accessible, accurate and understandable. Information about potential risks to a community must be honest, straightforward, and conversational. Absolute statements that exposure to hazardous waste incinerator emissions will or will not result in adverse health outcomes should be avoided. Messages pertaining to potential risks from exposure to hazardous waste incineration must be consistent and credible.

The Ross Environmental Services hazardous waste incinerator in Grafton, Ohio, which has been operating for more than 40 years, provides a good example of a comprehensive community involvement program.[57] Ross credits its success with programs such as (1) an open-door policy 24 hours per day, (2) scheduled open houses, (3) public meetings, (4) newsletters, and (5) telephone surveys every 2 years to gauge public sentiment to guide future actions appropriately. As a result of this comprehensive community monitoring system, Ross reports a high degree of community involvement and acceptance.

Summary and conclusions

The major goal of medical monitoring is to accurately collect and fully analyze clinical, epidemiologic, and other biomedical data for any evidence that the public and the environment are being, or could be, harmed by an existing or potential exposure scenario, such as from a hazardous waste incineration facility. This is based on the assessment of exposure and outcomes. Various exposure assessment methodologies have been utilized in the vicinity of incinerator facilities, including continuous stack monitoring for representative chemicals of concern, monitoring of environmental media (air, water, soil, plants, and animals), and biologic measurements in humans to assess exposure levels. The measurement of biologic markers may be a valuable tool for exposure assessment purposes. Mathematical models have been developed to estimate exposure levels in populations and many environmental exposure databases are available that could be expanded and modified to better accommodate public health surveillance purposes.

Surveillance strategies must utilize sound epidemiologic principles to characterize populations and monitoring techniques that allow for baseline measurements as well as long term follow-up for assessment of any changes in physical or mental health status in response to hazardous waste incinerator exposure. Applicable methods include sentinel event strategies, use of mapping techniques, monitoring vital statistics and reporting databases for outcome patterns of concern, and assessment of the perceptions of risk reported by community members.

Potential risks must be communicated to the public in such a way that will allow the public to participate effectively in the decision-making processes involved with the siting, permitting, and operation of a hazardous waste incineration facility. Only a comprehensive cooperative effort by government agencies, industrial representatives, unbiased academic institutions, public health officials and community representatives will ensure that the utilization of hazardous waste incineration facilities will not compromise the environment or the health of the public.

References

1. U.S. EPA/530-SW-88-018. Hazardous waste incineration: questions and answers. April 1988.
2. Hazardous Waste Treatment Council. Hazardous waste incineration: advanced technology to protect the environment. HWTC, Washington, DC; 1993.
3. Harrad SJ, Jones KC. A source inventory and budget for chlorinated dioxin and furans in the United Kingdom environment. *Science Total Environ* 1992;126:89–107.
4. Larsen EH, Moseholm L, Nielsen MM. Atmospheric deposition of trace elements around point sources and human health risk assessment. II. Uptake of arsenic and chromium by vegetables grown near a wood preservation factory. *Sci Total Environ* 1992;126(3):263–275.
5. Malkin R, Brandt-Rauf P, Graziano J, Paides M. Blood lead levels in incinerator workers. *Environ Res* 1992;59:265–270.
6. Travis CC, et al. A perspective on dioxin emissions from municipal waste incinerators. *Risk Anal* 1991;9(1):91–97.
7. Lisk DJ. Environmental implications of incineration of municipal solid waste and ash disposal. *Sci Total Environ* 1988;74:39–66.
8. Decker DW, Clark CS, Gia VJ, Kominski JR, Trapp JH. Worker exposure to organic vapors at a liquid chemical waste incinerator. *Am Ind Hyg Ass J* 1983;44:296.
9. Bloedner CD, Reiman EO, Schaller KH, Wettle D. Evaluation of internal cadmium, lead, and mercury exposure in workers of a modern waste combustion plant. *Zentraibl Arbitsmed* 1986;36:322.
10. NIOSH. Health Hazard Evaluation Report: The Caldwell Group, North Carolina, HETA 90-240-259, Cincinnati, Ohio, 1992.
11. Hallenbeck WH, Breen SP, Brenniman GR. Cancer risk assessment for the inhalation of metals from municipal solid waste incinerators impacting Chicago. *Bull Environ Contam and Toxicol* 1993;51:165–170.
12. Gustavsson P. Mortality among workers at a municipal waste incinerator. *Am J Ind Med* 1989;15:245–253.
13. Gustavsson P, Evanoff B, Hogstedt C. Increased risk of esophageal cancer among workers exposed to combustion products. *Arch Environ Health* 1993;48(4):243–245.
14. Watts RR, Lemieux PM, Grote RA, et al. Development of source testing, analytical, and mutagenicity bioassay procedures for evaluating emissions from municipal and hospital waste combustors. *Environ Health Perspec* 1992;98:227–234.
15. Ma XF, Babish JG, Scarlett JM, et al. Mutagens in urine sampled repetitively from municipal refuse incinerator workers and water treatment workers. *J Toxicol and Environ Health* 1992;37:483–494.
16. Bresnitz EA et al. Morbidity among municipal waste incinerator workers. *Am J Ind Med* 1992;22:363–378.

17. ATSDR. Study of symptom and disease prevalence at the Caldwell Systems Inc. Hazardous Waste Incinerator, Caldwell County, North Carolina, US Department of Health and Human Services, Atlanta, GA, February 1993.

18. Feigley CE, Hornung CA, Manera CA, Draheim CA, Weim, Oldenick R. Community study of health effects of hazardous waste incineration. In: *Hazardous Waste and Public Health*. Int. Congress on the Health Effects of Hazardous Waste. Andrews JS, Frumkin H et al. (Eds.) Princeton Scientific Publishing Co, Princeton. 1954, p. 765.

19. Rothenbacher D, Agenan D, Rhodes V, Shy C. Respiratory symptom prevalence in three North Carolina incinerator communities. In: *Hazardous Waste and Public Health*. Int. Congress on the Health Effects of Hazardous Waste. Andrews JS, Frumkin H et al. (Eds.) Princeton Scientific Publishing Co, Princeton. 1954, p. 757.

20. Zmirou D, Parent B, Porelon JL. Epidemiologic study of the health effects of atmospheric waste from an industrial and household refuse incineration plant. *Rev Epidemiol Sante Publique* 1984;32:391.

21. Taylor S, Elliot S, Eyles J, Frank J, Haight M, Sheiner D, Walter S, White N, Williams D. Psychosocial impacts in populations exposed to solid waste facilities. *Soc Sci Med* 1991;33:441.

22. Eyles J, Taylor SM, Johnson N, Baxter J. Worrying about waste: living close to solid waste disposal facilities in southern Ontario. *Soc Sci Med* 1993;37(6): 805–812.

23. Elliott SJ, Taylor SM, Walter S, et al. Modeling psychosocial effects of exposure to solid waste facilities. *Soc Sci Med* 1993;37(6):791–804.

24. Heath C W. Field epidemiologic studies of populations exposed to waste dumps. *Environ Health Persp* 1983;48:3–7.

25. Bradford Hill A. The environment and disease: association or causation? *Proc R Soc Med* 1965, 295–300.

26. GVRD. Burnaby incinerator: summary of soil and vegetation monitoring data; summary of ambient air quality monitoring data; summary of stack monitoring data. Greater Vancouver Regional District. September, 1992.

27. Smith WM, Weyman FO, Alberts MT. Pros and cons of including indirect exposure assessment in the human health risk assessments for permitting of boilers and industrial furnaces. Air Waste Association Annual Meeting, 1994;94-MP4.05.

28. Mason T, Hays HM. Disease among animals as sentinels of environmental exposures. In:(? - ed.)

29. Angerer J, Heinzow B, Reimann DO, et al. Internal exposure to organic substances in a municipal waste incinerator. *Int Arch Occup Environ Health* 1992;64(4):265–273.

30. Subcommittee of Pulmonary Toxicology Markers. Biologic markers in pulmonary toxicology. National Research Council, 1989.

31. Perera FP, Hemminki K, Gryzbowska E, et al. Molecular and genetic damage in humans from environmental pollution in Poland. *Nature* 1992; 360(6401): 256–258.

32. Lakhanisky T, Bazzoni D, Jadot P, et al. Cytogenetic monitoring of a village population exposed to a low level of environmental pollutants. Phase 1: SCE analysis. *Mutat Res* 1993;319(4):317–323.

33. Sexton K, Selevan SG, Wagener DK, Lybarger JA. Estimating human exposures to environmental pollutants: availability and utility of existing databases. Archives of Environmental Health. 1992;47960:398–407.
34. Corn M, ed. *Handbook of Hazardous Materials.* San Diego: Academic Press, 1993; Costner P, Thornton J. Playing with fire: hazardous waste incineration. Greenpeace, 1990.
35. Santoleri JJ, Lauber JD, Theodore L. Facts or myths: the burning issue of incineration. Air and Waste Management Association Annual Meeting, 1993;P-238-01.
36. Abelson PH. Health risk assessment. *Regul Toxicol Pharmacol* 1993; 17:219–223.
37. Stockwell J, et al. The U.S. EPA geographic information system for mapping environmental release of Toxic Chemical Release Inventory (TRI) chemicals. *Risk Anal* 1993;13:155–164.
38. Matanoski G, Selevan SG, Akiand G, et al. Role of exposure databases in epidemiology. *Arch Environ Health* 1992;47(6):439–446.
39. Goldman LR, Gomez M, Greenfield S, et al. Use of exposure databases for status and trends analysis. *Arch Environ Health* 1992;47(6):430–438.
40. Burke T, Anderson H, Beach N, et al. Role of exposure databases in risk management. *Arch Environ Health* 1993;47(6):421–429.
41. Graham J, Walker KD, Berry M, et al. Role of exposure databases in risk assessment. *Arch Environ Health* 1993;47(6):408–420.
42. Brooks SM. Bronchial asthma of occupational origin. In: Rom WN, ed. *Environmental and Occupational Medicine.* Boston: Little, Brown; 1992.
43. White J, et al. Respiratory illness in nonsmokers chronically exposed to tobacco smoke in the work place. *Chest* 1991;100:39–43.
44. Aldrich TE, Leaverton P. Sentinel event strategies in environmental health. *Ann Rev Public Health* 1993;14:205–217.
45. Williams FLR, Lawson AB, Lloyd OL. Low sex ratios of birth in areas at risk from air pollution from incinerators, as shown by geographical analysis and three-dimensional mapping. *Int J Epidemiol* 1992;21(2):311–319.
46. Selvin S, Merrill D, Schulman J, et al. Transformation of maps to investigate clusters of disease. *Soc Sci Med* 1988;26:215–221.
47. Knox E, Lancashire R. Epidemiology of congenital malformations. London:HMSO, 1990:126–128.
48. Mahoney M, Labrie D, Nascam P, et al. Population density and cancer mortality differentials in New York State, 1978–1982. *Int J Epidemiol* 1990:19: 483–490.
49. Felber E. Childhood cancers and congenital malformations in the state of Florida: an ecologic study [MSPH]. College of Public Health, University of South Florida, 1993.
50. ATSDR. Public Health Assessment Guidance Manual. Agency for Toxic Substances and Disease Registry, Atlanta, Georgia. U.S. Department of Health and Human Services, 1992.
51. Schofield WR, Herron JB. EPA's new policy on hazardous waste combustion: what does it mean to you? Air and Waste Management Association, 1994;94-WP1O1.02.
52. Brown P. Popular epidemiology and toxic waste contamination: lay and professional ways of knowing. *J Health Soc Behav Sci* 1992;33(3):267–281.

53. Ostry AS, Hertzman C, Teschke K. Risk perception differences in a community with a municipal solid waste incinerator. *Can J Public Health* 1993;84(5): 321–324.

54. Gilbert J. Translating technical information for the public and the media: a study of barriers and facilitators to effective communication. Air and Waste Management Association AnnualMeeting, 1994;94-TA34.01.

55. ASME. Hazardous waste incineration. A resource document. The American Society of Mechanical Engineers. New York, 1988.

56. Braile R. Is racism a factor in siting undesirable facilities? *Garbage* 1994;6(2):13–18.

57. Kelch MA. Personal letter and company literature, July 1994. Ross Environmental Services; Grafton, Ohio.

chapter thirteen

Quantitative uncertainty analysis

Stephen M. Roberts

Contents

Introduction

Risk assessments involving exposure to contaminants in the environment are inherently complex, typically incorporating numerous models to evaluate components ranging from contaminant concentrations at points of contact to dose-toxicity relationships. Risk assessments for hazardous waste incinerators are particularly complex because of the extensive environmental fate and transport modeling required, as well as the diversity of receptors and exposure pathways that must be considered. Risk assessments are not perfect, a fact that is readily acknowledged by those who conduct risk assessments

1-56670-250-X/99/$0.00+$.50
© 1999 by CRC Press LLC

and those who use them to make regulatory decisions. There is some degree of uncertainty associated with each of the models used in the risk assessment process, and in the values used for variables in these models. Accordingly, there is uncertainty associated with the estimates of risk that are derived. The existence of uncertainty at individual steps in the risk assessment process, and in the final outcome, does not make risk assessment a useless endeavor. There are, after all, no undertakings that do not entail some level of uncertainty. It is, however, vital that risk estimates be accompanied by some expression of the extent or degree of uncertainty regarding their accuracy if they are to be of value for risk management purposes. This is important, not only in indicating the confidence that can be placed in specific risk estimates, but also to determine whether a particular risk assessment, on the whole, offers sufficient confidence to permit a risk management decision to be made.

This section will describe concepts of uncertainty and identify sources of uncertainty in the risk assessment process. An overview of methods for estimating and expressing uncertainty in risk estimates is provided, along with a brief summary of current regulatory guidance on this issue, particularly as it relates to risk assessment of incinerators. Finally, practical issues and limitations in the evaluation of uncertainty are discussed.

Uncertainty vs. variability in risk assessments

It is important, when discussing uncertainty in risk assessments, to distinguish between uncertainty and variability. A good articulation of the distinction is provided by Bogen,[1] who refers to *uncertainty* as a lack of fundamental knowledge regarding a given risk-related characteristic. A *risk-related characteristic* is some quantity or relationship whose value or structure is at issue in a component of the risk assessment. According to Bogen, uncertainty can be either scientific or statistical in origin. *Variability*, in contrast, is used to refer to inter-individual differences or heterogeneity in regard to some risk-related characteristic that is distributed within a population.

The difference between uncertainty and variability can be illustrated in the following examples:

1. There is currently a variety of models with which to predict cancer risk at low-dose exposures based on data derived at high doses. Different models for this dose-response relationship can yield very different estimates of cancer risk for a given dose,[2] and it is unclear which of these models (if any) is correct. This ambiguity as to the best model for cancer risk extrapolation adds significant *uncertainty* to the risk assessment of carcinogens.
2. Calculated risk is dependent in part on the body weight of the exposed individual. Differences in risk as a function of body weight reflect *variability* among individuals, rather than scientific uncertainty.

All components of risk assessment possess some inherent uncertainty and variability to different degrees. For example, the choice of a chemical toxicity value such as a reference dose for use in a risk assessment may introduce uncertainty if it is based on animal dose-response relationships whose relevance to humans is unclear. In addition, the toxicity value is also variable in the sense that the susceptibility to toxicity from the chemical will vary somewhat among exposed individuals.

Both uncertainty and variability contribute to the range of possible risks that could arise from exposure to a contaminant in the environment. The distinction between uncertainty and variability is important, however, in part because the options available to address them and refine the estimates of risk are different. Uncertainty of statistical origin can be diminished through application of the appropriate statistical tests and perhaps through increasing sample size. Scientific uncertainty can often be reduced only through further research. Even if we had perfect knowledge and uncertainty was reduced to zero, projected risks would still fall within a range of values. For a number of parameters utilized in risk assessment, variability is an intrinsic property related to the exposed population and cannot be eliminated. Variability may be geographic, temporal, physiological, or behavioral in nature.[1] Improvement of the risk assessment in this case comes not from trying to reduce the variability (which may be impossible), but rather by developing increasingly accurate descriptions of the variability within the subject population.

Qualitative vs. quantitative uncertainty

The degree of uncertainty associated with a risk assessment can be expressed in qualitative or quantitative terms. The uncertainty analysis that accompanies most risk assessments performed for the EPA is qualitative, or at best, semiquantitative in nature. For example, guidance in conducting risk assessments for Superfund sites[3] suggests a listing and discussion of uncertainty that arose from:

1. *The definition of the physical setting of the site*, including assumptions regarding future land use and exposure pathways that exist now or in the future, and the selection of specific contaminants at the site to be included in the risk assessment;
2. *Model applicability and assumptions*, i.e., the validity of the models that are used to predict environmental fate and transport of contaminants and exposures;
3. *Selection of parameter values*, i.e., the accuracy of the values that are used as inputs in the fate, transport, and exposure models; and
4. *Tracking of uncertainty*, or the extent to which uncertainties at individual steps get carried through the risk assessment, and perhaps magnified.

Quantitative approaches are thought to be required only rarely, and qualitative treatment may be limited to listing key assumptions and characterizing the effect of the model or parameter assumption chosen as having the likelihood of potentially over- or underestimating the actual risk. The potential effect of the assumption on the final risk estimate may be described as "low," "moderate," or "high," depending on whether the estimate might create an error of one, two, or three or more orders of magnitude, respectively. Guidance from U.S. EPA[3] also suggests that uncertainties related to toxicity assessment (e.g., quality of the database upon which toxicity values were derived, the potential for synergistic or antagonistic reactions among multiple contaminants, possible health effects from contaminants omitted from the risk assessment, etc.) should also be discussed, although this is rarely done in more than a cursory fashion.

Qualitative expressions of uncertainty regarding a risk assessment are useful in identifying the various sources of uncertainty, and in conveying a broad sense of the relative precision of the risk estimates provided. In risk assessments conducted for regulatory purposes, uncertainty at individual steps is generally dealt with by adopting conservative assumptions, i.e., assumptions that tend to ensure that the actual risk is not underestimated. By identifying and summarizing these assumptions, and projecting roughly their effect on the risk estimate(s), a sense of the overall conservatism of the risk assessment can be gained. While a qualitative expression of uncertainty is important, an uncertainty analysis that is exclusively qualitative suffers from the absence of an explicit indication of the range of possible risk estimates consistent with the available information and the confidence that can be attached to specific risk estimates within that range. Some semiquantitative aspects may be included in these uncertainty analyses, such as order of magnitude estimates of influence on final risk values, but these estimates are not developed rigorously and their accuracy is often questionable. A further disadvantage is that qualitative uncertainty analyses do not lend themselves to identifying the most important source of uncertainty — those that contribute most to potential error in risk estimates.

Increasingly, the incorporation of a quantitative expression of uncertainty into risk assessments is being encouraged. The decision whether or not to include a quantitative expression of uncertainty is important, because it affects in a fundamental way the manner in which risk estimates are presented in the risk assessment. Most risk assessments currently performed use a *deterministic* approach in calculating risk; that is, for each of the input variables needed to calculate a risk estimate, a single value is selected from a range of possibilities. For example, dose estimates are typically derived using a single assumed body weight for exposed individuals (usually 70 kg for adults, or about 154 lbs.), despite the fact that exposed individuals come in a variety of body weights. Similarly, each of the individuals who fall into one of the exposure scenarios being evaluated in the risk assessment will be assumed to have the same exposure frequency and duration, the same rate of accidental ingestion of soil, the same area of skin available for dermal contact, etc. From this series of individual input values, a single risk estimate is derived for each exposure scenario under consideration. As discussed below, there can be uncertainty, variability, or both in the input values selected for risk calculation, and therefore variability and uncertainty in the final estimate of risk.

For regulatory purposes, to protect public health, it is important that the risk estimate upon which a risk management decision will be based not underestimate the actual risk. Accordingly, values to be used as input parameters are selected with the intent of producing a plausible, upper-bound, deterministic estimate of the risk associated with the site or activity being evaluated. This is not accomplished using worst-case assumptions for each and every model and parameter in the risk assessment; to do so would result in a theoretical maximum estimate of risk which would be completely unrealistic. Instead, the U.S. EPA has recommended that a combination of assumptions with varying degrees of conservatism be combined with the objective of deriving a point estimate of the highest risk that can be reasonably expected to occur from the exposure being evaluated. For Superfund sites, this has been termed "reasonable maximum exposure."[3] The same conceptual approach is embodied in the U.S. EPA guidance for risk assessment of combustor emissions,[4] although this specific terminology is not used.

One advantage of a deterministic approach is that the risk calculations are relatively simple. The bias with which models and input variables are selected ensures that the outcome is conservative, which makes it useful for regulatory purposes. Because it is conservative in nature, the meaning is clear if the calculated risk levels are low — less than those regarded as unacceptable. Interpretation becomes problematic, however, when point estimates of risk exceed what are regarded as *de minimis* (so low as to be insignificant) levels. The issue then becomes, do the *actual* risks exceed acceptable levels? The deterministic approach is not well equipped to address this issue. The presentation of a single point estimate of risk does little to convey the degree of uncertainty associated with its derivation, limiting its value as a basis for risk management decisions. Unfortunately, the presentation of a single risk estimate, e.g., 5.5×10^{-6}, also creates problems of perception in that it implies a much greater degree of precision in the risk estimation process than actually exists.

The difficulty lies in not knowing the degree of conservatism that has been imparted to the risk assessment through the selection of a combination of moderate, conservative, and worst-case assumptions. There is not explicit guidance as to how assumptions with varying degrees of conservatism should be combined to achieve the objective of a "reasonable maximum" or "high-end" exposure, or how the success or failure of the attempt can be assessed. By setting the bias high enough to swamp the uncertainty for many of the variables in the risk calculation, the outcome may be representative of exposure circumstances that rarely, if ever, happen.[5]

An alternative is to present risk estimates in probabilistic terms. While in a sense, cancer risks are presented in a probabilistic fashion (e.g., 1×10^{-6} risk, or one-in-a-million chance), they nonetheless represent point estimates of risk. What is meant here by probabilistic terms is the presentation of risk estimates as a probability distribution. This distribution would define the range of possible risk values, given the current available information, and the likelihood that individual risk estimates within this range represent the "true" risk. Probabilities can be expressed in terms of risk to one or more types of exposed individuals (e.g., individuals who, because of their place of residence or occupational activity, would be expected to have the greatest contaminant exposure), or the distribution of risk among a selected population at large, depending upon the objectives of the risk assessment.

Sources of uncertainty and variability in risk assessments

Risk assessments can involve extensive modeling of both physical and biological events and require a good understanding of how those models behave under conditions relevant to the facility or activity being assessed. The conceptual basis of the models, their functional validity, and the accuracy of inputs for these models are all potential sources of uncertainty in the risk assessment process. In this subsection, the nature and variety of sources of uncertainty are outlined.

There are several potential sources of uncertainty associated with risk assessments. These can be divided into three basic types:[6] (1) parameter uncertainty, (2) model uncertainty, and (3) decision-rule uncertainty. These are discussed in the following paragraphs.

Parameter uncertainty

There are three types of errors that can contribute to uncertainty regarding the values used for specific parameters within the risk assessment. One is *measurement error*. This arises because of practical or technical limitations in the ability to measure precisely many of the parameters needed to derive risk estimates. An example where measurement error might contribute to uncertainty in the risk assessment is in the determination of contaminant concentrations in environmental media (air, water, soil, etc.) to which individuals are exposed.

Random error can also contribute uncertainty to the accuracy of values selected for specific parameters in risk assessment. There is always the chance that samples taken, or observations made as part of an experiment, are not truly representative of what is being examined. The possibility of random error, also termed *sampling error*, is particularly great when the number of samples or observations is small. Using dose-toxicity information derived from only a limited number of clinical observations or experiments using few animals or dosages, for example, would contribute uncertainty due to the possibility of random error.

A third source of uncertainty in the generation of parameter values is *systematic errors*. Systematic errors arise from an inherent flaw in the data-gathering process and could arise, for example, from unintentional bias in sampling. An example of systematic error cited by Finkel[6] is the use of populations of healthy workers as being representative of the population at large. Due to a selection process associated with remaining employed in an active workplace, worker populations are generally more healthy than the general population and do not contain subgroups of individuals that may be particularly susceptible to toxicity or disease (e.g., children, the elderly, and the infirm). Uncertainties associated with extrapolating observations in workers to the general population, the so-called *healthy worker effect*, are well recognized by epidemiologists.

Model uncertainty

Risk assessments make extensive use of models in aspects ranging from environmental fate and transport to exposure assessment to the development of dose-toxicity relationships. Risk assessment models, and models in general, are devices used to express complex processes and relationships in manageable terms. Typically, models are created as simply as possible; added complexity that does not enhance the reliability of the model is considered undesirable. Occasionally, models are too simple and fail to capture all of the facets of the process or phenomenon that is being modeled. Models can also be conceptually incorrect, based on erroneous notions of the relationship between the component variables. Under these circumstances, an erroneous output will result no matter how precisely the input parameters have been determined.

There are a variety of ways in which uncertainty associated with models can be introduced. One is through the use of *surrogate variables*. Often, models contain one or more parameters that cannot be directly addressed. A common solution is to find by analogy a surrogate variable that can stand in its place. For example, to calculate the human cancer risk resulting from a specific dose of a particular toxicant, a numerical expression of the dose-cancer risk relationship in humans must be supplied as a parameter. For most toxicants, this is not available, so a number derived from a laboratory animal species must be used as a surrogate variable. Understandably, this adds uncertainty to the risk assessment.

Another source of uncertainty associated with models are *excluded variables*. Simply stated, these are the variables that are not included in the model for the sake of simplicity. If the impact on risk of excluding these variables is small, the level of uncertainty introduced by their omission is small. The difficulty, of course, is determining with confidence that their impact is small without actually considering them in the risk assessment. An example of an often-excluded variable in exposure assessments is one that accounts for the natural degradation of a contaminant in soil or water. Typically, it is assumed that contaminant concentrations remain unchanged over time, even for organic contaminants subject to well-described degradative processes in the environment.

A further source of uncertainty related to the use of models is the ability of the model to deal with *abnormal conditions* or unanticipated circumstances that might arise. As discussed above, models are designed to be as simple and manageable as possible. A given model may be well validated under generalized conditions, but its ability to effectively handle each and every circumstance that might reasonably be presented may be subject to uncertainty. For example, the model for the dose-toxicity relationship for a given toxicant employed in the risk assessment may account for the type(s) of toxicity anticipated in the general population, but may not account for exceptional sensitivity in some individuals due to genetic or other factors.

A final, and obvious source of uncertainty related to model use is *incorrect model form*. If the model is not structured correctly, there is the potential for significant error. An example of uncertainty related to model structure is provided by the array of models available to extrapolate cancer risk from high doses of carcinogens to low-dose exposures, as mentioned above. Each of the models works reasonably well in describing the relationship between carcinogen dose and cancer incidence at relatively high carcinogen doses, where virtually all of the dose-response data exist. When extrapolated to low carcinogen dosages, these models give widely disparate estimates of cancer incidence and risk. It is impossible to know which of these estimates is accurate, and consequently which model is best, due to the enormous practical difficulties in experimentally measuring low incidence phenomena (e.g., events occurring in 1 in 1,000 or 1 in 100,000 individuals similarly exposed). In the absence of empirical validation of the models, the selection of the model is based on judgment and therefore entails uncertainty.

Decision-rule uncertainty

This type of uncertainty arises from ambiguity or controversy as to how the risk assessment process should be applied to regulatory or social objectives. For example, it may be unclear what the best risk measure is in evaluating the consequences of contaminant exposure in a subject population. Should risk be expressed in terms of expected number of deaths, total expected life-years lost, or some other measure? Risk assessments can be directed to evaluate the probability of harm to an individual, or to a defined population at large. Which of these approaches is most appropriate, given the objectives of the risk assessment? What constitutes "acceptable risk," and in what terms should the acceptability be evaluated? These are but a few examples of decision-rule uncertainty.

Quantitative estimation of uncertainty

This section provides a brief, conceptual overview of the process for generating probabilistic estimates of risk and its accompanying uncertainty.

Detailed guidance in conducting a quantitative uncertainty analysis is beyond the scope of this document, but can be found elsewhere.[6-9]

Hammonds and co-workers[8] have prepared a concise listing of the steps required to perform a parameter uncertainty analysis. These are reproduced below:

1. Define the assessment end point.
2. List all uncertain parameters.
3. Specify maximum range of potential values relevant for unknown parameters with respect to the end point of the assessment.
4. Specify a subjective probability distribution for values occurring within this range.
5. Determine and account for correlations among parameters.
6. Using either analytical or numerical procedures, propagate the uncertainty in the model parameters to produce a (subjective) probability distribution of model predictions.
7. Derive quantitative statements of uncertainty in terms of a subjective confidence interval for the unknown value [representing the prediction end point (e.g., excess cancer risk or hazard index)].
8. Rank the parameters contributing most to uncertainty in the model prediction by performing a sensitivity analysis.
9. Obtain additional data for the most important model parameters and repeat steps 3 through 8.
10. Present and interpret the results of the analysis.

For the first step, it is necessary to determine the manner in which risk will be expressed. There are several possibilities. Examples include the number of excess deaths or occurrences of toxicity in a defined population, the total number of life-years lost in a population, the risk to a hypothetical average person, the risk to hypothetical maximally exposed individual, or the number of individuals in an exposed population who will have risk above an arbitrary "acceptable" level.[6] The decision regarding the desired expression of risk must be made first, since it shapes the way risk will be calculated and the kinds of uncertainties associated with that process.

The second step involves a careful delineation of all of the inputs and variables for which there is uncertainty. For complex risk assessments, such as those involving hazardous waste incinerators, this list will be substantial. For each of these parameters, the limits of potential values and a probability distribution within these limits must be specified (steps 3 and 4).

Assignment of probability distributions to uncertain parameters is a complex issue. In some cases there may be sufficient data that a distribution can be assigned empirically. Often, however, distributions and limits are assigned based on professional judgment, taking into consideration all available information concerning the parameter.

As indicated in step 5, it is important to account for correlations among variables. For example, the parameter of body weight is used in nearly all exposure calculations, and a parameter related to skin surface area is used when evaluating exposure from dermal contact. Body weight and skin surface area are logically related; in general, as one increases, so will the other.[10] Accordingly, it would not make sense in an uncertainty analysis to permit these parameters to vary independently. Interrelated parameters can be dealt with by specifying a correlation coefficient for the two parameters or by altering the model to reflect their interdependency.[8,10]

In step 6, model outputs are derived as a probability distribution, with the shape of the distribution dictated by the probability distributions of each of the component parameters. The uncertainty in the model output is generated by "propagating" the uncertainty and variability associated with the individual parameters. Propagation of uncertainty can be accomplished using analytical or numerical methods. For simple models described by simple equations, analytical methods developed for statistical error propagation may be sufficient. In a model in which parameters are added to derive an output, for example, the mean value for the output is equal to the sum of the means of the model parameters, and the variance of the output is equal to the sum of the parameter variances:[10]

$$\mu_o = \sum_{i=1}^{n} \mu_i \tag{1}$$

$$\sigma_o^2 = \sum_{i=1}^{n} \sigma_i^2 \tag{2}$$

where μ_o is the mean of the output, σ_o^2 the variance of the output, and n is the number of parameters in the model.

Variance can be propagated in models in which the parameters are multiplied by log-transforming the parameter values. In this way, the model is reduced to an additive model, since,

$$A = B \times C \text{ is equivalent to}$$

$$\ln(A) = \ln(B) + \ln(C) \tag{3}$$

and the output variance can be calculated in an analogous fashion. For models with any level of complexity, exact analytical approaches quickly become impractical. One option for dealing with this situation is to employ approximate analytical techniques based on Taylor series expansions of the function.[7] In this approach, the mean, variance, and occasionally higher order moments of the probability distribution are used to propagate the uncertainty.

Because of the relative complexity of many models used in the risk assessment process, and because of the sheer number of different models employed, numerical methods to propagate uncertainty are generally regarded as the most useful for risk assessment uncertainty analysis. Perhaps the best known of the numerical methods is the *Monte Carlo simulation*. A Monte Carlo simulation is a computer-aided, iterative process by which a probability distribution for model output is constructed. The computer randomly selects a value from within the probability distribution for each uncertain parameter. These parameter values are used to create a single output value according to the dictates of the model. The computer then calculates another, single output value, again based on randomly selected inputs from the various parameter distributions. This process is repeated over and over, with the computer selecting input values for each parameter randomly, but in a manner that is consistent with the probability distribution for that parameter. If the process is repeated a sufficient number of times, the collection of output values from all of the iterations will produce a probability distribution for model output. The number of runs, or iterations, required to complete a Monte Carlo simulation is dependent upon the precision needed with respect to the output distribution, but is independent of the number of inputs or parameters in the simulation. If, for some reason, very high precision is required, up to 100,000 iterations or more may be required; for most risk assessment applications, 10,000 iterations are usually sufficient.[7]

There are two different random sampling methods that can be employed with Monte Carlo simulations.[7] In the simple random sampling (SRS) approach, input values are selected randomly from the full probability distribution for each parameter. In an alternate approach, Latin Hypercube Sampling (LHS), probability distributions for each parameter are divided into sections of equal probability. Values are selected randomly from within each section, but each section is sampled only once. The LHS approach is more efficient than the SRS approach in that a stable distribution is reached more quickly, i.e., with fewer iterations. Details regarding the use of Monte Carlo simulation in risk assessments can be found in a variety of publications.[5,7-9,11]

In step 7, the probability distribution of the assessment end point is expressed in terms of a subjective confidence interval. Though quantitative, the confidence interval is termed subjective since it is typically based in part on parameter probability distortions that are derived through professional judgment (see step 4).

Steps 8 and 9 represent approaches to refining the uncertainty analysis. Different parameters rarely contribute equally to the overall uncertainty in the risk estimate. A variable may contribute disproportionately to overall uncertainty if the uncertainty associated with it is particularly large, or if the parameter itself is one of the prime determinants of calculated risk. Efforts to reduce uncertainty are most efficiently directed to its principal sources, and this can be determined using a sensitivity analysis. The general approach taken in such an analysis is to vary one parameter within the limits of its probability distribution and, holding all other parameters constant, evaluate the impact of these changes on risk output. Systematically evaluating each of the parameters can lead to a ranking of those with the greatest influence on overall uncertainty. To the extent possible, efforts should be made to reduce the uncertainty associated with these parameters, followed by the recalculation of the assessment end point (and its subjective confidence interval). The final step, step 10, is the presentation and interpretation of this final output.

Burmaster and Anderson[11] have published a list of "good practices" for the use of Monte Carlo analysis in conducting human health and ecological risk assessments. They emphasize the importance, when reporting the results of a Monte Carlo analysis, of providing the following information:

- A clear description of each of the critical inputs, e.g., formulas used to estimate exposure point concentrations, exposure doses, toxic potencies, hazard indices, and/or incremental lifetime cancer risks.
- Detailed information regarding the input distributions selected, including a graphical display of the distribution, descriptive parameters (mean, standard deviation, median, 95th percentile, etc., as applicable to the distribution type). A thorough justification should be provided for the selection of each distribution.
- The name and statistical quality of the random number generator used.
- Output distributions as graphs that show the allowable risk criterion (e.g., 10^{-6} cancer risk) and, for comparison, the point estimate that would be calculated using a deterministic approach. Also, a table is recommended which includes the mean, standard deviation, the minimum (if one exists), the 5th percentile, the median, the 95th percentile, and the maximum (if one exists) of the output distribution.
- A clear distinction between the effects of uncertainty vs. probability on risk estimates.
- The limitations of the methods and the interpretation of the results.

It is recommended that the analysis be focused on pathways and chemicals of greatest importance. Identification of these pathways and chemicals can be aided by the use of sensitivity analysis.

The procedures for addressing uncertainty in the selection of variables, as described above, are relatively straightforward. An analogous approach can be used to assess model uncertainty. If different models are available, the influence of model selection on outcome can be tested by substituting various models in the risk and uncertainty calculations. Alternately, parameters may be added to the uncertainty analysis expressly for the purpose of assessing model uncertainty. Professional judgment plays a major role in determining the probability distribution assigned to model uncertainty, and it may be difficult to get agreement on a distribution if there is controversy or intense debate over which model is most appropriate.

A quantitative assessment of decision-rule uncertainty is even more problematic. While it is theoretically possible, through expert elicitation, to derive probability distributions for uncertainty related to these issues, the value of these uncertainty estimates in the risk management process is less clear than those related to parameter or model uncertainty.

U.S. EPA guidance for uncertainty analysis for incinerators

U.S. EPA guidance for combustor emissions[4] addresses characterization of risk very briefly, essentially presenting as an example portions of an uncertainty analysis for terrestrial food chain and human soil ingestion pathways and concluding that:

> Uncertainty analysis, such as may be carried out by a stochastic sampling approach, is a useful tool for examining the influence of particular assumptions, input values, or pathways on the overall results, and for estimating overall ranges of uncertainty associated with the use of the methodology. Analyses of this type are particularly helpful whenever the interpretation and use of results derived from this methodology are important components of a risk assessment process.

Beyond this, no further guidance is provided in this document.

The more recently published addendum to this document[13] indicates that, in some cases, it may be useful to derive more detailed information regarding distribution of exposure and risk and presents limited, general guidance on how Monte Carlo simulation might be used. The following points are made:

- The addendum discusses how the impacts of uncertainty and variability can be systematically addressed using a "double loop" or "nested" simulation approach. The uncertain parameters are addressed in the outer loop, while population exposure variables are allowed to vary in the inner loop. The result is a distribution of population exposure or risk. The addendum points out that distributions for each of the uncertainties must be developed, and that this may be difficult. It recommends that since most of the distributions will be based on professional judgment, the document should clearly indicate that the outcome is largely dependent upon this judgment.
- The distinction between uncertainty and variability should be clearly made in the final display of a distribution generated by a Monte Carlo simulation. Graphically, this might be represented as uncertainty bands drawn around the distribution of variability.
- All variables used in a Monte Carlo simulation should be independent. A suggested approach for dealing with interdependent variables is to determine a distribution for one variable and use values from this distribution to compute the distribution for the second variable.

- It is cautioned that toxicity values used in risk assessments have embedded within them fixed exposure assumptions. This can lead to inconsistencies, since the same exposure assumptions elsewhere in the risk calculation may have different values assigned to them.
- General guidance on selection of distributions for use in Monte Carlo simulations is also provided. It is suggested that distributions should be based on local, rather than national, surveys whenever possible. Also, it is pointed out that values near the tail of a distribution are very sensitive to the type of distribution selected, and that the tails of ideal distributions extend to infinity. In order to truncate tails at reasonable bounding conditions, it is suggested that empirical histograms be considered when substantial empirical data are available.

Still more recently, the U.S. EPA has included a discussion of variability in the *Exposure Factors Handbook*,[13] and has provided draft guidance in conducting probabilistic risk assessment using Monte Carlo simulation.[9]

Limitations and practical considerations for quantitative uncertainty analysis

While quantitative uncertainty and variability analysis holds the promise of providing critical information to support a risk management decision, it should be recognized that there are some limitations to its application. A quantitative uncertainty analysis is a difficult task, and the difficulty increases with the thoroughness of the analysis. It is a significant undertaking, even for relatively simple risk assessments. For an inherently complex risk assessment, such as one for a hazardous waste incinerator, a full and comprehensive quantitative analysis involving each and every source of uncertainty is probably not feasible.

There are steps that can be taken to make a quantitative uncertainty analysis manageable. Perhaps the most important is to recognize that it is not necessary to conduct a full analysis of every possible source of uncertainty.[6] Screening approaches can be used to identify the contaminants and exposure pathways that are likely to dominate the overall risk,[8] and efforts regarding the uncertainty analysis can be focused on these without compromising the value of the analysis to the decision-making process.[8] A rudimentary sensitivity analysis can also be used to identify key assumptions and models for which a detailed uncertainty analysis is warranted. Finally, it is possible that there will be aspects within the risk assessment that are resistant to a meaningful quantitative uncertainty analysis. This would include nearly all decision-rule uncertainty and perhaps the uncertainty associated with some models. In these instances, the inability to address the uncertainty quantitatively should be acknowledged and a qualitative expression of uncertainty provided.

Summary of uncertainty and variability analysis for incinerator risk assessments

The uncertainty analysis is a critical aspect of any risk assessment, yet one that typically receives insufficient attention. In a sense, the uncertainty analysis will define the functional value of a completed risk assessment for regulatory decision making. Risk assessment methodology requires volumes of data and a significant amount of scientific and professional judgment, particularly in bridging data and knowledge gaps in facets important to the estimation of risk. The uncertainty contributed by less-than-perfect data and understanding of the physical and biological processes inherent in determining risk must be expressed, preferably in quantitative terms.

While quantitative uncertainty analyses are not yet a regular feature of risk assessments, the methodologies for conducting them exist. These methodologies require knowledge or estimation of the probability distributions for each uncertain factor, and most make use of computer technology to combine these individual probability distributions to derive a probabilistic estimate of risk. There are a number of advantages in expressing risk in probabilistic terms, with confidence intervals. One is that this approach readily lends itself to developing either population or individual risk estimates, depending on the objectives of the risk assessment. Another important advantage is that it portrays for the decision maker in a relatively clear form the limits within which the "true" risk is thought to lie, and its most probable value, given the current state of knowledge.

The difficulties in conducting an extensive quantitative uncertainty analysis should not be underestimated, however, particularly for risk assessments as complex as those for hazardous waste incinerators. A full quantitative uncertainty analysis is probably not feasible, at least presently, for large complex risk assessments. On the other hand, a full analysis is probably not necessary for practical purposes. Through the use of sensitivity analysis techniques, the scope of the quantitative uncertainty analysis can be managed to focus on the most important sources of uncertainty.

References

1. K. T. Bogen, *Uncertainty in Environmental Health Risk Assessment*. Garland, New York, 1990.
2. U.S. EPA, *A Descriptive Guide to Risk Assessment Methodologies for Toxic Air Pollutants*, Environmental Protection Agency, Research Triangle Park, NC, 1993.
3. U.S. EPA, *Risk Assessment Guidance for Superfund Volume I Human Health Evaluation Manual (Part A) Interim Final*, Environmental Protection Agency, Washington, DC, 1989.
4. U. S. EPA, *Methodology for Assessing Health Risks Associated with Indirect Exposure to Combustor Emissions*, Environmental Protection Agency, Cincinnati, 1990.

5. K. M. Thompson, D.E. Burmaster, A.C. Crouch, Monte Carlo techniques for quantitative uncertainty analysis in public health risk assessments, *Risk Analysis*, 12:53–63 (1992).

6. A. M. Finkel, *Confronting Uncertainty in Risk Management: A Guide for Decisionmakers*. Resources for the Future, Washington, 1990.

7. M. G. Morgan, M. Henrion, *Uncertainty: A Guide to Dealing with Uncertainty in Quantitative Risk and Policy Analysis*, Cambridge University Press, New York, 1990.

8. J. S. Hammonds, F. O. Hoffman, *An Introductory Guide to Uncertainty Analysis in Environmental and Health Risk Assessment*, U.S. Department of Energy, 1992.

9. U.S. EPA. Supplemental guidance to RAGS: The use of probabilistic analysis in risk assessment. Part E. OSWER, 1998.

10. D.E. Burmaster, Lognormal distributions for skin area as a function of body weight. *Risk Analysis* 18:27–32 (1998).

11. D. E. Burmaster, P. D. Anderson, Principles of good practice for the use of Monte Carlo techniques in human health and ecological risk assessments, *Risk Analysis* 14:477–481 (1994).

12. U.S. EPA, *Addendum to Methodology for Assessing Health Risks Associated with Indirect Exposure to Combustor Emissions*, Environmental Protection Agency, Washington, 1993.

13. U.S. EPA. *Exposure Factors Handbook*. Vol. 1. EPA/600/p.95/002Ba, 1996.

chapter fourteen

Regulatory compliance within the U.S. commercial hazardous waste incineration industry

John Schert

In May 1994, the Florida Department of Environmental Protection (FDEP) contracted with the Florida Center for Solid and Hazardous Waste Management (Center) to conduct a comprehensive study of the health effects of commercial hazardous waste incinerators. The research effort focused on four questions:

1. What is known about existing commercial hazardous waste incinerators and their impacts on human health?
2. Can the impacts of a proposed commercial hazardous waste incinerator be evaluated before it is built, and, if so, how?
3. What is the regulatory compliance record of existing commercial hazardous waste incinerators?
4. What methods can be used to monitor a commercial hazardous waste incinerator's impacts after it is built?

At the time that the FDEP–Center research project contract was signed, two permit applications had been filed with the FDEP to build commercial hazardous waste incinerators in Florida. One facility was proposed to be built in central Florida near the town of Bartow, and another facility was proposed to be built in northern Florida near the state capital, Tallahassee. A great deal of public opposition had developed in response to these proposed plants, and the Florida legislature had directed the FDEP to conduct a study regarding the need for these facilities.

The Center assembled a team of health care professionals, public health experts, epidemiologists, and other scientists and engineers to conduct this research. The team included faculty from Florida State University, the University of Florida, the University of Central Florida, the University of South Florida, and the University of Miami. The team was responsible for providing answers to questions 1, 2, and 4. Center staff were responsible for answering question 3.

At the time this research was conducted, there were approximately 20 commercial hazardous waste incinerators operating in the United States. A comprehensive study of the compliance records of all commercial hazardous waste incinerators was beyond the scope and budget for this project. As such, Center staff made site visits to several hazardous waste incinerators and developed compliance summaries for a selected number of these facilities using the contract services of two former U.S. Environmental Protection Agency (EPA) employees who had a great deal of experience in this area.

On-site record searches were conducted in the U.S. EPA offices in Chicago, Dallas, and New York and in state environmental program offices in New Jersey and Ohio. Other records used in this effort were obtained from states through extensive correspondence with state environmental agencies and the hazardous waste industry. In many cases, compliance records were not readily available from the agency responsible for the inspections of the facility and a site visit had to be made to the regulatory agency responsible for the records in order to inspect the records.

The regulatory records associated with these facilities are usually quite extensive (one facility may take up more than a whole file cabinet), and great volumes of information had to be reviewed and copied so that the materials could be closely studied and summarized at a later point in time. The process usually followed was to first contact the federal or state Freedom of Information (FOI) Act officer responsible for the agency records and make an official request to inspect the files. Following this, an appointment was made with the FOI officer to visit the regulatory office, and the Center staff traveled to the office to inspect and view the facility files that had been gathered and assembled by the FOI officer. The relevant portions of the files were then marked for copying. The copying was paid for and the copied files were mailed to the Center. The process of obtaining relevant documents was quite laborious and time consuming.

The former U.S. EPA employees from Region II and Region V who reviewed the compliance files and developed the compliance summaries for the incinerators that were studied had extensive experience in the permitting of hazardous waste incinerators. Their expertise was quite valuable, and this portion of the research project could not have been accomplished without their help. It was necessary to hire staff with this very special expertise, as the information contained in the compliance and enforcement files of these facilities was quite voluminous and highly technical.

The facilities that were the subject of this research are located in various regions of the United States and represent a cross-section of the commercial hazardous waste industry, including one of the oldest incinerators in the industry (Chemical Waste Management, Chicago, Illinois) and one of the newest incinerators (WTI, East Liverpool, Ohio). The Chemical Waste Management incinerator is no longer in service and was not in service at the time this research was conducted.

Center staff visited several hazardous waste incineration facilities in connection with this study, including six commercial hazardous waste incinerators (the Chemical Waste Management incinerator, now inactive, in Chicago; the Chemical Waste Management incinerator in Port Arthur, Texas; the Rollins Environmental Services incinerator in Deer Park, Texas; the Rollins Environmental Services incinerator in Bridgeport, New Jersey; the USPCI incinerator west of Salt Lake City, Utah; and the Westinghouse-Aptus incinerator in Aragonite, Utah). (The Westinghouse-Aptus incinerator in Utah has since been acquired by Rollins Environmental Services.) Also visited were a boiler/industrial furnace (BIF) in North Carolina (Carolina Solite) and an industrial hazardous waste incinerator operated by DuPont in Port Orange, Texas.

In examining the compliance records of these facilities, it was readily apparent that the regulatory compliance histories varied considerably from one facility to another. Compliance summaries were developed for the following 10 facilities.

Aptus, Inc., Aragonite, Utah. This facility is located in an extremely remote section of Utah. The nearest residential dwelling is 34 miles away. Another commercial hazardous waste incinerator (USPCI) is located several miles from the Aptus facility. Both the Aptus incinerator and the USPCI incinerator are in a section of Utah that was specifically designated by the state government for this particular type of industrial land use. Both facilities are surrounded by large federal land holdings. The Aptus facility has a portable building on-site owned by the State of Utah, Bureau of Solid and Hazardous Waste.

CWM Chemical Services, Inc., Chicago Incinerator, Chicago, Illinois. The incinerator at this facility is no longer operating and has not operated commercially since February 1991, except for the purpose of stack emissions testing from August 1991 to January 1992. The facility has reportedly been sold to another company that operates an industrial wastewater treatment plant on an adjoining piece of property. The facility is located in a heavily industrial area which includes municipal solid waste landfills, an inactive solid waste incinerator, and properties designated as Superfund sites. The incinerator, in its current configuration, was operated by a number of owners from 1980 to 1992. Previous operations on the property during the period of 1970 to 1980 consisted of wastewater treatment, chemical stabilization, and biological treatment lagoons.

ENSCO (Environmental Systems Company) Inc., El Dorado, Arkansas. This facility is located on the site of a former petroleum refinery that was operated from 1921 to 1972. The incinerator at this facility has been operated since 1974. This facility reportedly processed more than 65,000 tons of waste in 1992.

Norlite Corporation, Cohoes, New York. This facility burns waste oil, used machining lubricants, and solvents in a rotary kiln producing lightweight aggregate for use in the production of light weight concrete. The facility also has a strip mine and a quarry.

Rhone-Poulenc Basic Chemicals Company, Houston. This facility is located in a very industrial setting and is surrounded by a cement plant, an oil storage facility, a shipyard, and a cemetery. Industrial use of the property dates back to 1917–1919 when the property was first used as a charcoal production facility and a sulfuric acid production facility. The incinerator is used primarily in the regeneration of spent sulfuric acid generated by the chemical and petroleum industry. The facility has operated under a series of owners.

Rollins Environmental Services (LA), Baton Rouge, Louisiana. This facility is a treatment (incinerator) and disposal facility (landfill) that has been in operation since 1970. This facility reportedly processed over 32,000 tons of waste material in 1992.

Rollins Environmental Services Inc., Bridgeport, New Jersey. This facility is located in a coastal rural area with very few neighbors. Commercial operations began at this facility in 1969. It is the only full service commercial incinerator operating on the East Coast. The New Jersey Environmental Protection Agency inspects this facility weekly.

Rollins Environmental Services, Deer Park, Texas. Operations at this facility began in the early 1970s prior to the passage of RCRA. The site was originally used as a truck transfer facility and consisted of a series of holding ponds and lagoons used by the Rollins transport fleet to clean transport trucks and other vehicles. A number of large petrochemical refineries/manufacturing facilities are located in close proximity to this facility. The site is currently used as both a treatment facility (incinerator) and as a disposal facility (landfill). With an annual hazardous waste capacity of approximately 205,000 tons per year, this facility is reportedly one of the largest of its kind in the country and incinerates large amounts of material for the Texas and Louisiana petrochemical industry. This facility reportedly processed more than 85,000 tons of waste material in 1992.

Waste Technologies Industries (WTI), East Liverpool, Ohio. WTI is one of the newest commercial hazardous waste incinerators in the United States. This facility has been the focus of a great deal of criticism due to the siting of the facility near a residential neighborhood and a public school and the fact that the facility has been located in a river valley which allegedly has poor air pollution dispersion characteristics. The

Ohio EPA has stationed three full-time employees at the WTI facility to allow inspections at any time.

Florida Solite Company, Russell, Florida. This facility began operating in Clay County, Florida, just north of the city of Green Cove Springs, in 1959. The facility began burning liquid burnable material (LBM) in 1973. (LBM can consist of either hazardous materials, such as spent solvents, or nonhazardous materials, such as used oil and liquid organic waste.) The facility produced a lightweight aggregate product used in numerous construction applications where high strength-to-weight, insulating, fire-resistant, and weather-resistant properties are desired. Clay was mined on site and "fired" in three kilns. In 1995 Florida Solite Company informed the Florida Department of Environmental Protection that it planned to cease its Florida operations. The facility is no longer in operation.

These compliance summaries are highly detailed and were published in their entirety in the appendix of a report entitled, "Evaluation of the Health Impacts Associated with Commercial Hazardous Waste Incinerators," July 1995. This report is available from the Florida Center for Solid and Hazardous Waste Management at the University of Florida, Gainesville, Florida.

The compliance summaries in this report were provided to the staff of the respective incinerators prior to publishing the report in order to provide the companies an opportunity to correct any errors or to provide additional information prior to publication. This was done in an effort to ensure that the information that was presented was accurate.

Compliance summaries are usually of great interest to prospective customers of commercial hazardous waste incinerators. Customers who send hazardous waste to these incinerators are often very concerned about the liability they incur via their hazardous waste disposal contractor because of the federal policy of making generators of hazardous waste responsible for the actions of the company/facility that takes care of disposal or treatment of the waste. If the disposal/treatment facility disposes or treats the waste in a manner that causes an environmental contamination problem, the company that originally generated the waste can be held responsible for helping to pay to clean up the problem.

An interesting thing happened when the compliance summaries were provided to the companies that operate the incinerators for review. At one company, the staff was very appreciative of the fact that a university research project had summarized the compliance history of their facility. It seems that the compliance officer for this facility receives a large number of inquiries for compliance information from prospective clients. The compliance summary produced as part of this report provided all the information that this person needed for his prospective clients. In addition, the compliance summary was developed by a third party, which typically has high credibility.

A particularly relevant and informative U.S. EPA–OSHA report on the issue of environmental compliance by the hazardous waste industry is available from the U.S. EPA. The report, "Evaluation of Compliance With On-Site Health and Safety Requirements at Hazardous Waste Incinerators," was released on May 23, 1991, and indicates U.S. EPA's concerns with the "apparent overuse of waste feed cut-offs and emergency by-passes at some facilities." The companies which were studied as part of this U.S. EPA–OSHA research project are not specifically identified in the report. Because of the sensitive nature of the information which was gathered, the U.S. EPA did not want to publicly identify the companies they studied.

The Executive Summary of the U.S. EPA–OSHA compliance report notes that:

> The U.S. EPA identified a total of 75 violations of its standards at the 29 facilities inspected. These violations included 14 for failure to provide adequate information and/or training to employees; 16 for non-compliance with the contingency plans and emergency response requirements; 29 for non-compliance with general inspections and preparedness and prevention requirements; and 16 for failure to comply with operational procedures requirements. Of these 16 violations, only 5 related specifically to incinerator operations.

The U.S. EPA also noted a significant number of waste feed cut-offs and emergency by-pass openings. The waste feed cut-off system is intended to stop waste entering the incinerator combustion unit when certain operating conditions are exceeded. Emergency by-passes are intended to prevent ground level fugitive emissions and possible explosions from excessive pressures in the combustion unit. While both devices are designed for safety purposes, the frequent use of these devices at some facilities probably indicates a need to improve the operating procedures at these facilities.

It should be noted that a "Notice of Violation" indicated in the compliance summary does not necessarily mean that there was, in fact, a violation of a regulation or permit condition. In some cases, the "Notice of Violation" was subsequently withdrawn by the government agency that issued it.

One state, North Carolina, actually has regulatory inspectors stationed full time at incinerators to observe the operation of the incinerator first hand. One criticism of this practice is the potential for a familiar relationship to develop between the inspector and the personnel at the regulated facility. This sense of familiarity might make it difficult for the inspector to take regulatory action against the facility.

At some facilities, the state government and the regulated facility have developed cooperative agreements where stack gas emission monitors that monitor the concentrations of different pollutants in the air pollution emissions of the facility have also been installed in the offices of the regulatory agency so that regulatory staff can know exactly what the emission concentrations are from a facility at any point in time on an ongoing and real-time basis. In addition, at some facilities, the state government has installed a portable building on the grounds of the facility complete with telephones and desks to provide support for state inspectors when they are at the facility.

Index